AF615309

Polyelectrolyte Gels

ACS SYMPOSIUM SERIES 480

Polyelectrolyte Gels

Properties, Preparation, and Applications

Ronald S. Harland, EDITOR
KV Pharmaceutical Company

Robert K. Prud'homme, EDITOR
Princeton University

Developed from a symposium sponsored
by the American Institute of Chemical Engineers
at the American Institute of Chemical Engineers Annual Meeting
Chicago, Illinois,
November 11–16, 1990

American Chemical Society, Washington, DC 1992

Library of Congress Cataloging-in-Publication Data

Polyelectrolyte gels: properties, preparation, and applications / Ronald S. Harland, editor, Robert K. Prud'homme, editor.

p. cm.—(ACS symposium series; 480)

"Developed from a symposium held at the National Meeting of the American Institute of Chemical Engineers, Chicago, Illinois, November 11–16, 1990."

Includes bibliographical references and indexes.

ISBN 0–8412–2176–6

1. Polyelectrolytes—Congresses. 2. Colloids—Congresses.

I. Harland, Ronald S., 1962– . II. Prud'homme, Robert K., 1948– . III. National Meeting of AIChE (1990: Chicago, Ill.) IV. Series.

QD382.P64P63 1992
547.7—dc20 91–37617
CIP

The paper used in this publication meets the minimum requirements of American National Standard for Information Sciences—Permanence of Paper for Printed Library Materials, ANSI Z39.48–1984. ♾

Copyright © 1992

American Chemical Society

All Rights Reserved. The appearance of the code at the bottom of the first page of each chapter in this volume indicates the copyright owner's consent that reprographic copies of the chapter may be made for personal or internal use or for the personal or internal use of specific clients. This consent is given on the condition, however, that the copier pay the stated per-copy fee through the Copyright Clearance Center, Inc., 27 Congress Street, Salem, MA 01970, for copying beyond that permitted by Sections 107 or 108 of the U.S. Copyright Law. This consent does not extend to copying or transmission by any means—graphic or electronic—for any other purpose, such as for general distribution, for advertising or promotional purposes, for creating a new collective work, for resale, or for information storage and retrieval systems. The copying fee for each chapter is indicated in the code at the bottom of the first page of the chapter.

The citation of trade names and/or names of manufacturers in this publication is not to be construed as an endorsement or as approval by ACS of the commercial products or services referenced herein; nor should the mere reference herein to any drawing, specification, chemical process, or other data be regarded as a license or as a conveyance of any right or permission to the holder, reader, or any other person or corporation, to manufacture, reproduce, use, or sell any patented invention or copyrighted work that may in any way be related thereto. Registered names, trademarks, etc., used in this publication, even without specific indication thereof, are not to be considered unprotected by law.

PRINTED IN THE UNITED STATES OF AMERICA

Chem
QD
382
.P64P5
1992

ACS Symposium Series

M. Joan Comstock, *Series Editor*

1992 ACS Books Advisory Board

V. Dean Adams
Tennessee Technological University

Alexis T. Bell
University of California—Berkeley

Dennis W. Hess
Lehigh University

Mary A. Kaiser
E. I. du Pont de Nemours and Company

Gretchen S. Kohl
Dow-Corning Corporation

Bonnie Lawlor
Institute for Scientific Information

John L. Massingill
Dow Chemical Company

Robert McGorrin
Kraft General Foods

Julius J. Menn
Plant Sciences Institute, U.S. Department of Agriculture

Marshall Phillips
Phillips Consulting

A. Truman Schwartz
Macalaster College

Stephen A. Szabo
Conoco Inc.

Robert A. Weiss
University of Connecticut

Foreword

THE ACS SYMPOSIUM SERIES was founded in 1974 to provide a medium for publishing symposia quickly in book form. The format of the Series parallels that of the continuing ADVANCES IN CHEMISTRY SERIES except that, in order to save time, the papers are not typeset, but are reproduced as they are submitted by the authors in camera-ready form. Papers are reviewed under the supervision of the editors with the assistance of the Advisory Board and are selected to maintain the integrity of the symposia. Both reviews and reports of research are acceptable, because symposia may embrace both types of presentation. However, verbatim reproductions of previously published papers are not accepted.

Contents

THE UNIVERSITY
WITHDRAWN
OF IOWA LIBRARIES
The University of Iowa

THE UNIVERSITY OF IOWA LIBRARIES
WITHDRAWN

Preface

POLYELECTROLYTE GELS ARE THE BASES for multibillion dollar businesses that affect all our lives: personal care products (e.g., diapers, feminine care products, and incontinence products); industrial separations (e.g., sewage treatment, membrane processes, and protein and biological synthesis processes); and pharmaceuticals (e.g., controlled-release technology, bioadhesives, and enteric dosage forms). Developing these applications of gels has required a fundamental understanding of the relationships between synthesis, characterization, and end-use performance.

The development of materials, instruments, and products requires a judicious selection of chemical and physical characteristics of the gel. The final properties of the gelled system are determined by varying the type, degree, and distribution of ionizable and nonionizable functional groups, cross-linkers, and reaction conditions. Therefore, in addition to an understanding of polymer synthesis, a knowledge of polymer structural characterization, thermodynamics, diffusion theory, and physical chemistry is required to achieve the desired polyelectrolyte gel.

This volume is presented to provide an overview of polymer gels with particular emphasis on the unique characteristics of polyelectrolyte gels. The authors include participants from the symposium on which this book is based, and additional contributions were made by other leading researchers in the field. These chapters provide the reader with a comprehensive book on polyelectrolyte gels.

An introduction is provided by Toyoichi Tanaka of the Massachusetts Institute of Technology covering the state of the art and the potential of polyelectrolyte gels. The following chapters provide in-depth descriptions of gel synthesis, characterization, thermodynamics, mass transport, and novel applications. Each chapter provides descriptions of current research and a review of noteworthy contributions in the specific area by leading experts.

Each chapter has been reviewed by one (and in most cases two) independent experts, both editors, and a proofreader. But in the end, it is the strength of the individual researchers who contributed chapters that ensures the strength and impact of the volume.

We thank all the contributors for their responses to our requests and to those of the American Chemical Society. In addition, we thank the staffs of KV Pharmaceutical Company and Princeton University for their support.

RONALD S. HARLAND
KV Pharmaceutical Company
St. Louis, MO 63144

ROBERT K. PRUD'HOMME
Princeton University
Princeton, NJ 08544

August 29, 1991

Chapter 1

Phase Transitions of Gels

Toyoichi Tanaka

Department of Physics and Center for Materials Science and Engineering, Massachusetts Institute of Technology, Cambridge, MA 02139

Polymer gels are known to exist in two distinct phases, swollen and collapsed. Volume transition occurs between the phases either continuously or discontinuously in response to chemical and physical stimuli such as temperature, solvent composition, pH, ionic composition, electric field, light, and particular molecules. For a gel to undergo the phase transition, it is necessary that polymers interact with each other through both repulsive and attractive interactions and the balance of competing interactions has to be modified by various stimuli. The phase behavior of a gel, therefore, crucially depends on the nature of interactions between polymers. Recently new phases and volume transitions between them have been discovered in some gels. Detailed examination of the gel phase behavior provides a deep insight into the polymer-polymer interactions and configurations of polymers. The knowledge on physical and chemical fundamentals of gel phase transition will play a role as guiding principles for a wide variety of technological applications of gels as functional elements.

During the last decade gel research has made remarkable progress owing to the strenuous efforts by researchers from many different fields. Since ancient times gels have been closely affiliated with our daily life. From the early 1940's numerous synthetic gels have been designed and developed for various usages *(1,2)*. What, then, has made the gel research very popular again? The purpose of this article is to provide an answer to this question *(3)*.

Gels are found everywhere. In our bodies, the cornea, the vitreous, and the connective tissues are gels. The surfaces of the internal tracts such as the stomach and lung are covered with gels. The basement membranes for the kidney and blood vessels are also gels. These membranes are believed to play a fundamental role in the transport of water and solute molecules.

Disposable diapers and sanitary napkins use gels as super water absorbents.

0097–6156/92/0480–0001$06.25/0
© 1992 American Chemical Society

For these industries it is crucial to understand the physical and chemical principles that govern the degree and speed of gel swelling in the physiological conditions.

Sheets of gels are developed that tightly wrap fresh fishes and meats. The gels keep them moist but absorb unnecessary excess water and are useful for efficient transportations and storages.

Gels are used for agricultural purposes as retainers of water and solutes.

Gels are indispensable materials as molecular sieves to separate molecules according to size in gel permeation chromatography and electrophoresis.

Soft contact lenses, artificial lenses, artificial vitreous, and materials used in plastic surgeries are made of gels. Gels are also used as control delivery systems for drugs and perfumes.

Builders mix a concrete powder with water. It is difficult to mix a powder and a liquid homogeneously, and it is necessary to put an excess amount of water for a thorough homogeneous mixing. Recently a method has been developed where gels rather than water are used. Mixing of two powders is much easier than mixing powder and water and requires much less of water. After thorough mixing the gel gradually shrinks due to the pH change squeezing out water into the concrete powder uniformly, thus a concrete with an extremely high density can be obtained. Its hardness and strength are similar to those of rock.

The gels play a vital role in the fields of medicine, foods, chemical, agricultural, and other industries. The list of gels and their usages would be too long to be fully described in this chapter. Readers may refer to some recently published books on applications *(1-2,8-9)*.

Structure of gels

The unique structure allows gels to be useful in various applications *(10,11)*. Gels are cross-linked networks of polymers swollen with a liquid. For example, two thirds of the human body consists of water, and almost all the biological reactions and other activities occur in water. Water needs a container to retain its shape as an entity; nature invented two kinds of containers, cells and gels. A cell membrane provides a boundary to water, whereas a polymer network incorporates water in its intersticious space with its affinity due to interaction energy and polymer entropy. These containers allow water to retain its shape.

The history of the science and technology of membranes is long. Recent revival of the technology of the Langmuir-Blogget films allowed remarkable progress in the field. Fruitful outcome in the membrane technology is expected to flourish during the next decade. Incidentally gel research is also expected to grow rapidly during the next decade.

As our society becomes richer and more sophisticated, and as we increasingly recognize that the natural resources are not unlimited, materials with better quality and higher functional performance become more wanted and needed. Soft and gentle materials are beginning to replace some of the hard mechanical materials in various industries. Recent progress in biology and polymer sciences that is unveiling the mystery of marvelous functions of biological molecules and organelles promises new development in gel technologies. All these factors bring us to realize the importance and urgent needs of establishing gel sciences and technologies *(1)*.

The principles of phase transitions and critical phenomena in polymer gels are essential for the understanding of gels and of polymer systems in general *(12)*. These phenomena provide an experimental tool with which to explore the fundamentals underlying the molecular interactions and recognitions in natural and synthetic polymers.

Historical Background

Theoretical works

Study of the phase transition of gels can be traced back to the theoretical work by Dusek and Patterson in 1968, who suggested the possibility of a discontinuous volume change of a gel when an external stress is imposed upon it *(13)*. Using the Flory-Huggins equation of state, these researchers realized that with the presence of an external force, Maxwell's loop appears in the isobar of the gel. A similar phenomenon in a single polymer chain, known as the coil-globule transition, has been theoretically studied by Ptitsen and Eizner *(14)*, and Lifshitz, Grosberg, and Khokhlov *(15)*, and deGennes *(16)*. In a sense a gel phase transition is a macroscopic manifestation of a coil-globule transition.

Collective diffusion of gels

On the experimental side the first relevant event may be the observation of the so-called collective diffusion, which was found in 1973 using quasielastic light scattering on acrylamide gels *(17)*. Light is scattered by the density fluctuations of polymer network or the phonons in the polymer network. The phonon would propagate without water, but in water the phonon is completely overdamped and the motion becomes a diffusion process. The equation for the displacement **u** of a point in the gel whose average location was x is given by:

$$\rho\frac{\partial^2\vec{u}(\vec{x},t)}{\partial t^2} = K\frac{\partial^2\vec{u}}{\partial\vec{x}^2} - f\frac{\partial\vec{u}}{\partial t} \tag{1}$$

The left hand side represents the density ρ times acceleration, the first in the rhs is the elastic restoring force. K denotes the longitudinal elastic modulus. These two terms constitute the sound wave equation within an elastic medium, which, in the case of a gel, is the polymer network . The last term is the viscous friction between the network and water. In most gels the inertia term is much smaller than the friction and elastic terms and can be neglected, and the equation becomes that of a diffusion.

$$\rho\frac{\partial^2\vec{u}}{\partial t^2} = K\frac{\partial^2\vec{u}}{\partial\vec{x}^2} - f\frac{\partial\vec{u}}{\partial t} \cong 0 \quad \Rightarrow \quad \frac{\partial\vec{u}}{\partial t} = \frac{K}{f}\frac{\partial^2\vec{u}}{\partial\vec{x}^2} \tag{2}$$

It is interesting that polymers undergo a diffusion process even though they are all cross-linked into a network. This process is, therefore, called the collective diffusion of a gel.

However, the notion that an elastic medium obeys the diffusion equation when it moves in a viscous fluid is more general and universal than the diffusion of isolated molecules in water. For example, the diffusion of ink in water obeys the diffusion equation. Without water the ink solute can be considered as a gas whose elastic constant and friction is given by using Stokes formula:

$$K = nkT \quad \text{and} \quad f = n6\pi\eta r \tag{3}$$

where n is the number density of the ink molecules, η is the water viscosity, and r is the radius of the ink molecule. From these two equations the Stokes-Einstein formula is derived for the diffusion coefficient of particle:

$$D = \frac{K}{f} = \frac{nkT}{n6\pi\eta r} = \frac{kT}{6\pi\eta r} \tag{4}$$

Thus the diffusion of ink molecules can be derived from the equation (2).

The light scattering study confirmed the diffusion process in gels. For example, the mode decayed in proportion to the square of the wave-vector (~ q^2) as theoretically predicted from the equation (2). The collective diffusion mode has been confirmed in various gel systems. Munch and colleagues showed an interesting demonstration on the collective nature of the diffusion of a polymer network. These researchers measured a solution containing polymers of a same chemical structure but with two different molecular weights *(18)*. The two diffusion coefficients were observed, each corresponding to each molecular weight. When the polymers were cross-linked, the correlation function of scattered light became a single exponential and the diffusion coefficient became a single value. This observation revealed the essence of nature of collectivity of the diffusion process of all the connected polymers.

Viscoelastic parameters were determined using light scattering technique on various gels *(19)*. The collective diffusion theory has been improved by taking into account of the counter flow of water, the effect of shear modulus, and shape dependence *(19-22)*.

Critical behavior of gels

In 1977 the critical phenomena were discovered while the light intensity scattered from a gel was being measured as a function of temperature *(23)*. As the temperature was lowered, both the intensity and the fluctuation time of scattered light increased and appeared to diverge at a special temperature. The phenomenon was explained as the critical density fluctuations of polymer network, although the polymers were cross-linked *(24-25)*. The critical behavior was not along the critical isobar, but the spinodal line was reached under super cooled conditions.

Phase transition of gels

The light scattering study of the acrylamide gels in water showed that the fluctuations of the polymer network diverged at minus 17°C. This finding raised a question of ice formation although such possibility was carefully checked and eliminated by the measurement of refractive index of the gel. Such a question could be answered if the spinodal temperature were raised to much above the freezing temperature. The gels were placed in acetone-water mixtures with different concentrations to find a proper solvent in which the opacification of the gel occurred at room temperature. The next day half of the gels were swollen, and the remaining gels were collapsed, meaning the gel volume changed discontinuously as a function of acetone concentration.

The experiments were repeated but were not reproducible: new acrylamide gels were made with various recipes and their swelling curves were determined as a function of acetone concentration, but they were all continuous. It was later recognized that the gels that showed the discontinuous transition were old ones, that is, gels made a month before and left within tubes in which they were prepared. Repeated experiments were all carried out on "new" gels, and thus they underwent a continuous transition.

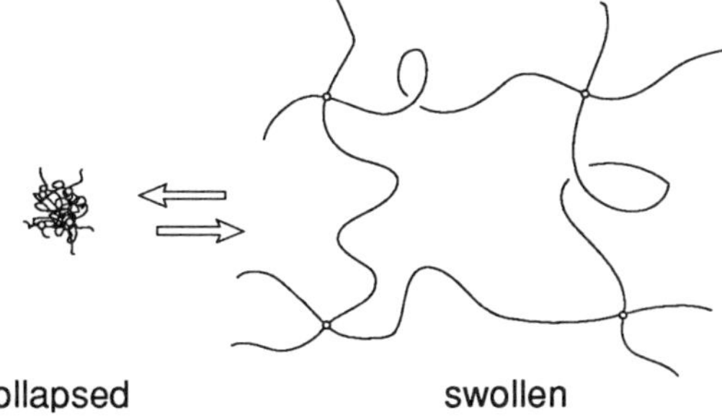

Figure 1. The phase transition of a gel is a reversible and discontinuous folding and unfolding of the polymer network.

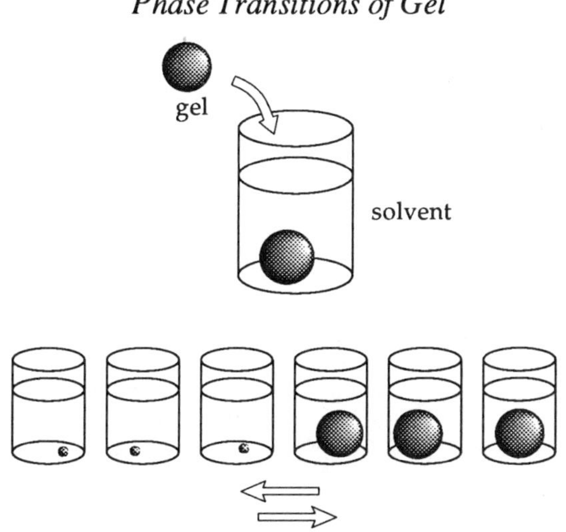

Figure 2. The phase transition of a gel is observed as a reversible and discontinuous volume change in response to temperature and other environmental variables.

Later the difference of new and old gels was identified as ionization which induced an excess osmotic pressure within the gels leading to the discontinuous transition. Hydrolysis was gradually taking place in the gel in a mildly high pH solution. This explanation was experimentally proven by artificially hydrolyzing the gel and observing the increase in the discontinuity of the volume transition *(26)*. It was also shown theoretically that when the osmotic pressure due to counter ions was added to the equation of state of a gel, Maxwell's loop appeared in the swelling curves (temperature vs volume) and a discontinuous transition emerged. As the ionization increased the transition temperature was predicted to increase and the volume change at the transition was found to be larger.

Figure 3 below shows the swelling curves of acrylamide gels immersed in acetone water mixtures with different volume compositions *(26)*. The gel is swollen in water and gradually shrinks as acetone is added to the solution. The days shown in the graphs indicate the length of hydrolysis of the acrylamide gels in the pH 10 solution of tetramethylethylenediamine, the accelerator of the gelation of acrylamide.

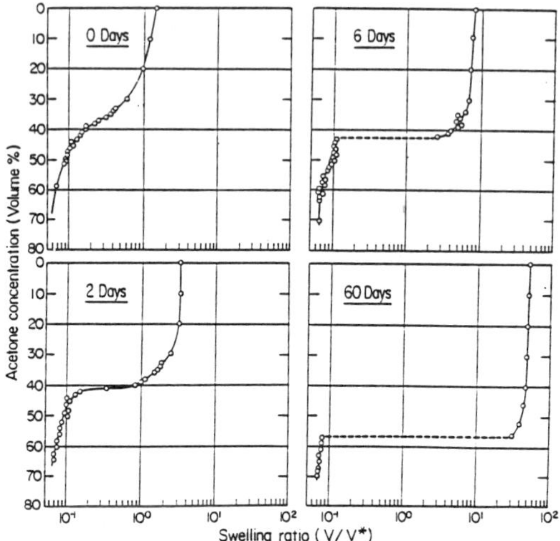

Figure 3. The equilibrium swelling volumes of acrylamide gels immersed in acetone–water mixtures with different volume compositions. (Reproduced with permission from reference (*26*). Copyright 1980 American Physical Society)

As the hydrolysis proceeds the transition acetone concentration becomes higher and the volume change at the transition increases. The volume change at the transition reaches 500 times at 60 days hydrolysis. A special hydrolysis duration exists at which the transition becomes discontinuous for the first time. The special condition of temperature, solvent composition and gel composition, at which the swollen phase and collapsed phase merge corresponds to the critical point.

Phase transition is universally observed in gels

The set of the swelling curves shown in the figure has a strong resemblance to the set of volume-temperature curves of water. Water becomes vapor at 100°C and changes its volume seventeen hundred times at one atmosphere. As the pressure decreases the vaporization temperature is lowered and the volume change at the transition becomes large. (This phenomenon is why a mountain climber needs a pressurized cooker to boil rice at high altitudes.) The ionization causes a similar effect on the gel phase transition. In gels the ionization adds an internal pressure, which is opposite to an external pressure. Ionization of a gel has a similar effect as decreasing the external pressure in water-vapor transition. Theoretically the swelling curve is given by minimizing the free energy per polymer segment with respect to polymer density or equivalently by the zero osmotic pressure on the gel. The Flory-Huggins formula gives the free energy in the form *(10,11,23)*:

$$F = F_{\text{rubber}} + F_{\text{counter-ion}} + F_{\text{virial}}$$

$$\frac{F_{\text{rubber}}}{VkT} \sim \nu_e \left[\left(\frac{\rho}{\rho_o}\right)^{1/3} + \left(\frac{\rho}{\rho_o}\right) \ln\left(\frac{\rho}{\rho_o}\right) \right]$$

$$\frac{F_{\text{counter-ion}}}{VkT} \sim f \nu_e \left(\frac{\rho}{\rho_o}\right) \ln\left(\frac{\rho}{\rho_o}\right)$$

$$\frac{F_{\text{virial}}}{VkT} \sim \left[v + \left(\frac{f}{\rho_o}\right)^2 \right] \rho^2 + w\rho^3 + x\rho^4 + \cdots \tag{5}$$

where F_{rubber} represents the free energy of rubber elasticity, $F_{\text{counter-ion}}$ represents the counter ion osmotic pressure, and F_{virial} denotes the set of virial terms including the charge-charge repulsion. The curves calculated using this formula satisfactorily describes the phenomena qualitatively. Thus, the phase transition of a gel may be considered a phenomenon similar to the gas-liquid phase transition. The latter occurs in the group of separated molecules, whereas in the former all the molecules are connected in the form of a network.

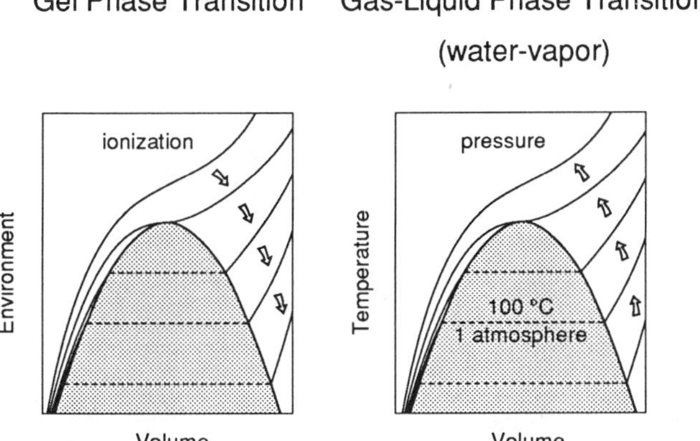

Figure 4. Equation of states of a gas-liquid system and a gel.

The analogy between the water-vapor transition and gel transition indicates that the gel transition should be observed in any gel, since the gas-liquid transition is also universal to any materials. The universality of the gel phase transition was confirmed by the observations of phase transition in gels of various synthetic polymers. Most of the synthetic polymers in which the transition was found had the vinyl chain backbone.

As is known only three types of polymers exist in nature: polypeptides, polysaccharides, and polynucleotides. These polymers have backbone structures totally different from each other and from vinyl polymers. Amiya chose a representative from each one of these biopolymer groups and cross-linked them by covalent bonds. Gels made of gelatin, agarose, and DNA exhibited a phase transition *(27)*. Many synthetic gels have also been studied and the universality of the phase transition in gels seems to have been well established *(28-32)*. It may now be concluded that the phase transition of gels should be universal to any gel.

Theories of phase transition of gels have been extensively improved. Readers should refer to references such as *(33-36)*.

Examples of gel phase transition in the biological world

Now that the phase transition is universally observed in gels, it is natural to ask whether nature makes use of the phenomenon in biological activities. Verdugo found a fascinating example where the gel phase transition is indeed utilized in biological activity *(37)*. Slug mucin, which is stored in the body in an extremely compact form, swells more than 1000 times when secreted out of the body and absorbs water in its neighborhood. By doing so the slugs are able to keep water and maintain the moist environment necessary for survival. The idea is also used for recently available disposable diapers. The question has been raised by some biologists concerning how slugs can create such highly dense mucin for storage in the body full of water. A slug might need a significant osmotic pressure, thus a large energy to do so.

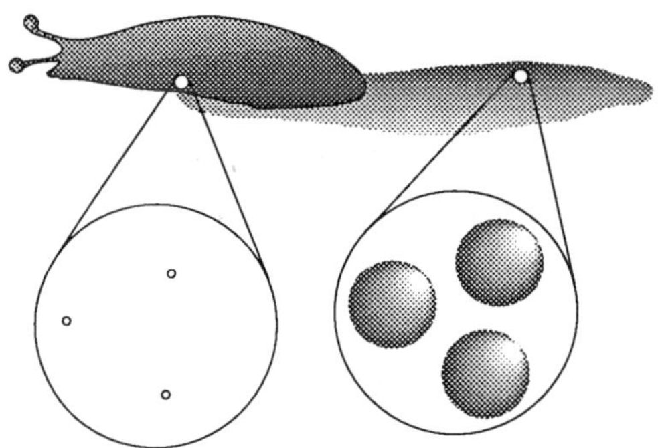

Figure 5. Slugs release compactly packed gel beads to absorb water to maintain a moist environment. The gel beads are shrunken in spite of the aqueous body fluid thanks to the phase transition induced by calcium ions.

Verdugo and his colleagues found that the mucin undergoes phase transition in acetone-water mixtures, and by adding calcium ions. Namely slug does not need much energy to squeeze the mucins, but only have to increase calcium concentration around mucins. Such mechanisms seem to be used for other animals such as the eight-eyed eels in Japan, and probably for the mucins that cover our internal ducts. This finding is not only amusing but seems to have a profound implication for biological activities. It is interesting to note how sophisticated are the slug's method that utilize gels near the phase transition threshold and thus in the reversible condition, compared to the super water absorbents in diapers which are used only in the irreversible processes.

Verdugo further found that the delivery of histamines is carried out by phase transition of granular mucus gels *(38)*. The granules contain highly concentrated histamine molecules, which are released upon swelling transition after the granules are secreted out of a mast cell. The gel granules are collapsed with the presence of high concentration of histamine for recycling.

Some researchers consider that the phase transition of gels plays a vital role in muscular contraction and relaxation *(39)*, and also in neural excitation and signal transmission *(40)*. It is known that a cytoskeletal gel is needed for membrane excitations and the gel changes its volume upon polarization and repolarization *(41)*.

Universality Class of Gels

Many systems undergo phase transitions. Water becomes ice below 0°C and boils at 100°C. Helium becomes superfluid below 2.7°K. A liquid crystal changes its structure among nematic, smectic, cholesteric, and isotropic phases. Polymers precipitate when solvent becomes poor. Associated to these phase transitions are the critical fluctuations in density, magnetization, composition, and orientations. One of the greatest achievements of modern physics is the understanding of such phase transitions and critical phenomena of various systems in a general way. In particular, phase transitions can be classified into a small number of groups called universality classes *(42)*. The classes are characterized by the set of critical exponents indicating the power relations among deviations of variables from the critical values *(43)*. Therefore, a good quesstion is to which universality class do gels belong.

Yong Li determined the various critical exponents on the poly(N-isopropylacrylamide) gel that undergoes the phase transition driven by hydrophobic interaction. The exponents were determined along the critical isobar.

measured quantity	critical exponent
specific heat	$\alpha = -0.05 \pm 0.15$
critical isobar	$\delta = 4.2 \pm 0.5$
coexistence curve	$\beta = 0.40 \pm 0.10$
density fluctuations	$\gamma = 0.90 \pm 0.25$
relaxation rate	$\nu = 0.45 \pm 0.07$
swelling or shrinking rate	$\nu = 0.45$

These critical exponents are consistent with the 3-D Ising system *(44)*. It is important to note that this question is not settled yet, since the critical exponents measured so far are all related to the volume or density of gels, but there are other order parameters such as off-diagonal strain components and also unresolved experimental studies of effects of shear modulus on the volume phase transition. Onuki provided a theoretical consideration on this problem *(45-46)*.

Diminishing of gel-solvent friction

According to the study of dynamic light scattering of acrylamide gel, the friction between the water and the polymer network decreases reversibly and diminishes as the gel approaches to a critical point or spinodal line *(23)*. The physical picture is that the density fluctuations become large as the critical point is approached, which create effectively large pores as determined by the correlation length. The correlation length diverges and the friction diminishes at the critical point. Tokita found the decrease of gel-solvent friction by three orders of magnitude in poly(N-isopropylacrylamide) gel in water *(47-48)*.

Fundamental biological interactions and gel phase transitions

Four fundamental interactions play an essential role in biological nature in determining the structures and specific functions of macromolecules and their assemblies: hydrogen bonding; hydrophobic interactions; van der Waals interactions; and ionic interactions *(49)*. Molecular recognition that occurs between DNA strands, antigen-antibody, receptor-hormone, enzyme-substrate, and actin-myosin filaments are all achieved by exquisitely coordinating these forces in space. The magnitude, temperature dependence, and behavior in aqueous environment are totally different among the fundamental forces, and it is those differences that allow the variety in the biological functions created by these forces. Understanding how these forces determine the phase behaviors of polymers is of fundamental importance.

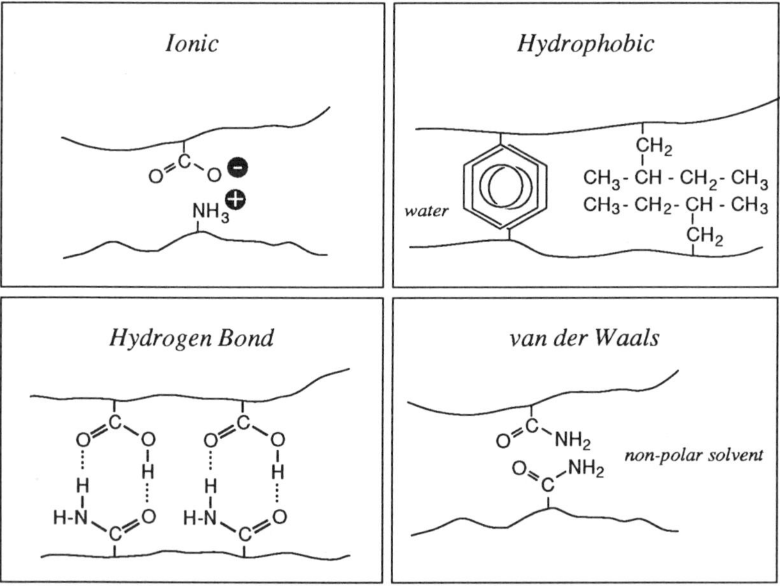

Figure 6. Four fundamental attractive interactions in the biological nature.

Gel phase transition is a result of a competitive balance between a repulsive force that acts to expand the polymer network and an attractive force that acts to shrink the network. The most effective repulsive force is the electrostatic interaction between the polymer charges of the same kind, which can be imposed upon a gel by introducing ionization into the network. The osmotic pressure by counter ions adds to the expanding pressure. The attractive interactions can be van der Waals, hydrophobic interaction, ion-ion with opposite kinds, and hydrogen bonding. The phase transition was discovered in gels induced by all the fundamental forces.

van der Waals interaction

The partially hydrolyzed acrylamide gel undergoes phase transition in acetone water mixtures *(26)*. The main polymer-polymer affinity is due to the van der Waals interaction. It was necessary to add acetone, a nonpolar poor solvent, to water in order to increase the attraction to a sufficiently large value to induce the transition. The transition is also observed when the temperature is varied while the solvent composition is fixed near the transition threshold at room temperature. The gel swells at higher temperatures and shrinks at lower temperatures.

Hydrophobic interaction

In an attempt to find a gel that undergoes a volume phase transition in pure water, rather than acetone-water mixture, Hirokawa studied gels which had side groups with more hydrophobicity than that of acrylamide *(50)*. In the figure below the volumes of copolymer gels of N-isopropylacrylamide and sodium acrylate in water are plotted as a function of temperature as determined by Hirotsu, Hirokawa, and Tanaka *(51)*. This gel is unique in that the solvent is a simple liquid in contrast to the previous cases where mixtures of solvents were necessary to observe the phase transition. Without ionizable sodium acrylate, the gel undergoes a discontinuous volume change by eight times at T=33.0°C. As the sodium acrylate concentration increases, the transition temperature rises and the volume change becomes larger.

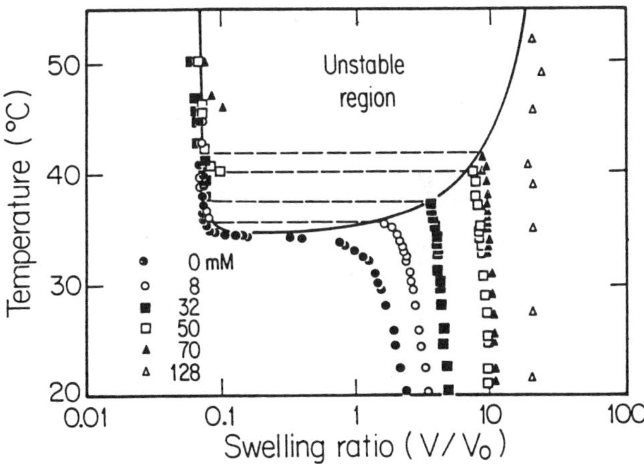

Figure 7. The volumes of ionized N-isopropylacrylamide gels in water as a function of temperature. As the ionization increases the volume change at the transition becomes larger and the transition temperature rises. (Reproduced with permission from reference *(51)*. Copyright 1987 American Physical Society)

It is interesting to observe that the gel swells at lower temperatures and collapses at higher temperatures. This temperature dependence, which is opposite to the transition induced by van der Waals interaction, is due to the hydrophobic interaction of the gel and water. The water molecules in the vicinity of hydrophobic polymer chains have more ordered structures, and thus a lower entropy, than those away from the polymers. At higher temperatures the polymer network shrinks and becomes more ordered, but the water molecules excluded from the polymer network become less ordered. As a whole the gel collapses amount to a higher entropy of the entire gel system as should be. More detailed theory and experiments have been carried out *(52-54)*.

Hydrogen bonding

We now look for a gel that undergoes phase transition in pure water, but in the temperature dependence opposite to that of hydrophobic interaction. Such a phase transition has been found in an interpenetrating polymer network (IPN) consisting of two independent networks intermingled each other *(55)*. One network is poly(acrylic acid) and the other poly(acrylamide). The gel was originally designed and developed by Okano and his colleagues *(56-59)*, who found that the gel is shrunken at low temperatures in water, and the volume increased as temperature rose. There was a sharp but continuous volume change at about 30°C. These researchers identified the main interaction to be hydrogen bonding and also pointed out the importance of the so-called "zipper" effect, which described the cooperative nature of the interaction between two polymers *(56)*. Such polycomplexation phenomena were studied extensively in solutions of various polymer pairs *(60-64)*.

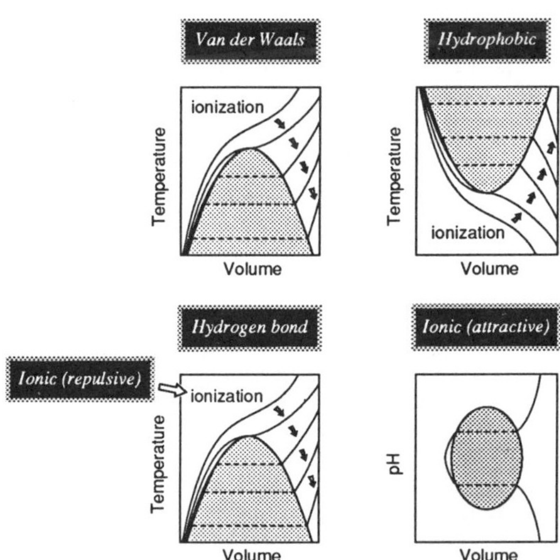

Figure 8. The phase transitions of gels induced by one of the fundamental attractive interactions in the biological nature. (Reproduced with permission from reference *(55)*. Copyright 1991 Macmillan Magazines Ltd)

By slightly ionizing the gel, Ilmain, Tanaka, and Kokufuta succeeded in inducing the discontinuous volume transition of the IPN in pure water *(55)*. The transition temperature was approximately 20°C and a large hysteresis was observed.

Electrostatic attraction
Polymers having both cationic and anionic groups are called polyampholytes. These polymers can be either positively or negatively charged, and repel each other over short ranges, but attract over long ranges *(65-69)*. Myoga and Katayama studied such a gel and observed that at neutral pH's the gel is indeed shrunken and at both higher and lower pH's *(70)* it is swollen. In the neutral pH's both cations and anions are ionized, and they attract each other, thus the gel shrinks. Otherwise the one of the ionizable groups is neutralized, while the other remains ionized, and the gel swells. The volume change was gradual and continuous.

Recently Annaka found a phase transition in a polyampholyte gel of copolymers of acrylic acid and methacrylamidopropylammonium chloride *(71)*. Phase transitions at higher pH and at lower pH are both accompanied by large pH hystereses.

These observations of phase transitions driven mainly by each one of the fundamental biological interactions allow us to draw a general picture of how polymers interact with each other through these interactions. It is important to note that each interaction is strong enough to induce a collapse but weak enough to allow swelling transition near the body temperature. The differences in the nature of each interaction are clearly revealed in the phase transition, particularly in the temperature dependence and in the size of hysteresis. Nature uses those differences to create extremely effective and specific molecular recognition mechanisms. It will be interesting to study the phase behavior of gels where combinations of these fundamental interactions are introduced and their balance is varied by changing variables such as temperature, solvent composition, and pH. We shall show in the next section that such a study is now revealing a fascinating new aspect in the phase behavior of gels.

New phases

We have seen that each of the four fundamental interactions induces a transition between two different phases in water near body temperature. What will happen if two or more interactions are combined within a gel? Would a gel, and in general a polymer system, have new phases other than just swollen or collapsed phases?

Such phases are known to exist in the biological world. For example, every protein has a unique and stable structure. Configuration of a natural protein must be at its free energy minimum separated by free energy barriers from other possible configurations. By thermodynamic definition the protein configuration should be a phase. Similarly an antibody forms a pair only with its antigen. Any other combination does not lead to a pair. Configuration of such a pair should thus be separated by the free energy barrier from other possible combinations. Again by definition the antibody-antigen should be a phase.

However, no such phases were observed in synthetic polymer systems. Only biopolymers, not synthetic polymers, may have such phases. It took billions of years for nature to select some 10^6 proteins for biological activities, whereas theoretically 20^{400} variety is possible for polypeptides assuming each consists of 400 amino acids of 20 kinds. The number of actual proteins is negligible compared to the theoretically possible number of polypeptides. Proteins could be an exception. However, new phases have been recently found by Annaka in copolymer gels consisting of cationic and anionic groups that form interpolymer hydrogen bondings *(71)*. As temperature or pH is varied, the gel changes its volume discontinuously in water among many phases distinguished by different volumes. The number of phases and the transition thresholds depend on the ratio of the cationic and anionic monomers in the polymer network. Each phase can be reached by following a different way of changing pH. New phases have also been discovered by Ishii in a gel made of homogeneous polymers where polymers are interacting through hydrogen bonding and hydrophobic interactions *(72)*.

The internal structures of these new phases are not yet understood. The finding demonstrated the possibility that the marvelous functions and structures that were considered to be available only to biopolymers, may be achieved by ordinary synthetic polymers. It is a democratization of polymers.

On the other side, the concept and principles of specificity, individuality, and diversity will be introduced into the polymer science. Macroscopic behaviors of polymers are far richer than one used to think and crucially depend on how the fundamental interactions are coordinated within the polymers.

Various stimuli induce phase transition

Gels undergo volume phase transition in response to different kinds of stimuli. Before the phase transition was found in gels, various researchers have developed gels that change their swelling degrees when a stimulus was applied to the gels. This article will describe only the systems that use the phase transition phenomenon.

Light sensitive gels

Gels have been designed and synthesized that undergo phase transition in response to light *(73-74)*. Two such reports have each demonstratied a different mechanism. One uses the photo-ionization effect. Mamada and colleagues synthesized copolymer gels of N-isopropylacrylamide (NIPA) and the photo-sensitive molecule, Bis(4-dimethylaminophenyl)-4'-vinylphenyl-methane-leucocyanide *(73)*. NIPA gels are thermosensitive gels and undergo phase transition in pure water in response to temperature. Without ultraviolet irradiation the gels underwent a sharp, yet continuous volume change, whereas upon ultraviolet irradiation they showed a discontinuous volume phase transition. For fixed appropriate temperatures the gels discontinuously swelled in response to irradiation of ultraviolet light and shrank when the light was removed. The phenomena were caused by osmotic pressure of cyanide ions created by ultraviolet irradiation.

The second example uses visible light absorption that increases the temperature locally within the thermosensitive gel *(74)*. The gel consists of a covalently cross-linked copolymer network of N-isopropylacrylamide and chlorophyllin, a combination of thermo-sensitive gel and chromophore. In the absence of light, the gel volume changes sharply but continuously as the temperature is varied. Upon irradiation the transition temperature is lowered, and beyond a certain irradiation threshold the volume transition becomes discontinuous. The phase transition is presumably induced by local heating of polymer chains due to the absorption and subsequent thermal dissipation of light energy by the chromophore. The results are theoretically interpreted using the Flory-Huggins equation of state of gels, assuming that the local temperature increment is proportional to the chlorophyllin density, thus proportional to the polymer density.

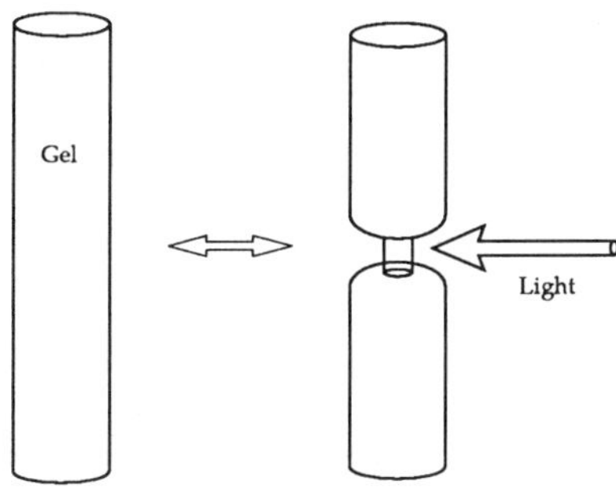

Figure 9. The phase transition of a gel induced by light. (Reproduced with permission from reference *(74)*. Copyright 1990 Macmillan Magazines Ltd)

Thermo-sensitive gels

Ito and colleagues synthesized various polymers and gels that collapse as temperature is increased using hydrophobic polymers *(75)*. This study allows production of a gel with any desired transition temperature. It is necessary to develop gels with various transition temperatures that swell upon increasing temperature, rather than decreasing temperature *(76-78)*. The gels with polymers interacting with hydrogen bonds and van der Waals interactions should have such a temperature dependence.

Thermosensitive gels are widely used for delivery systems *(79-82, 56-59)* and actuators *(83-86)*.

Solvent sensitive gels

Some gels have been synthesized that undergo the phase transition twice as the solvent composition is monotonically varied from 0% to 100%. Such reentrant behavior has also been observed when temperature or pH were used as variables.

Ion and pH sensitive gels

Gels are developed that respond to ionic and pH changes *(87-88)*. Siegel and colleagues have been extensively studying the pH sensitive gels, and opened the door to medical applications *(89-91)*.

Electric field sensitive gels

Some gels have been synthesized that undergo phase transition under an electric field. Before the phase transition was found the shrinking and swelling effect of an electric field was recognized and studied by several researchers *(92-95)*. Tanaka and colleagues found the phase transition in hydrolyzed acrylamide gel in 50% acetone water mixtures. Their original interpretation that the electrophoretic of polymer network might be responsible for the phase transition does not seem correct *(96)*. The most important effect seems to be the migration and redistribution of counter and added ions within the gel *(97)*.

Biochemically sensitive gels

Recently various schemes have been developed where a gel can undergo phase transition when a particular kind of molecule is present. This transition is achieved by embedding biochemically active elements such as enzymes or receptors within a gel that is placed near the transition threshold. When target molecules enter the gel, the active element converts them into other molecules or form a complex, which disturbs the equilibrium of the gel inducing the swelling or collapsing phase transition. Such a scheme has been extensively explored by various researchers *(56-59,79-82,98-100)*.

Kokufuta, Zhang and Tanaka developed a gel system that undergoes reversible swelling and collapsing changes in response to saccharides, sodium salt of dextran sulfate (DSS) and α–methyl-D-mannopyranoside (MP) *(101)*. The gel consists of a covalently cross-linked polymer network of N-isopropylacrylamide into which concanavalin A (ConA) is immobilized. At a certain temperature the gel swells five times when DSS ions bind to ConA due to the excess ionic pressure created by DSS. The replacement of DSS by non-ionic MP brings about a collapse of the gel.

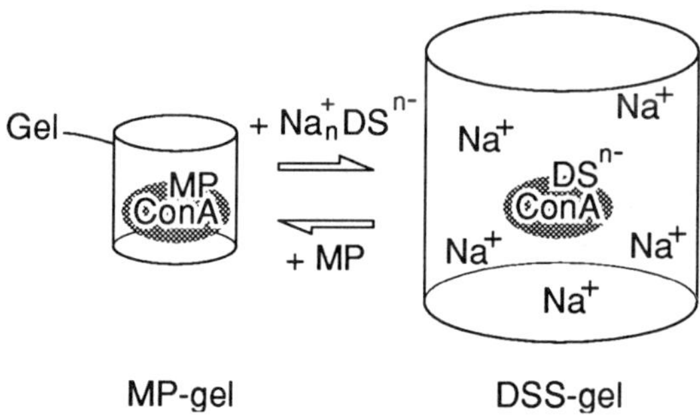

Figure 10. The phase transition of a gel in response to saccharides. (Reproduced with permission from reference *(101)*. Copyright 1991, AAAS)

The transition can be repeated with excellent reproducibility. In the next page Figure 11 shows how the swelling curve changes in the presence and absence of these saccharides.

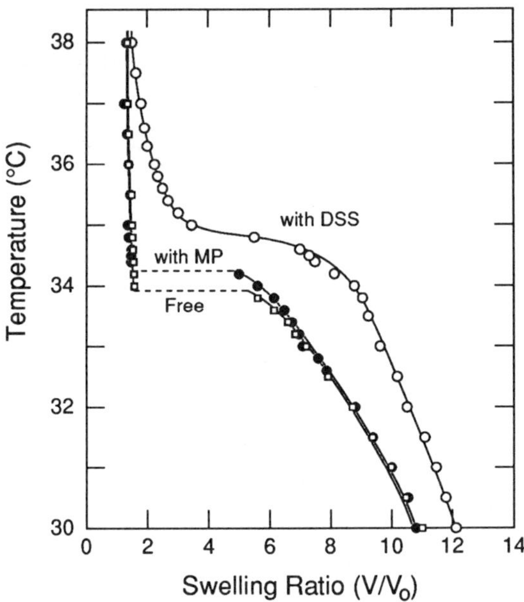

Figure 11. The swelling curves of the saccharide-sensitive gel in response to DSS and MP. (Reproduced with permission from reference *(101)*. Copyright 1991, AAAS)

We will soon be in a position to design at the molecular level a gel that responds to a desired stimulus under a specified condition.

Stress sensitive gels

Onuki *(40-41)*, Hirotsu and Onuki *(96)*, and Suzuki *(97)* have shown that the transition temperature is stress dependent.

Some future directions

Gel research is wide and deep. We have learned a great deal from the analogy of gels and other systems such as magnetic systems and gas-liquid systems. During the earlier stage of research, the detailed characteristics of gels were purposefully ignored to advance gel research into a rigorous physics. Such efforts indeed allowed a better and more general understanding of various aspects of gels.

However, gels have their own unique features. For example, gels have shear elasticity, that polymer solutions do not have. Because of the shear elasticity, for example, the kinetics of gel shrinking and diffusion of ink have a fundamental difference. It was mentioned earlier in this chapter that a gel swelling and shrinking obeys the diffusion equation. But it is actually necessary to add one more equation to fully describe the kinetic process, namely, the overall energy due to shear deformation should always be minimized, because the relaxation of shear deformation does not require relative motion between the network and water. This additional condition makes the diffusion process dependent not only on the size of a gel but on its shape. For example, Li predicted *(104)* that the relative values of effective diffusion coefficient of sphere : cylinder : slab = 1 : 2/3 : 1/3. This prediction was confirmed experimentally and interpreted as follows. A spherical gel diffuses in all the three dimensions. A cylindrical gel can diffuse only in the radial direction, since there should be no relative motion between water and network along the axis. The diffusion along the radius is shared by the axial shear relaxation and thus becomes two-thirds that of a spherical gel. A slab of gel can diffuse only along the thickness, not within the plain. The diffusion should be shared by two other dimensions, and the overall diffusion becomes one third of the sphere.

Because of the shear elasticity, a gel can have a wide variety of patterns during an extensive swelling or shrinking processes *(105)*. Permanent structural inhomogeneities are also specific to gels, and to identify their origin is an important task.

Universality classes of the phase transition of gels may well be dependent on the attractive interaction that drives the transition. For example, the hydrogen bond requires specific orientation for the two molecules, and thus new order parameters such as hydrogen bond density and orientation will be needed to describe the equation of state. Then the universality class of gels having hydrogen bonding as a key attractive interaction may be different from that of the hydrophobic gel, which is the only gel system whose universality class has been identified.

Finding principles associated with the uniqueness of gels is thus one of the key directions.

A second important direction for future research is to establish the principles underlying the individuality or specificity of gels. Namely, the principles should be established that predict the phase behavior of gels from the knowledge of their chemical structures. As shown in the recent studies of new phases of gels, it is clear that small changes in chemical composition induces a drastic change in the macroscopic phase properties of gels. This area has to be explored through combined efforts of experiments and theory *(107-109)*.

Finally, a rapid progress has been underway in technological applications of polymer gels, including chemical design, synthesis, fabrication in various length scales from μm to cm, and development of devices *(110-111)*. These applications are described in detail in the other portion of this book.

Literature Cited.

(1) DeRossi, D., Kajiwara, K., Osada, Y., and Yamauchi A., *Polymer Gels* Plenum Press, New York, 1991
(2) Yamauchi, A., (Editor), *Organic Polymer Gels (Japanese)*, Gakkai Publisher, Tokyo, 1990
(3) In this article only gels cross-linked by permanent chemical bondings will be considered. There are numerous reviews for physically cross-linked gels. Refer, for example *(4-7)*.
(4) Kramer, O. (Editor), *Biological and Synthetic Polymer Networks*, Elsevier Appl. Sci., London, 1988
(5) Clark, A. H. and Ross-Murphy, S. B., Adv. Polym. Sci., **1987**, *83*, 60
(6) Stepto, R.F.T. and Haward, R. N., *Development in Polymerization - 3. Network Formation and Cyclization*, Appl. Sci., London and New York, 1982

(7) Stauffer, D., *Introduction to Percolation Theory*, Taylor & Francis, London, 1985
(8) Masuda, F., *Superabsorbent Polymers (in Japanese)*, Kyoritsu Press, Tokyo, 1987
(9) Fushimi, T., *Ideas of Applications of Superabsorbent Polymers (in Japanese)*, Kogyo Chosakai, Tokyo, 1990
(10) Flory, P. J., *Principles of Polymer Chemistry* Cornell University Press, Ithaca, NY, 1957
(11) DeGennes, P. G., *Scaling Concepts in Polymer Physics*, Cornell University, Ithaca, New York, 1979
(12) Tanaka, T., *Sci. Am.* , **1981**, *244*, 110.
(13) Dusek, K. and Patterson, D., J. Polym. Sci., **1968**, *Part A-2, 6*, 1209.
(14) Ptitsen, O. B.. and Eizner, Yu. E., Biofizika, **1965**, *10*, 3
(15) Lifshitz, I. M., Grosberg, A. Yu., and Khokhlov, A., R., Rev. Modern Phys., **1978**, *50*, 683-713
(16) DeGennes, P. G., Phys. Lett., **1972**, *A38*, 339
(17) Tanaka, T., Hocker, L. O., Benedek, G. B., J. Chem. Phys., **1973**, *59*, 5151
(18) Munch, J. P., Candau, S. J., Duplessix, R., Picot, C., Herz, J., and Benoit, R., J. Polym. Sci. **1976**, *14*, 1097
(19) Geissler, E. and Hecht, A. M., Macromolecules, **1980**, *13*, 1276, , **1981**, *14*, 185
(20) Doi, M. and Edwards, S., *The Theory of Polymer Dynamics*, Oxford Univ. Press, Clarendon, 1980
(21) Onuki, A., Phys. Rev. A, **1989**, *39*, 2308
(22) Peters, A. and Candau, S. J., Macromolecules, **1986**, *19*, 1952, **1988**, *21*, 2278
(23) Tanaka, T., Ishiwata, S., and Ishimoto, C., Phys. Rev. Lett., **1977**, *39*, 474
(24) Tanaka, T.; Phys. Rev. Lett., **1978**, *40*, 820
(25) Hochberg, A., Tanaka, T., and Nicoli, D., Phys. Rev. Lett., **1979**, *43*, 217
(26) Tanaka, T., Fillmore, D.J., Sun, S.-T., Nishio, I., Swislow, G., and Shah, A., *Phys. Rev. Lett.*, **1980**, *45*, 1636
(27) Amiya, T. and Tanaka, T., Macromolecules, **1987**, *20* 1162
(28) Katayama, T. and Ohta, A., Macromolecules, **1985**, *18*, 2781
(29) Siegel, R. A. and Firestone, B. A., Macromolecules, **1988**, *21*, 3254
(30) Ilavsky, M., Macromolecules, **1978**, *59*, 5151
(31) Hrouz, J., Ilavsky, M., Ulbrich, K., and Kopecek, J., Europ. Polym. J., **1981**, *17* 361
(32) Ilavsky, M., Hrouz, J., and Ulbrich, K., Polym. Bull., **1982**, *7* 107
(33) Khokhlov, A. R., Polymer **1980**, *21*, 376
(34) Grosberg, A. Yu. and Khokhlov, A. R., *Statistical Physics of Macromolecules*, Nauka, Moscow, 1989
(35) Grosberg, A. Yu., Nechaev, S. K., Shakhnovich, E. I., J. de Phys. (Paris), **1988**, *49*, 2095
(36) Khokhlov, A. R. and Nechaev, S. K., Phys. Lett., **1985**, *112-A*, 156
(37) Verdugo, P., Biophys. J., **1986**, *49*, 231
(38) Verdugo, P., Annu. Rev. Physiol., **1990**, *52*, 157
(39) Shay, J. W., *Cell and Muscle Mobility*, Plenum Press, New York, 1984
(40) Tasaki, I., *Physiology and Electrochemistry of Nerve Fibers*, Academic Press, New York, 1982
(41) Arai, T. and Matsumoto, G., J. Neurochem., **1988**, *51*, 1825, **1989**, *52*, 93
(42) Wilson, K. G. and Kogut, J. B., Physics Reports, **1974**, *12c*, 77
(43) Stanley, H. E., *Introduction to Phase Transition and Critical Phenomena*, Oxford Univ. Press, Oxford, 1971

(44) Li, Y. and Tanaka, T., J. Chem. Phys., **1989**, *90*, 5161
(45) Onuki, A., Phys, Rev. A., **1988**, *38*, 2192
(46) Onuki, A., J. Phys. Soc. Japan, **1988**, *56*, 699 and 1868
(47) Tokita, M., Tanaka, T., J. Chem. Phys., **1991**, *in press*
(48) Tokita, M., Tanaka, T., Science, **1991**, *in press*
(49) Lehninger, A. L., I., *Principles of Biochemistry,* Worth Publishers, New York, 1982
(50) Hirokawa, Y. and Tanaka, T., J. Chem. Phys., **1984**, *81*, 6379
(51) Hirotsu, S., Hirokawa, Y., and Tanaka, T. J. Chem. Phys., **1987**, *87*, 1392
(52) Otake, K., Tsuji, T., Konno, M., and Saito, S., J. Chem. Eng. Japan, **1988**, *21*, 443
(53) Inomata, H., Yagi, Y., and Saito, S., Macromolecules, **1990**, *29*, 4887
(54) Ichita, H., Miyano, Y., Kiyota, y., and Nakano, Y., *Proc. Gel Conference in Tokyo*, Polymer Soc. Japan, pp.92-93, 1991
(55) Ilmain, F., Tanaka, T., and Kokufuta, E., Nature, **1991**, *349*, 400
(56) Okano, T., Bae, Y. H., and Kim, S.W., in *Modulated Control Release System, (Ed. by Kost, J.)*, CRC Press, *(in press)*
(57) Okano, T., Bae, Y. H., Jacobs, H., and Kim, S.W., J. Controlled Release, **1990**, *11*, 255
(58) Bae, Y. H., Okano, T., and Kim, S.W., Makromol. Chem. Rapid Commun., **1987**, *8*, 481, **1988**, *9*, 185
(59) Bae, Y. H., Okano, T., and Kim, S.W., J. Controlled Release, **1989**, *9*, 271
(60) Baranovsky, Yu. V., Litmanovich, A. A., Papisov, I. M., Kabanov, V.A., Europ. Polym. J., **1981**, *17*, 969
(61) Eustace, D.J., Siano, D.b., and Drake, E.N., J. Appl. Polym. Sci., **1988**, *35,* 707
(62) Osada, Y., J. Polym. Sci. Polym. Chem. Ed. **1979**, *17*, 3485
(63) Tsuchida, E. and Abe, K., *Interaction between Macromolecules in Solution*, Springer-Verlag, Berlin (1982)
(64) Abe, K. and Koide, M., Macromolecules, **1977**, *10*, 1259
(65) Edwards, S., King, P. R., and Pincus, P., Ferroelectrics, **1980**, *30*, 3
(66) Qian, C. and Kholodenko, A. L., J. Chem. Phys., **1988**, *89*, 5273
(67) Khokhlov. A. R., and Kachaturian, K. A., Polymer, **1982**, *23*, 1742
(68) Higgs, P., and Joanny, J-F., J. Chem. Phys., **1991**, *94*, 1543
(69) Kantor, Y. and Kardar, M., (to be published) show that a polyampholyte has a repulsive interaction in the short range.
(70) Myoga, A., and Katayama, S., Polym. Prep. Japan, **1987**, ***36***, 2852
(71) Annaka, M. and Tanaka, T., (to be published)
(72) Ishii, K. and Tanaka, T. (to be published)
(73) Mamada, A., Tanaka, T., Kungwatchakun, D., and Irie, M., Macromolecules, **1990**, *24*, 1605
(74) Suzuki, A. and Tanaka, T., Nature, **1990**, *346*, 345
(75) Ito, S., Collected papers of Kobunshi (in Japanese), **1989**, *46*, 437
(76) Hirasa, O., Koubunshi, **1987**, *35*, 1100,
(77) Otake, K., Inomata, H., Konno, M., and Saito, S., Macromolecules, **1990**, *23*, 283
(78) Kudo, S., Kosaka, K., Konno, M, and Saito, S., Polym. Prep. Japan, **1990**, *39*, 3239

(79) Hoffman, A. S., J. Controlled Release, **1987**, *6*, 297
(80) Hoffman, A. S., and Ratner, B. D., ACS Symposium Series, **1976**, *31*, 1
(81) Hoffman, A. S., and Afrasiabi, A., and Dong, L. C.,
J. Controlled Release, **1986**, *4*, 213
(82) Hanson, S. R., Harker, L. A., Ratner, B. D., and Hoffman, A. S.,
Biomaterials 1980, J. Wiley & Sons, London, pp 95-126, 1982
(83) Suzuki, M., and Sawada, Y., J. Appl. Phys., **1980**, *51*, 5667
(84) Suzuki, M., Tateishi, T., Ushida, T., and Fujishige, S., Biorheology, **1986**, *23*, 274
(85) Grodzinsky, A. J. and Melcher, J. R., IEEE Trans. Biomed. Eng., **1976**, *23*, 421,
(86) Huang, X., Unno, H., Akehata, T., and Hirasa, O.,
J. Chem. Eng. Japan, **1987**, *20*, 123, **1988**, *21*, 10, **1987**, *21*, 651
(87) Hirokawa, Y. and Tanaka, T., in *Microbial Adhesion and Aggregation*,
Edited by Marshall, K.C., Springer Verlag, Berlin, 1984
(88) Ricka, J. and Tanaka, T., Macromolecules, **1984**, *17*, 2916
(89) Ohmine, I. and Tanaka, T., J. Chem. Phys., **1982**, *77*, 5725
(90) Siegel, R. A., Faramalzian, M., and Firestone, B. A., and Moxley, B. C.,
J. Controlled Release, **1988**, *8*, 179
(91) Siegel, R. A. and Firestone, B. A., J. Controlled Release, **1990**, *(in press)*
(92) Osada, Y., *Advances in Polymer Science*, Springer-Verlag, Berlin,
1987, *82*, 1
(93) Osada, Y., Umezawa, K., and Yamauchi, A., Makromol. Chem., **1988**, *189*, 3859
(94) DeRossi, D., et al, Trans. Am. Soc. Artif. Inter. Organs, XXXII,
Makromol. Chem., **1988**, *189*, 3859
(95) Kurauchi, T., Shiga, T., Hirose, Y., and Okada, A., *Polymer Gels,* pp237-246,
Plenum Press, New York (1991)
(96) Tanaka, T., Nishio, I., Sun, S-T., and Ueno-Nishio, S., Science, **1982**, *218*, 467
(97) Giannetti, G., Hirose, Y., Hirokawa, Y., and Tanaka, T., in *Molecular Electronic Devices (Carter, F. L. et al Editors)*, Elsevier Appl. Sci., London (1988)
(98) Albin, G. W., Horbett, T. A., Miller, S. R., and Ricker, N. L.,
J. Controlled Release, **1987**, *6*, 267
(99) Dong, L. C.. and Hoffman, A. S., J. Controlled Release, **1986**, *4*, 213
(100) Park, T. G.. and Hoffman, A. S., Appl. Biochem. Biotech., **1988**, *19*, 1
(101) Kokufuta, E., Zhang, Y-Q., and Tanaka, T., Nature, **1991**, *351,* 302
(102) Hirotsu, S. and Onuki, A., **1989**, *58,* 1508
(103) Suzuki, A., (unpublished)
(104) Li, Y. and Tanaka, T., J. Chem. Phys., **1990**, *92,* 1365
(105) Onuki, A., *Formation, Dynamics, and Statistics of Patterns, (Edited by Kawasaki, K.)*, World Sci. Singapore, 1989
Sekimoto, K. and Kawasaki, K., J. Phys. Soc. Japan, **1987**, *56*, 2997
(106) Kantor, Y. and Kardar, M., (to be published) show that a polyampholyte has a repulsive interaction in the short range.
(107) Golubovic, L., and Lubensky, T., Phys. Rev. Lett., **1989**, *63*, 1082
(108) Shakhnovich, E. I., and Gutin, A. M., Europhys. Lett., **1989**, *8,* 327
(109) Goldbart, P., and Goldfeld, N., Phys. Rev. Lett., **1987**, *58*, 2676
(110) Umemoto, S., Itoh, Y., Okui, N., and Sakai, T.,
Reports on Progress in Polym. Phys. in Japan, **1988**, *31*, 295
(111) Umemoto, S., Okui, N., and Sakai, T., *Polymer Gels,* pp257-270,
Plenum Press, New York (1991)

RECEIVED September 12, 1991

SYNTHESIS AND CHARACTERIZATION

Chapter 2

Heterophase Synthesis of Acrylic Water-Soluble Polymers

D. Hunkeler[1] and A. E. Hamielec

Institute for Polymer Production Technology, Department of Chemical Engineering, McMaster University, Hamilton, Ontario L8S 4L7, Canada

A mechanism has been developed for the inverse-microsuspension polymerization of acrylic water soluble monomers. This free radical reaction scheme includes elementary reactions for polymerization in the aqueous and organic phases, nucleation in the monomer droplets and heterophase oligoradical precipitation, the latter being the dominant initiation process. A unimolecular termination reaction with interfacial species has also been elucidated and has been found to compete with and often dominate over the conventional bimolecular reaction. A kinetic model is developed which includes the influence of ionogenic monomers and polyelectrolytes and is able to predict the rate, molecular weight and composition data well for polymerizations of acrylamide and copolymerizations with quaternary ammonium cationic monomers. A categorization and systematic nomenclature for heterophase water-in-oil and oil-in-water polymerizations is also developed based on physical, chemical and colloidal criteria.

Acrylic water soluble polymers gain their utility from their large macromolecular size, ionic substitution and expanded configuration in aqueous solution. As such they are used primarily for water modification purposes and are typically employed as drag reduction agents (1), superabsorbents and in microemulsion form as pushing fluids (2).

Polyacrylamide is the ideal backbone for these supermolecular structures since it has the highest $k_p/k_t^{1/2}$ for any commercial olefinic monomer. These polymers are generally synthesized by free radical methods which yield large linear chain lengths, up to 10^7 daltons, without altering the side chain acrylate functionality (3). Anionic and cationic comonomers, such as sodium acrylate and quaternary ammoniums, are often incorporated at low compositions to yield polyelectrolyte flocculants (4). These ionogenic polymers are applied to the clarification of aqueous solid-liquid suspensions including mining wastes, potable water and paper suspensions.

[1]Current address: Department of Chemical Engineering, Vanderbilt University, Nashville, TN 37235

0097–6156/92/0480–0024$06.00/0
© 1992 American Chemical Society

The objective of this research is the development of a mechanism and kinetic model to predict the molecular properties, as a function of the synthesis conditions, for the heterophase polymerization of acrylic water soluble monomers.

Results and Discussion

The synthesis of acrylic water soluble polymers in aqueous solutions is limited by their high molecular weight and the large exothermicity of polymerization. These respectively lead to high viscosities and thermal instabilities. To overcome this problem we must employ either dilute solutions with an associated economic tradeoff in terms of inefficient reactor utilization or a heterophase water-in-oil polymerization process.

In this chapter the working definition of a heterophase polymerization will include only processes where differences in solubility of various phases exist at the outset of the reaction. That is to say only "bulk" or "macroscopic" effects will be considered. Microheterogeneities such as stereochemistry, chain composition and chain architecture, which are inherent in all free radically synthesized polymers due to the stochastic nature of propagation and termination reactions, will be neglected. This restriction allows for the identification of four independent polymerization regimes, as is illustrated in Figure 1. These are: I: Macroemulsion Polymerization; II: Inverse-Macroemulsion Polymerization; III: Microemulsion Polymerization; and IV: Inverse- Microemulsion Polymerization.

There are two primary criteria for process distinction. The first is a surface tension driving force, ΔY, where Y_{A-H} is the surface tension between the aqueous phase and hydrophilic moiety of the emulsifier and Y_{L-0} the surface tension between the lipophile and the organic phase. This surface tension driving force delineates the dispersion structure. If ΔY is less than zero an oil-in-water dispersion forms, whereas if ΔY is larger than zero a water-in-oil dispersion is produced. The second is a threshold emulsifier concentration also exists above which, for a given organic and aqueous phase, a thermodynamically stable "microemulsion" spontaneously forms. Below this threshold, which will also depend on temperature, monomer concentration and the chemistry of the emulsifier, thermodynamically unstable (kinetically stable) "macroemulsions" are produced. However, the prefix "macro" is often omitted for brevity which can lead to ambiguities in nomenclature. The difficulties in assigning a proper nomenclature will be demonstrated later in this chapter. The prefix "inverse" has historically been used for water-in-oil emulsions (5) in contrast to "direct" or "conventional" oil-in-water dispersions for which the prefix is implied but is not often explicitly stated.

Macroemulsions and Inverse-Macroemulsions can be further sub-categorized according to a secondary criterion - the level of surfactant with respect to the critical micelle concentration At surfactant concentrations below the cmc, a suspension of a dispersed monomer phase in a continuous media exists. Nucleation occurs predominantly in the monomer droplets, and each polymer particle therefore behaves as an isolated batch polymerization reactor. The continuous phase serves primarily to reduce viscosity and dissipate heat, although under certain circumstances it can contain the initiating species. The nomenclature "Suspension Polymerization" and "Inverse-Suspension" polymerization (6) have been appropriately employed. By contrast, at emulsifier levels above the micellar threshold, micelles or inverse-micelles are formed which may have a significant role in nucleation. These "emulsion" or "inverse-emulsion" polymerizations are distinguished by two phenomena: an average number of radicals per polymer particle on the order of 1 (suspensions have 10^{2-4} radicals/particle) and nucleation proceeding outside the monomer droplets. The heterophase polymerization domains can then be extended, as shown in Figure 2.

In suspension and inverse-suspension polymerizations, the monomer droplet size is inversely related to the emulsifier level. At high emulsifier concentrations the particle sizes are reduced to such an extent that the total interfacial area is large relative to the droplet volume. Under such conditions radical reactions within the interfacial layer

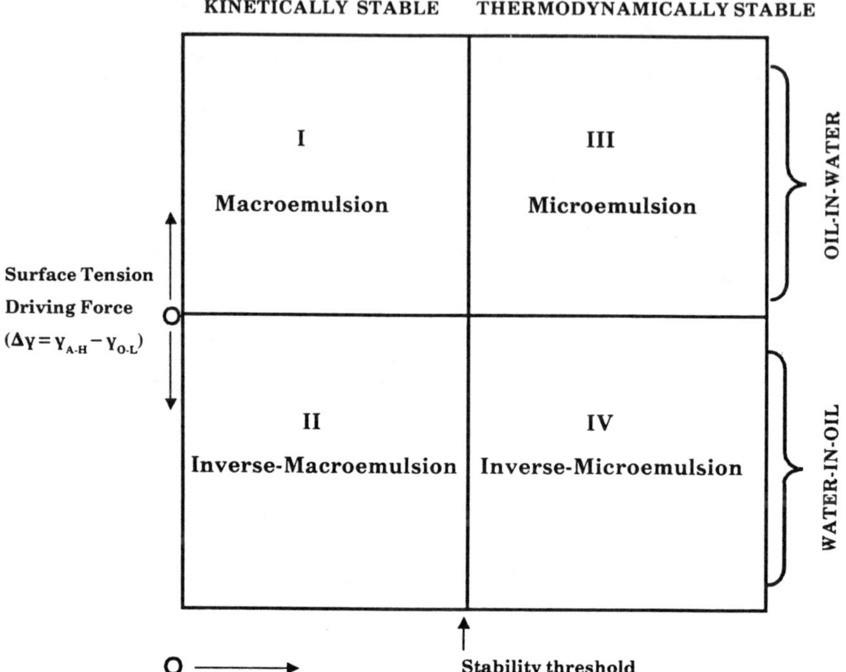

Figure 1. Categorization of heterophase polymerization domains.

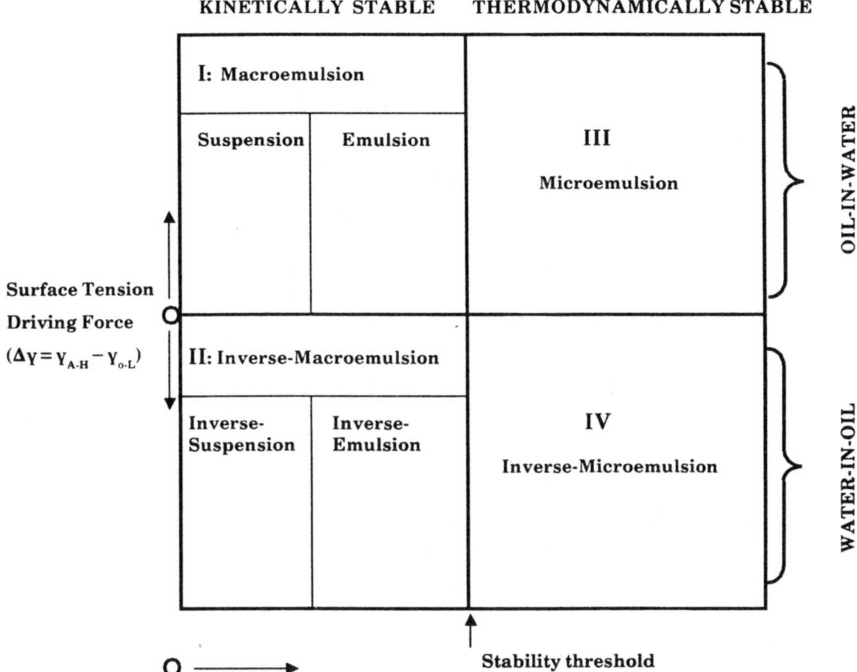

Figure 2. Categorization of heterophase polymerizations, including the macroemulsion subdomains.

become competitive with the propagation, termination and transfer reactions occurring within the monomer droplet. The onset of such phenomena is for diameters of 1-10 microns. To distinguish these processes from suspensions, which are characterized by radically inert interfaces, the nomenclature "microsuspension" has been proposed (7). The prefix "micro" designates the nominal particle size, and the suffix "suspension" confers the dominant characteristics of the synthesis. For water-in-oil polymerizations, the analogous process is "Inverse-microsuspension". As this is the most common synthesis method for high molecular weight acrylic water soluble polymers, it serves as a good basis for continued discussion.

Inverse-Microsuspension Polymerization

Inverse-microsuspension involves the dispersion of a water soluble monomer, in solution, in a continuous organic phase. The hydrocarbon can be either aliphatic or aromatic. These polymerizations are thermodynamically unstable "Inverse-Macroemulsions" (regime IIa by the preceding scheme), and to prevent coalescence require both continuous, vigorous, agitation and the addition of a low HLB steric stabilizer The monomer droplets are typically on the order of 1-10 microns and are heterodisperse in size due to their formation via a breakup-coalescence mechanism. A schematic of the polymerization is shown in Figure 3. The average particle size is governed by the interfacial tension which is controlled by the temperature and agitation conditions. Each droplet contains monomer, water, primary and macroradicals, and dead polymer chains. Inverse-micelles have not been detected with nucleation and polymerization proceeding within the monomer droplets. Initiation is most often carried out chemically with the free radical azo or peroxide species. The initiator can be located in either the dispersed (aqueous) or continuous (organic) phases. If water soluble initiators are employed, all the reactive species are contained in the aqueous phase and each monomer droplet behaves as an isolated microbatch polymerization reactor. Kinetically these processes resemble solution reactions (8). The localization of monomer and initiator in the same phase can, however, lead to secondary initiation reactions through monomer enhanced decomposition (9).

When the initiator and monomer are segregated (oil soluble initiator), we must consider the reactions in the organic and aqueous phases, as well as the interfacial layer.

Mechanism

The mechanism, which will be presented as an elementary reaction scheme in the following section, can be described as follows (7, 11): a chemical initiator decomposes in the continuous phase through thermal bond rupture and liberates two active radicals (step 1). These radicals can deactivate by reacting with either emulsifier, if radically active groups are present, or hydrocarbon (steps 2, 3). Since low efficiencies of initiation have been reported ($f = 0.02\text{-}0.2$; Ref. 7), these radical annihilation reactions are more probable than propagation with monomer which is marginally soluble in the organic media (step 4). A typical water soluble monomer such as acrylamide will, however, have a solubility in organic media of approximately 1% (step 8), and the solubility increases in the presence of emulsifier. Although this concentration is negligible from a mass balance perspective, this level is sufficient to have important kinetic consequences. Oligoradicals are formed (step 5) and as their chain length increases, their solubility limit in the organic phase is exceeded and precipitate across the interfacial boundary into the monomer droplets. Experimental evidence (7) has shown that the oligoradial precipitation is mass transfer controlled reaction (steps 6, 7). Once in the interior of the monomer droplet, the oligoradicals will undergo the normal chain addition, transfer and termination reactions (steps 9-12). Figure 4 illustrates these reactions.

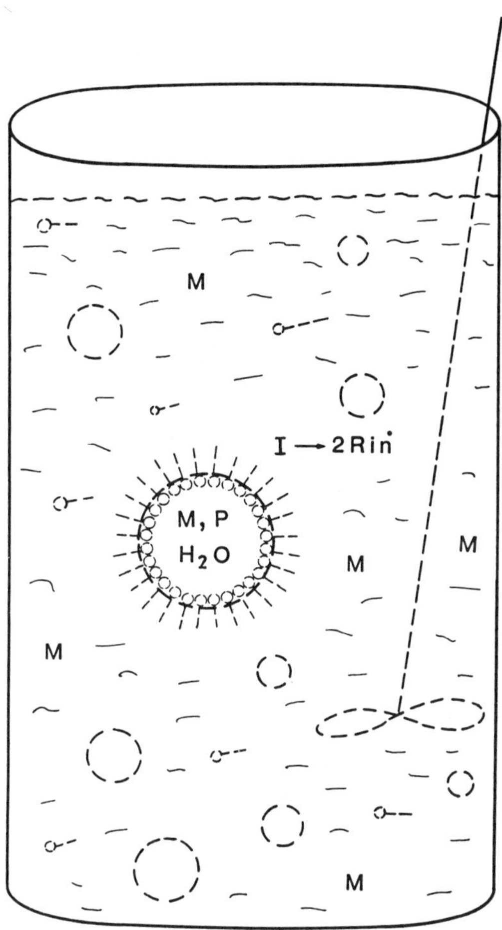

Figure 3. Schematic diagram of an inverse-microsuspension polymerization reactor. The circles designate monomer droplets, O--- is an emulsifier molecule, M a monomeric species, I an initiator, $R_{iN}^{\bullet}$ a primary radical, and P a dead polymer chain. An impeller is added at the right of the reactor to designate the thermodynamic instability of the system and the need for energy input to maintain emulsification.

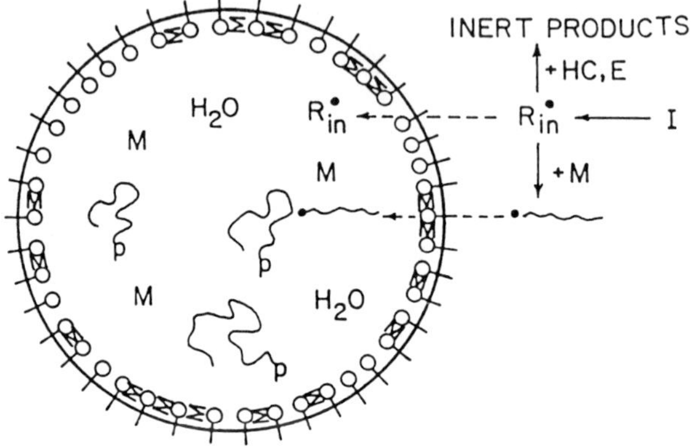

Figure 4. Elementary radical reactions in inverse-microsuspension polymerization. I, $R_{iN}^\bullet$, HC, E, M and P designate initiator, primary radicals, hydrocarbon, emulsifier, monomer and dead polymer respectively. The symbol O--- is an interfacial emulsifier molecule. A small fraction of the monomer is also observed within the interfacial layer. The dashed arrows indicate mass transfer between the continuous (organic) and dispersed (aqueous) phases

Until this stage the mechanistic discussion has been general for inverse-microsuspension polymerizations. However, there are several unique phenomena which have been observed for polymerizations in paraffinic media with fatty acid esters of sorbitan as emulsifiers. First, the transfer to monomer constant is lower for water-in-oil polymerizations than in solution. The magnitude is dependent on the type of organic phase and monomer, with certain acrylic monomers acting as cosurfactants (12, 13). In such an interfacial configuration, the olefinic groups are oriented toward the organic phase, and the α, β carbons are less accessible to radical attack. Secondly, unimolecular termination with the hydrophilic head of the emulsifier occurs if the water soluble moiety contains extractable groups. For example, sorbitanmonooleate has five labile hydroxy groups on the hydrophilic level (step 13). This interfacial reaction generates a dead polymer molecule and an emulsifier radical. The latter has two courses of reaction. The emulsifier radical can either heterodiffuse into the organic phase and be consumed by hydrocarbon radical scavenging species (step 15), or it can propagate with interfacial monomer (step 14). Experimental evidence has shown that the latter reaction consumes approximately twenty percent of emulsifier radicals, with the remainder deactivated in the oil phase (7). If propagation of the emulsifier radical occurs, the chain length of the macroradical will increase rapidly and the polymer will become predominantly water soluble. The macroradical will then be absorbed into the interior of the droplet where it can undergo the classical free radical reactions and eventually terminate. However, the macromolecule contains a terminal emulsifier unit. If the emulsifier contains greater than one radically active functional group, the molecule can undergo long chain branching reactions (step 16). These reactions are shown in Figure 5.

Based on the preceding discussion, the following elementary reaction scheme has been elucidated for the inverse-microsuspension polymerization of acrylic water soluble monomers with oil soluble initiators (10, 11).

Organic Phase Reactions:

1. $$I \xrightarrow{k_d} 2R^{\bullet}_{in,o}$$

2. $$R^{\bullet}_{in,o} + E \xrightarrow{k_1} \text{inert products}$$

3. $$R^{\bullet}_{in,o} + HC \xrightarrow{k_4} \text{inert products}$$

4. $$R^{\bullet}_{in,o} + M_o \xrightarrow{k_p} R^{\bullet}_{1,o}$$

5. $$R^{\bullet}_{r,o} + M_o \xrightarrow{k_p} R^{\bullet}_{r+1,o}$$

where I, $R^{\bullet}_{in}$, HC, E and $R^{\bullet}_{r}$ are the symbols for initiator, primary radicals, hydrocarbon, emulsifier, monomer and macroradicals, respectively. The subscript "o" designates an oil phase species.

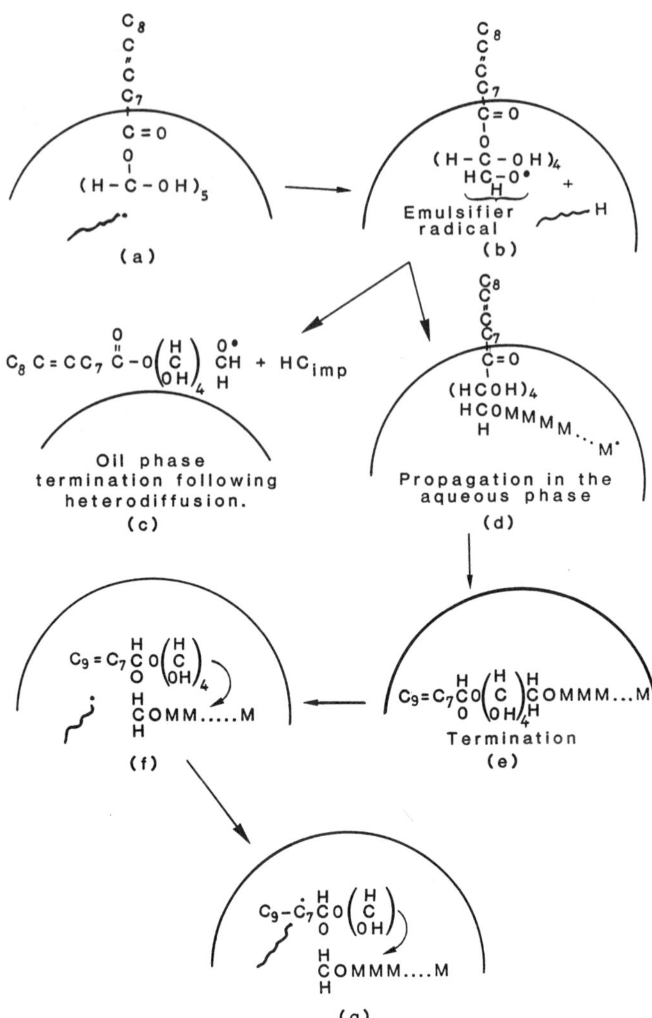

Figure 5. Interfacial reactions occurring within an inverse-microsuspension polymerization particle. Each figure shows a section of a particle, the region of higher curvature being the aqueous phase with a sorbitanmonooleate molecule situated at the interface. In figure 5a, a macroradical approaches the interfacial emulsifier and abstracts a hydrogen from the labile hydroxy group on the hydrophilic head of the emulsifier generating a dead polymer chain an emulsifier radical (5b). This emulsifier radical has two courses of reaction: heterodiffusion to the organic phase followed by termination with hydrocarbon (HC) radical scavenging species (5c) or propagation with interfacial monomer (5d). For the latter, as the chain length increases the polymer becomes predominantly water soluble and is absorbed into the interior of the droplet. There it undergoes the classical free radical propagation and transfer reactions and will eventually terminate (5e). If the terminal emulsifier group on the macromolecule contains additional radically active groups, such as unsaturated carbons or hydroxy groups, it will become a candidate for long chain branching reactions (5f). In such a reaction two linear chains join at the terminal group (5g). If all chains would undergo such a reaction it would be possible to double the average molecular weight of the final polymer.

Mass Transfer Between Phases:

6.

$$R^{\bullet}_{in,o} \xrightarrow{k_r} R^{\bullet}_{in,w}$$

7.

$$R^{\bullet}_{r,o} \xrightarrow{k_{r,r}} R^{\bullet}_{r,w}$$

8.

$$M_o \rightleftharpoons M_w$$

where the subscript "w" denotes a water phase concentration. $k_{r,r}$ is the mass transfer constant for an oligoradical of length r. $k_{r,r}$ will tend to k_r at small chain lengths.

Aqueous Phase Reactions:

9.

$$R^{\bullet}_{in,w} + M_w \xrightarrow{k_p} R^{\bullet}_{1,w}$$

10.

$$R^{\bullet}_{r,w} + M_w \xrightarrow{k_p} R^{\bullet}_{r+1,w}$$

11.

$$R^{\bullet}_{r,w} + M_w \xrightarrow{k_{fm}} P_r + R^{\bullet}_{1,w}$$

12.

$$R^{\bullet}_{r,w} + R^{\bullet}_{s,w} \xrightarrow{k_{td}} P_r + P_s$$

where P_r represents a dead polymer chain of length r.

Interfacial Reactions:

13.

$$R^{\bullet}_{r,w} + \Gamma_E \xrightarrow{k_{fE}} P_r + E^{\bullet}$$

14.

$$E^{\bullet} + M_w \xrightarrow{k_p} {}''R^{\bullet}_{1,w}$$

15.

$$E^{\bullet} + HC_{imp} \xrightarrow{k_5} \text{inert products}$$

where Γ_E is the surface emulsifier concentration, $E^{\bullet}$ is an emulsifier radical, HC_{imp} represents a hydrocarbon impurity or radical scavenging species and $''R^{\bullet}$ is the symbol for a macroradical with a terminal emulsifier group. Terminally unsaturated macromolecules are generated if the emulsifier contains a labile group(s) on the hydrophilic head so it may act as a unimolecular terminating agent for the macroradical (step 13). The latter are incorporated onto the terminal unit of the polymer chain through the propagation of emulsifier radicals (step 14). It is obvious that both chemically initiated macroradicals (R•) and those with terminal emulsifier groups ("R•) can undergo the classical free radical polymerization reactions (steps 9-12). Additionally, the emulsifier containing a macroradical can participate in long chain branching reactions, provided it has additional radically active functional groups (the first is consumed in the transfer to emulsifier reaction). These groups would include labile hydrogens and unsaturated carbons on either the hydrophilic or hydrophobic moiety of the molecule

Long Chain Branch Formation:

16. $"R^\bullet_{r,w} + "R^\bullet_{s,w} \xrightarrow{k^*} "R^\bullet_{r+s,w}$

$R^\bullet_{s,w} \qquad R^\bullet_{r+s,w}$

Kinetic Model

Initiation

Balances on primary radicals in the organic and aqueous phases:

$$\frac{d[R^\bullet_{iN,o}]}{dt} \simeq 0 = 2k_d[I] - k_p[R^\bullet_{iN,o}][M_o] - k_4[R^\bullet_{iN,o}][HC] - k_1[R_{iN,o}][E_o] - \frac{k_r}{V_o}\left(\frac{[R^\bullet_{iN,o}]}{\phi_r} - [R^\bullet_{iN,w}]\right) \quad (1)$$

$$\frac{d[R^\bullet_{iN,w}]}{dt} \simeq 0 = \frac{k_r}{V_w}\left(\frac{[R^\bullet_{iN,o}]}{\phi_r} - [R^\bullet_{iN,w}]\right) - k_p[R^\bullet_{iN,w}][M_w] \quad (2)$$

Where I_1, $R^\bullet_{iN}$, M and HC designate the initiator, primary radicals, monomer and hydrocarbon, respectively. The subscripts "o" and "w" designate organic and aqueous species, V_o is the volume of the organic phase, k_r a mass transfer constant of primary radicals between the organic and aqueous phases and ϕ_r the primary radical partition coefficient.

The partitioning of monomer between the continuous and dispersed phases is modelled by the following equation:

$$[M_o] = \phi_M [M_w]$$

where M_w is the monomer partition coefficient.

The rate of initiation (R_I) can be defined as the propagation rate of primary radicals with monomer (equation 3) or in terms of the efficiency of initiation and decomposition of the initiator (equation 4):

$$R_I = k_p[M_w][R^\bullet_{iN,w}] \quad (3)$$

$$\text{or } R_I = 2fk_d[I] \quad (4)$$

By combining equations (1) to (4) we can solve for the efficiency of initiation (f):

$$f = \frac{1}{1 + \frac{\phi_r}{k_r^*}\left[\frac{k_4[HC]}{a_{sp}} + \frac{k_1[E_o]}{a_{sp}}\right]} \cdot \frac{V_o}{V_w} \tag{5}$$

where Asp is the specific interfacial area, related to the total interfacial area (Ar) by:

$$Asp = A_T/V_o$$

and k_r is the overall primary radical mass transfer constant:

$$k_r = k_r^* A_T$$

Rate of Polymerization: Overall Macroradical Balance

$$\frac{d[R^\bullet]}{dt} \simeq 0 = R_I - k_{td}[R^\bullet]([R^\bullet] + [''R^\bullet]) - k_{fE}[R^\bullet]\Gamma_E + k_{fm}[M_w][''R^\bullet] \tag{6}$$

where $''R^\bullet$ is a macroradical with terminal unsaturation (terminal emulsifier group) and $R^\bullet$ is a macroradical with no terminal unsaturation. Γ_E is the interfacial emulsifier concentration.

Macroradical Balance on Chains with Terminal Emulsifier Groups

$$\frac{d[''R^\bullet]}{dt} \simeq 0 = k_p[E^\bullet][M] - k_{td}[''R^\bullet]([R^\bullet] + [''R^\bullet]) - k_{fE}[''R^\bullet]\Gamma_E - k_{fm}[M_w][''R^\bullet] \tag{7}$$

where

$$[R^\bullet] = \sum_{r=1}^{\infty} [R_r^\bullet] \tag{8}$$

$$[''R^\bullet] = \sum_{r=1}^{\infty} [''R^\bullet_r] \tag{9}$$

$$[R^\bullet_T] = [R^\bullet] + [''R^\bullet] \tag{10}$$

Emulsifier Radical Balance

$$\frac{d[E^\bullet]}{dt} \simeq 0 = k_{fE}[R_T^\bullet]\Gamma_E - k_p[E^\bullet][M_w] - k_5[HC_{imp}][E^\bullet] \tag{11}$$

Combining equations (6) to (11), a quadratic equation can be derived for the total radical concentration:

$$R_I + k_p[E^\bullet][M_w] - k_{td}[R^\bullet_T]^2 - k_{fE}[R_T^\bullet]\Gamma_E = 0 \tag{12}$$

If unimolecular termination with interfacial emulsifier is considered as the dominant macroradical consumption reaction, the following equation for the rate of polymerization (R_p) can be derived:

$$R_p = \frac{2 f k_d k_p [I][M]}{(1 - f_e) k_p \Gamma_E} \tag{13}$$

where f_e is the efficiency of initiation of emulsifier radicals and is given by:

$$f_e = \frac{1}{1 + \dfrac{k_5 [HC_{imp}]}{k_p [M_W]}} \tag{14}$$

where k_5 is a rate constant describing a reaction between emulsifier radicals and hydrocarbon impurities (HC_{imp}).

By comparison, experimental investigations have found the following rate dependence (Table 1):

$$R_p \propto [I]^1 [M]^1 [E]^{-a}$$

where a depends on the emulsifier.

Figure 6 shows a conversion-time plot for an acrylamide polymerization at different levels of AIBN (Azobisisobutryonitrile) initiator. The agreement between the kinetic model and experimental data are good for both the initial rate region, where the first order dependence is predicted, and the high conversion region. Similar agreement is observed in Figure 7 for the copolymerization of acrylamide and dimethylaminoethyl methacrylate, a quaternary ammonium monomer.

In Figure 8 the measured composition drift for an inverse-microsuspension copolymerization of acrylamide and dimethylaminoethyl methacrylate is compared with predictions based on reactivity ratios measured in solution polymerization. The data lie within the 95% confidence envelope, and we can conclude that inverse-microsuspension and solution polymerization are kinetically equivalent with respect to propagation. These data supports the hypothesis of nucleation and polymerization in monomer droplets.

Figure 9 shows the dependence of the rate of polymerization on the emulsifier concentration. As the emulsifier saturates the interface at low levels of sorbitanmonooleate, the addition of further emulsifier increases the concentration in the continuous phase. The excess emulsifier increases the consumption of primary radicals and lowers the efficiency of initiation (equation 5). A prior investigation has also shown that increasing the level of agitation generates a larger interfacial area (Asp) increasing both the efficiency (equation 5) and rate of polymerization (7).

The efficiency of initiation has also been observed to increase as the solubility of the acrylic water soluble monomer rises (18). In such cases a larger fraction of primary radicals are consumed by propagation (step 4) and a smaller amount by organic phase radical scavengers (steps 2, 3). The preceeding evidence is strong support for the oligoradical nucleation postulate.

The elucidation of monomolecular termination with emulsifier or the "reactive interface" is significant in two regards; it explains the kinetic observations of four independent laboratories who have reported a first order dependence of the polymerization rate on initiator concentration (Table 1) and has practical applications. For example, the kinetic model predicts that the linear molecular weight of a polymer can be increased up to 100% without reducing the latex stability by appropriately selecting the interfacial recipe. Normally, modifications to the recipe which improve stability, such as the addition of a cosurfactant, reduce molecular weight. Figures 10 and 11 show an increase in molecular weight with conversion when sorbitanmonooleate is used as an emulsifier. When sorbitanmonostearate is employed molecular weights are 30% lower. Sorbitanmonooleate

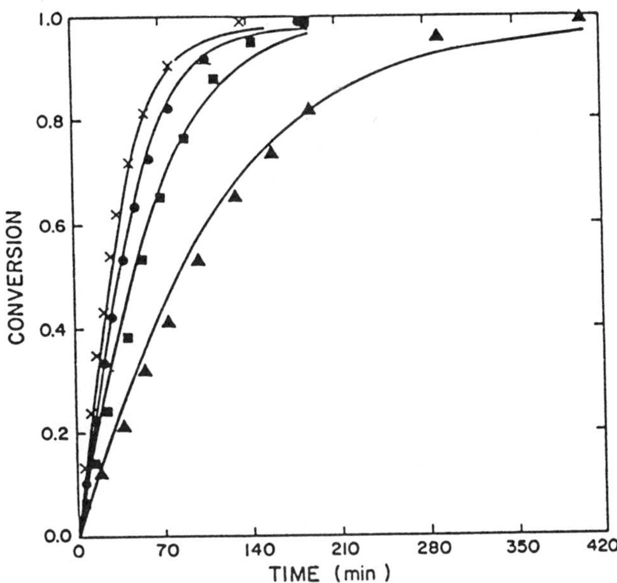

Figure 6. Conversion time data and kinetic model predictions (—) for an acrylamide polymerization with azobisisobutyronitrile as an initiator. The temperature was 47 °C, monomer concentration 5.75 mol L^{-1} of water, emulsifier concentration (sorbitanmonooleate) 0.103 mol L^{-1} of oil, aqueous to organic phase ratio 1:1, and agitation 1000 rpm. Initiator concentration 1.92 (▲), 4.02 (■), 6.03 (●), and 7.92 (×).

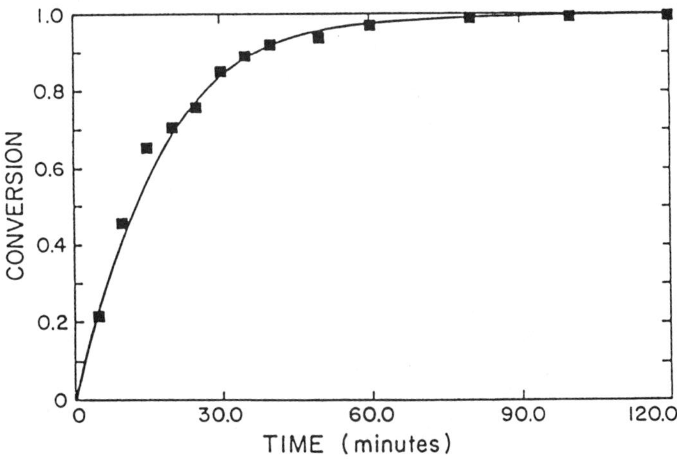

Figure 7. Conversion time data (■) and kinetic model predictions (–) for an acrylamide/dimethylaminoethyl methacrylate copolymerization at 60 °C. Experimental conditions [Monomer] = 6.90 mol/L_w, [Azobisisobutyronitrile] = 3.329 · 10^{-3} mol/L_0, f_{10} = 0.875, V_W/V_0 – 0.74 (10).

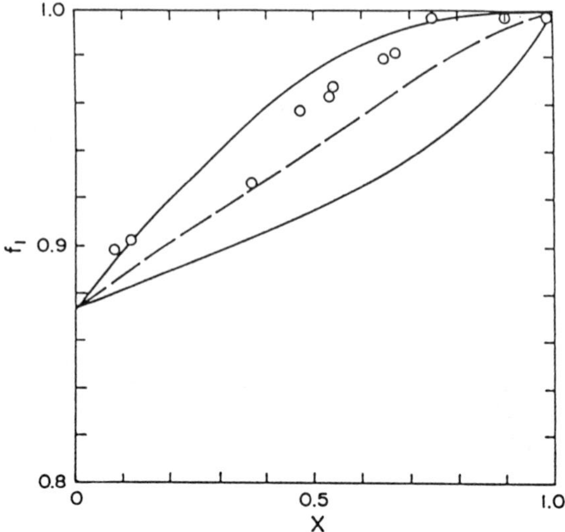

Figure 8. Drift in monomer composition with conversion for an acrylamide-dimethlaminoethyl methacrylate copolymerization at 50°C. (o): Experimental data measured by HPLC (17), (- - -): Predicted composition based on the reactivity ratios measured in solution, (—): 95% confidence limits based on the reactivity ratios measured in solution (10).

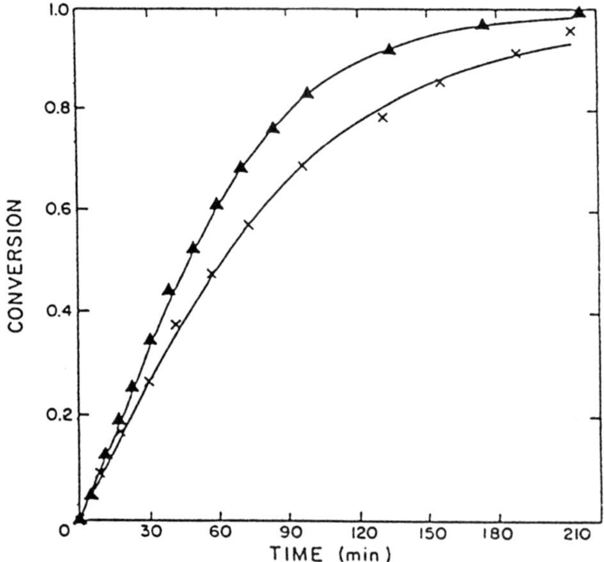

Figure 9. Conversion-time data and kinetic model predictions (—) for an acrylamide polymerization with sorbitanmonooleate at 0.103 (▲) and 0.211 (X) mol L^{-1} of oil. Other conditions are the same as in Figure 6, with an initiator level of 4.02 mmol L^{-1} of oil.

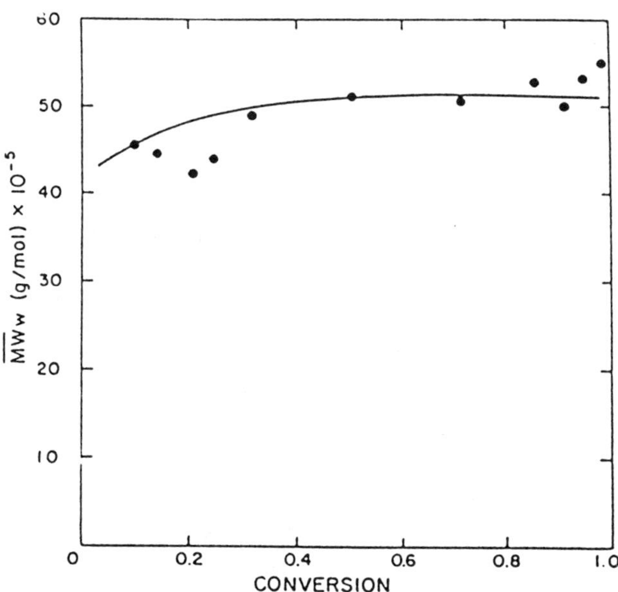

Figure 10. Experimental with average molecular weight (MW_w) data (●) and model predictions for a reaction under the following conditions: [Acrylamide] = 5.75 mol/L_w, [Sorbitanmonooleate] = 0.103 mol/L_0 [Azobisisobutyronitrile] = 4.02 mmol/L_0, V_W/V_0 = 1:1, Temperature 47 °C, Agitation 1000 RPM (7).

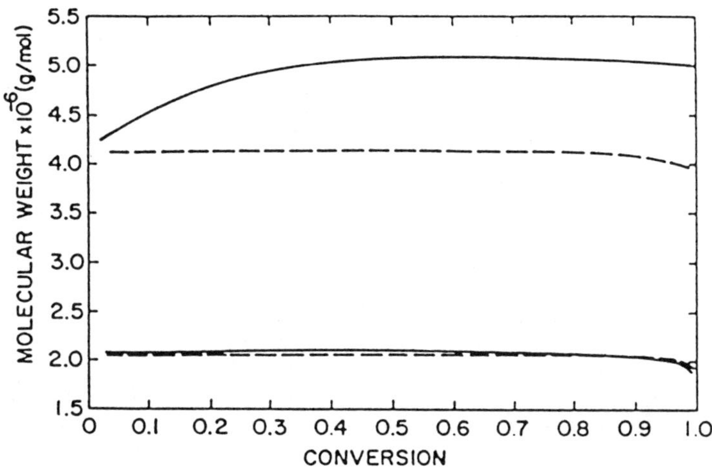

Figure 11. Number and weight average molecular weights for polymerizations with sorbitanmonooleate (—) and sorbitanmonostearate (- - -). Conditions are identical to those in Figure 6 (7), with an initiator concentration of 4.02 mmol/L^{-1} of oil.

Table 1. Reports of a first order dependence of the polymerization rate on the initiator concentration. All investigations were heterophase water-in-oil polymerizations of acrylamide

Investigator	Reference	"a" in $R_p \propto [I]^a$
Kurenkov	14	1.0
Baade	15	1.0
Hunkeler	7	0.95→0.99
Candau	16	0.99

and sorbitanmonostearate are chemically identical emulsifiers with the exception that the stearate does not contain an unsaturated carbon. This double bond is radically active and allows macromolecules with terminal oleate groups to undergo long chain branching reactions, whereas the stearate terminated molecules are inert. Such behavior is a strong indication that the monomolecular termination reaction, which manifests itself as a first order dependence of polymerization rate on initiator concentration, is indeed caused by transfer to emulsifier.

In summary, the monomolecular termination reaction influences the polymerization mechanism in two manners. A second termination step (13) exists which can compete with, and often dominate over, the conventional bimolecular radical annihilation reaction (step 12). Furthermore, the transfer product, an emulsifier radical, acts as a secondary source of initiation. This reaction generates up to 20% of the polymer chains and the effect is most pronounced at the outset of the reaction (7). Therefore, the first order rate dependence on initiator concentration is a consequence of modifications to both the chain initiation and termination mechanisms.

Conclusions

A mechanism has been developed for the inverse-microsuspension polymerization of acrylic water soluble monomers. This reaction scheme includes three previously unidentified reactions: the heterophase transfer of primary and oligoradicals; unimolecular termination of macroradicals with interfacial emulsifier; and a secondary initiation from emulsifier radicals which generates macromolecules which are candidates for long chain branching reactions. It has been demonstrated that by appropriately selecting the interfacial composition it is possible not only to influence latex stability but also the molecular properties of the macromolecule.

A mechanism or kinetic model for heterophase water-in-oil polymerizations must be specific to the polymerization domain. General models will have limited utility, and Inverse-Suspension, Inverse-Emulsion and Inverse-Microemulsion should be developed independently as has been the useful precedent for "direct" oil-in-water polymerizations. Furthermore, the physical and colloidal phenomena cannot be described without explicitly considering the unique chemistry of the reagents, particularily the type of organic phase and emulsifier.

Literature Cited

(1) Kulicke, W.M.; Kotter, M.; Grujer,M. *Advances in Polymer Science*, **1989**, *89*, 1.
(2) Meltzer, X.L. *Water Soluble Polymers*, Noyes Data Corporation, Park Ridge, New Jersey, 1979.
(3) Halverson, F.; Lancaster, J.E.; O'Connor, M.N. *Macromolecules*, **1985**, *18*, 1139.
(4) Klenina, O.V.; Fomina, V.I.; Klenin, V.I.; Avetisyan, P.K.; Medvedev, G.P.; Klenin, S.I.; Bykova, YE.N.; Milovskaya, YE.B. *Vysokomol. soyed.*, **1984**, *A26*, 271.
(5) Vanderhoff, J.W.; Bradford, E.B.; Tarkowski, H.L.; Shaffer, J.B.; Wiley, R.M. *Adv. Chem. Ser.*, **1962**, *34*, 32.
(6) Dimonie, M.V.; Boghina, C.M.; Marinescu, N.N.; Marinescu, M.M.; Cincu, C.J.; Oprescu, C.G. *Eur. Polym. J.*, **1982**, *18*, 693.
(7) Hunkeler,D.; Hamielec, A.E.; Baade, W. *Polymer*, **1989**, *30*, 127.
(8) Hunkeler, D. *Macromolecules* (accepted for publication).
(9) Hunkeler, D. "Mechanism and Kinetics of the Persulfate Initiated Polymerization of Acrylamide", *Water Soluble Polymers*, S. Shalaby, G.B. Butler, C.L. McCormick, eds., American Chemical Society, Washington, D.C., 1991.
(10) Hunkeler,D.; Hamielec, A.E. *Polymer* (accepted for publication).
(11) Hunkeler, D. *British Polymer Journal* (accepted for publication).
(12) Holtzcherer, C.; Candau, F. *Colloids and Surfaces*, **1988**, *29*, 411.
(13) Holtzcherer, C.; Durand, J.P.; Candau, F. *Colloid and Polymer Science*, **1987**, *265*, 1067.
(14) Kurenkov, V.F.; Verizhnikova, A.S.; Myagchenkov, V.A. *Doklady Akademii Nauk SSR*, **1984**, *278*, 1173.
(15) Baade, W.; Reichert, K.H. *Eur. Polym. J.*, **1984**, *20*, 505.
(16) Carver, M.T.; Dreyer, U.; Knoesel, R.; Candau, F.; Fitch, R.M. *J. Polym. Sci. Pt. A: Poly. Chem.*, **1989**, *27*, 2179.
(17) Hunkeler, D.; Hamielec, A.E.; Baade, W. "Polymerization of Cationic Monomers with Acrylamide", in *Polymers in Aqueous Media*, J.E. Glass, ed., American Chemical Society, Washington, D.C., 1989.
(18) Hunkeler, D. Ph.D. Dissertation, McMaster University, Hamilton, Canada 1990.

RECEIVED August 29, 1991

Chapter 3

Structure and Properties of Surfactant-Bridged Viscoelastic Hydrogels

Abdulkadir J. Dualeh[1] and Carol A. Steiner

Department of Chemical Engineering, City College of the City University of New York, 140th Street and Convent Avenue, New York, NY 10031

Surface-active graft copolymers form viscoelastic hydrogels in the presence of unmicellized surfactant. These gels have been characterized with respect to their volume, rheological properties, and the nature and structure of the intermolecular linkages as a function of composition and temperature. The linkages are found to resemble comicelles composed of surfactant molecules and hydrocarbon side chains from the polymer. The composition of the comicelles varies in a predictable manner with that of the bulk solution, by analogy with comicelles of monomeric surfactants. This correlation in turn governs the bulk properties of the network.

Interest in the solution properties of surface-active graft copolymers has grown in recent years with the synthesis of new such polymers and the recognition that these polymers may interact with other components of a solution, such as surfactants and dispersed particles, in a unique and potentially useful manner. On their own in solution, water-soluble surface-active graft copolymers may aggregate either inter- or intramolecularly, either increasing or reducing the apparent intrinsic viscosity of the macromolecules (*1–11*) depending on the nature of the solvent and whether or not the critical overlap concentration has been reached. In the presence of suspended particles, these polymers adsorb at the solid/liquid interface, promoting steric stabilization of the suspension (*12*). In surfactant solutions, the hydrophobic moieties of surface-active graft copolymers act as nucleation sites for adsorption of surfactant aggregates. This adsorption can result in either the break-up or enhancement of intermolecular hydrophobic interactions, depending on the composition of the solution and hence of the adsorbed aggregates (*12–16*).

[1]Current address: Ciba–Geigy Corporation, MR/2, Saw Mill River Road, Ardsley, NY 10502

0097–6156/92/0480–0042$06.00/0
© 1992 American Chemical Society

Besides having unique solution properties, surface-active graft copolymers also have the ability to form stiff, water-swollen hydrogels at polymer levels as low 0.6% by weight (*13, 14*). These gels may be either charged or nonionic, depending on the structure of the starting polymer and the composition of the solution from which the gels form. The synthesis, structure, and properties of a group of hydrogels formed from one particular surface-active graft copolymer will be the subject of this report.

Hydrogels in general, and polyelectrolyte gels in particular, are valuable in many industrially important applications, including superabsorbent technology, chromatographic separations, and controlled drug delivery devices. The various properties exploited in these applications are, respectively, their high swelling ratios, preference for retaining specific solutes, and ability to act as reservoirs for drugs. Any improvement in the method of preparation or the discovery of other polymer structures which behave like polyelectrolyte gels is of fundamental interest.

Polyelectrolyte gels are generally made by covalently crosslinking the polymer chains with bifunctional monomers. Alternatively, and of greater relevance to the present study, polymeric hydrogels can be made without covalently linking the polymer chains. These networks are held together physically by secondary forces (van der Waals, hydrogen bonding, etc.) among some segments of the polymer chain. The linkages are weak compared to covalently linked gels and can thus be made and broken relatively easily by gentle changes in the environment. At least two conditions are required in order to form gels of this type from aqueous solutions. First, some segments of the polymer must prefer to associate with themselves more than with water molecules or other segments of the polymer. Second, the bonds formed must be strong enough to hold the chains together despite thermal motion of the chains. In some cases, however, favorably interacting segments may be prevented from doing so by physical limitations imposed by the structure of the chain. Factors such as chain flexibility and tacticity contribute strongly. It is known, for example that tacticity plays an important role in the ability of polymethacrylic acid to form gels. All of these factors must be taken into consideration in selecting a starting polymer for the formation of non-covalently bonded hydrogels.

We have made mechanically stable hydrogels by contacting hydrophobically modified hydroxyethyl cellulose (HMHEC), an amphiphilic nonionic copolymer, and the surfactant sodium dodecyl sulfate (SDS) in aqueous solution. These hydrogels form spontaneously at certain specific concentrations of copolymer and surfactant, without any chemical reaction to crosslink the polymer chains. The materials exhibit rheological behavior similar to that of covalently crosslinked polyelectrolyte gels; however they respond differently to changes in ionic strength and temperature. These properties, coupled with the mechanical stability of the gels, make these new materials useful for industrial exploitation

Experimental

The polymer hydrophobically modified hydroxyethylcellulose (HMHEC), MW 10^6, was obtained from Aqualon Co., Wilmington, DE. Its main chain, hydroxyethyl cellulose (HEC), is composed of anhydroglucose units rendered water soluble by addition of ethylene oxide (EO) groups. The HEC is then modified with C_{12} alkyl grafts at sufficiently high level to render the polymer insoluble in water. Sodium dodecyl sulfate (SDS), 99% pure, was obtained from Fluka and used as received. The critical micelle concentration was found by conductivity to be 8×10^{-3} M. Sample preparation and methods are detailed elsewhere (*14*).

The bulk properties of the gels were investigated using dynamic rheological measurements and observation. Microscopic investigation of the gel structure was carried out using two different fluorescence techniques, described elsewhere and below. The first, fluorescence of pyrene, makes use of the fact that certain characteristics of the fluorescence spectrum of pyrene provide a quantitative measure of the hydrophobicity of the pyrene's environment (*17*). Using this method we were able to establish that our hydrogels are characterized by distinct hydrophobic regions within an aqueous network. The second method involves measuring the extent of interaction between a fluorescence probe and quencher which are chosen because of their relative insolubility in water, so that quenching only takes place within the hydrophobic regions of the gel. The resulting fluorescence spectra provide a measure of the size, or aggregation number, of these regions, if they exist, according to the following relation (*18*):

$$\ln (I_o/I) = [Q]/[M] \quad (1)$$

where I_o and I are the fluorescence emission intensity of the probe without and with quencher, respectively, [Q] is the quencher concentration and [M] is the concentration of aggregates in the sample. The slope of this plot is thus the reciprocal of the molar concentration of aggregates, and the aggregation number is given by [M] divided by the total moles of aggregated species, e.g. surfactant molecules and hydrophobic side chains. The number of alkyl grafts and surfactant molecules making up each aggregate can be obtained from:

$$N_s = ([surf]_o - [surf]_{free})/[M] \quad (2)$$

$$N_a = ([alkyl]_o - [alkyl]_{free})/[M] \quad (3)$$

where $[surf]_o$ and $[alkyl]_o$ are the total surfactant and alkyl graft concentrations in the gel, respectively, and $[surf]_{free}$ and $[alkyl]_{free}$ are, respectively, the surfactant and alkyl concentrations in the gel that are not part of an aggregate, and can be estimated as detailed elsewhere (*14*).

Results and Discussion

A disagreement often occurs about what is and is not a "true gel." Some researchers insist that a true gel must be chemically crosslinked in a permanent network, others look for a long plateau in the storage modulus vs frequency spectrum, and still others consider a gel to be a mass of polymer that swells isotropically without dispersing in an excess of solvent. Our gels fall into the last category. The gels which we will be discussing here are one piece, viscoelastic, flexible, clear and macroscopically homogeneous. These gels fill the container in which they are formed, but retain the shape of the container (typically a right circular cylinder) for at least several days after removal from the vessel. The swollen volume of the gels depends on the composition of the solvent; however in some solvents their volume is remarkably stable, even upon immersion for a period of several days. The rheological properties of the gels are shear rate dependent, but a long frequency range exists over which the storage modulus, while not exactly constant, varies only slightly.

Effect of Solution Composition. HMHEC gels may or may not form depending on the initial composition of the solution. Upon addition of HMHEC to pure water, in the absence of surfactant, the copolymer precipitates out of the solution in the form of individual swollen globules. Above 1×10^{-3} M SDS and below the critical micelle concentration of the surfactant, mixtures of HMHEC and SDS form a polymer-rich gel phase and a supernatant phase composed primarily of surfactant and water. At surfactant concentrations near the cmc, the solutions are one phase but exhibit recoil on pouring. The solutions remain clear on further addition of SDS but cease to exhibit recoil on pouring and can pass through a filter. At higher surfactant levels no gel phase forms; all the polymer is soluble in the surfactant solution. In contrast, HEC is quite soluble in both water and SDS, forming one- phase viscous or non-Newtonian solutions at concentrations comparable to those at which a gel formed with HMHEC. Thus it is clear that the HMHEC/SDS gels are characterized by linkages involving both the hydrophobic side chains and the surfactant. The pyrene fluorescence experiments demonstrated that these linkages, or some other component of the network, are significantly more hydrophobic than if the network were homogeneous. It was our aim to determine the nature of polymer and surfactant interactions that give rise to these hydrophobic regions and to network formation.

Typical rheological behavior of the gels is shown in Figure 1, showing the storage (G′) and loss (G″) moduli as a function of frequency for a HMHEC gel made from a solution containing 5.6 mM SDS and 0.04 weight percent polymer at 25 °C. In contrast to chemically cross-linked systems, which are characterized by a true plateau over a long range of frequency in the G′ vs frequency spectrum, we observe a shallow but finite slope in G′ over the frequency range 10–60 rad/s. It can also be seen that over the frequency range studied, G′ is everywhere greater than G″, i.e. no terminal zone is reached. The behavior seen here is characteristic of a temporary network of interactions whose stability depends on the stresses to which the interactions are subjected.

In chemically cross-linked systems, the plateau value of the storage modulus (G_n) is inversely proportional to the molecular weight between crosslinks (M_x) (*19*). By analogy, we can consider the molecular weight between our temporary linkages, $M_l(\omega)$, to observe the same inverse relationship with the value of G′ at that frequency, $G'(\omega)$. $G'(\omega)$ was obtained from our data at 30 and 50 rad/s by averaging the 6 values of G′ spanning each of those frequencies. This procedure was used to minimize the effect of random scatter, which became considerable at the high end of the frequency scale. At constant polymer concentration and applied frequency, the storage modulus decreases with increasing surfactant concentration as shown in Figure 2; thus some of the intermolecular linkages must disappear as surfactant is added to the system.

Based on all of the above methods of analysis and observation of HMHEC gels made with sodium dodecyl sulfate (SDS), we conclude that the structural element holding the gel together takes the form of adsorbed aggregates resembling comicelles consisting of both SDS and hydrocarbon side chains from the polymer. When HMHEC is placed in contact with SDS and water, the side chains act as nucleation sites for the adsorption of surfactant, thereby shielding themselves from unfavorable interactions with the water. While we have not as yet been able to characterize the distribution of comicelle sizes which obtain at a particular bulk composition, the average size is within the expected range for comicelles of SDS and monomeric surfactants such as poly(oxyethylene) lauryl ethers, with structures similar to that of the side chains on our polymer (*20*).

The data present a consistent picture. Above the critical micelle concentration (cmc) of the surfactant, when the bulk SDS concentration is high enough to form pure SDS micelles, the polymer is readily solubilized and no gel forms. Below the cmc, a stiff gel forms whose storage modulus is an inverse function of the total surfactant concentration, so the extent of intermolecular interaction increases as the surfactant concentration goes down. This relationship is also reflected in the fact that the total volume of the gel phase increases with SDS concentration below the cmc, up to the limit where the polymer forms a one-phase solution at or just above the cmc. Thus far below the cmc, where pure SDS micelles cannot form, side chains from the polymer become incorporated into comicelle-like aggregates which bridge macromolecular chains, resulting in the formation of a network. At higher surfactant concentrations, more total aggregates can form, so the average number of side chains per aggregate goes down. This reduction results in a higher molecular weight between linkage points and the observed lower G′ and higher gel volume (Figure 3). We have been able to verify using the fluorescence quenching

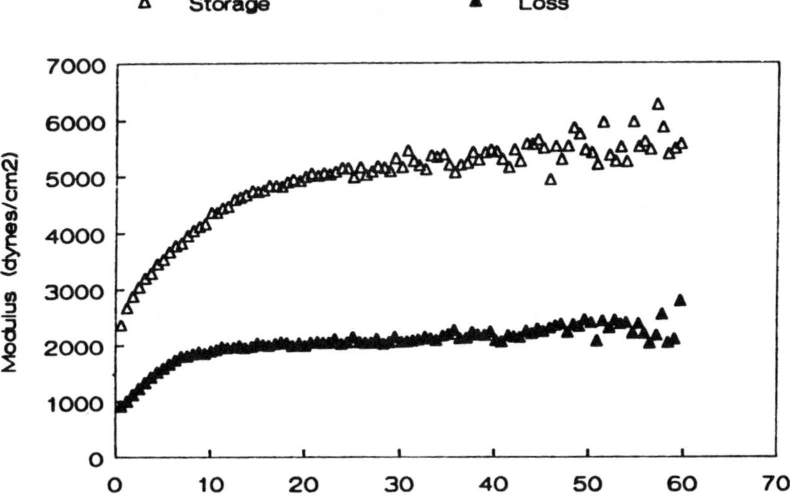

Figure 1. Typical dynamic response of hydrogels. Δ, storage modulus, G′; ▲, loss modulus, G″.

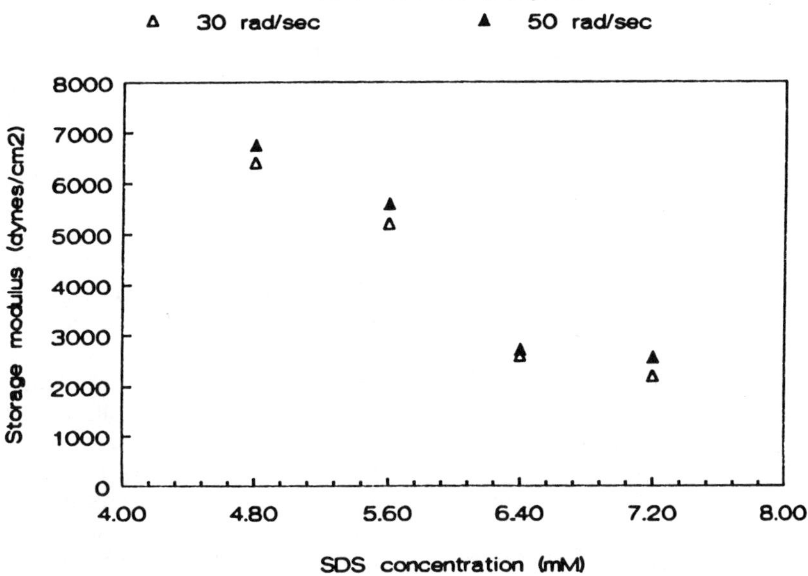

Figure 2. Effect of SDS concentration on the storage modulus (G′) of the hydrogels. Δ, G′(30); ▲, G′(50).

technique described above that the number of alkyl grafts in an aggregate lies on a smooth curve when plotted against the storage modulus evaluated at 30 rad/sec (Figure 4). Thus, strong evidence exists of the intimate relationship between the structure of the network and the structure of the aggregates.

Effect of Temperature. Preliminary experiments have been conducted to determine the effect of temperature on gel volume. As noted above, the gel volume gives a good indication of the extent of intermolecular interaction in the system, and comparison of the results with what is known about the behavior of monomeric surfactant comicelles will provide independent confirmation of our hypothesis on the nature of these interactions. We used two methods of preparing the gels to study the effect. In method 1, a graduated cylinder containing a preformed gel (gel made at 25 °C) in contact with its supernatant was immersed in a constant temperature bath at 50 °C and the gel volume monitored. It took about 5 hours for the gels to attain a fixed, final volume. In method 2, the polymer was added to a preheated (50 °C) surfactant solution and maintained at 50 °C. The solutions were mixed for the same period as those made at room temperature and then allowed to stand and equilibrate in the constant temperature bath, leading to the formation of gels. The final gel volume was generally reached in less than 1 hour, which is also the typical equilibration time for gels mixed up and equilibrated at 25 °C.

At each of two different solution compositions studied, we found repeatedly that the final gel volume was independent of the method of preparation. This finding points to the existence of a thermodynamically determined final gel volume. It also demonstrates that the polymer chains in the preformed gel are mobile, albeit constrained, and can rearrange to their equilibrium configuration at ambient conditions given sufficient time. The discrepancy in equilibration times for the two methods of preparation is a reflection of the degree to which the chains are constrained in the gel.

The effect of raising the temperature of HMHEC hydrogels in 4.8 mM SDS or 5.6 mM SDS from 25 °C to 50 °C is to decrease the equilibrium gel volume by 7.9% or 7.5%, respectively. The preformed gels visibly shrink when heated, and the final volume is stable for a period of at least several days.

We have seen that the gel volume is inversely related to the number of crosslinking hydrophobic aggregates in the network. Recall, too, that the number of linkages goes down with increasing bulk surfactant concentration, so that not all surfactant aggregates act as linkages. Put another way, as the amount of surfactant available to form aggregates is increased, both the total number of aggregates and the ratio of surfactant to side chain in each aggregate increases, giving rise to fewer crosslinking hydrophobic aggregates overall in the network and a higher gel volume. In light of this, and given the known effect of temperature on the cmc of ionic surfactants, we offer the following explanation for the observed decrease in gel volume with temperature.

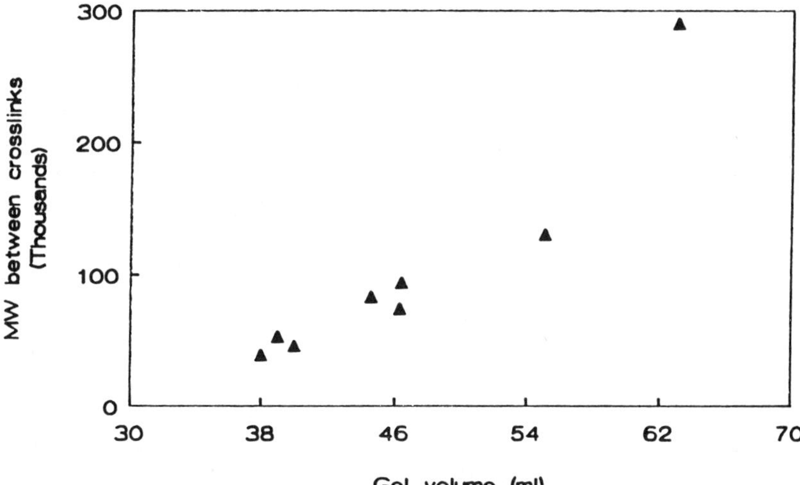

Figure 3. Correlation between rheological properties and gel volume.

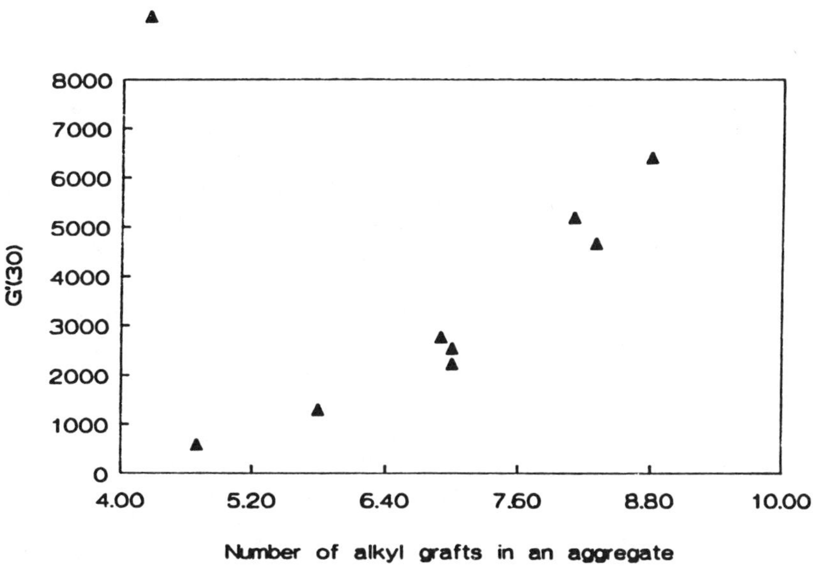

Figure 4. Effect of aggregate composition on modulus.

We postulate that the lower gel volumes at high temperature are due to a lower number of micellizable surfactant molecules, hence aggregates, available to act as intermolecular linkages between polymer chains. The cmc of ionic surfactants increases by as much as a factor of 10 with an increase in temperature of about of 70 ° C (*21*). By analogy, the effective concentration of adsorbed (i.e. aggregated) surfactant in our HMHEC/SDS systems goes down with increasing temperature. It is true that the hydrophobic grafts will themselves become more soluble in water as the "hydrophobic effect" is diminished and water acts more like a "normal fluid" with increasing temperature. The effect will be fewer alkyl grafts needing to avoid water, creating a tendency towards formation of gels with higher volumes than the 25 ° C controls at the same composition. Further, the solubility of the hydrophobes needs to increase by only ~1/N (where N is the total aggregation number; most of the molecules comprising the aggregate are surfactant) that of surfactant molecules to compensate for the tendency towards lower gel volume caused by the increase in surfactant solubility. Nevertheless, lower gel volumes are observed in our experiments so apparently the thermal solubility increase of the hydrophobes is low compared to that of the surfactant. We are currently investigating the total number of aggregates in the gel as well as the effect of temperature on the solubility of POE surfactants in water in an effort to permit better quantitative analysis of our results.

Effect of Salt. The effect of salt on the final gel volume was investigated by adding 0.1M NaCl to the HMHEC/SDS solutions during mixing. All solutions were studied at 25 ° C. We saw no effect of NaCl on the final gel volume at either of 2 bulk compositions studied. This result might once again be explained by looking at the effect of NaCl on the surfactant cmc and aggregation number. In general, salts are known to reduce the cmc of ionic surfactants. SDS is well characterized in this respect. In the absence of a cosurfactant, NaCl concentrations of up to 0.4 M reduce the cmc of SDS by an order of magnitude (*22*). However, the aggregation number of SDS micelles also rises from 55 to 162 on addition of salt (*23*), diminishing the effect of the lower cmc on the total number of aggregates that form. We are not aware of any study of the effect of salt on the cmc and aggregation number in solutions containing both SDS and a cosurfactant. It appears that in our system, the opposing effects of reducing the cmc and increasing the aggregation number balance each other in such a way as to leave the number of crosslinking aggregates in the network unaltered.

Conclusions

We have seen that hydrogel networks may be made using surface-active graft copolymers, and that the microstructure and bulk properties of these networks may be easily manipulated. We are now conducting further temperature and salt studies, as well as neutron scattering studies to determine the average size of the aggregates, in an effort to characterize these networks more completely. It remains now to seek out practical applications for these new materials which exploit their unique two-phase structure. The fact that pyrene partitions into the hydrophobic microdomains suggests that these materials may be used to advantage in separations or controlled release applications, where solutes may be sequestered in or released from the microdomains in a manner dependent on their relative solubilities in the two phases of the gel. Investigations into these areas are currently under way.

References

1. Landoll, L. M. *J. Polym. Sci., Polym. Chem. Ed.* **1982,** *20,* 443.
2. Gelman, R. A.; Barth, H. G. *Adv. Chem. Ser.* No. 213, 1986, 101.
3. Schulz, D. N.; Kaladas, J. J.; Maurer, J. J.; Bock, J.; Pace, S. J.; Schulz, W. W. *Polymer* **1987,** *27,* 2110.
4. Valint, P. L., Jr.; Bock, J. *Macromolecules* **1988,** *21,* 175.
5. Bock, J.; Valint, P. L., Jr.; Pace, S. J.; Siano, D. B.; Schulz, D. N.; Turner, S. R. In *Water-Soluble Polymers for Petroleum Recovery,* Stahl, G. A.; Schulz, D. N.; Plenum: New York, 1988; p 147.
6. Bock, J.; Siano, D. B.; Valint, P. L., Jr.; Pace, S. J. In *Polymers in Aqueous Media*, Advances in Chemistry Series 223; Glass, J. E., Ed.; American Chemical Society: Washington, DC, 1989; p 411.
7. Siano, D. B.; Bock, J.; Myer, P.; Valint, P. L., Jr. *ibid.* p 425.
8. Selb, J.; Gallot, Y. *Makromol. Chem.* **1980,** *181,* 809.
9. Selb, J.; Gallot, Y. *Makromol. Chem.* **1980,** *181,* 2605.
10. Varelas, C. G.; Steiner, C. A. In *Absorbent Polymer Technology,* Brannon-Peppas L.; Harland, R., Eds.; Elsevier: New York, 1990; p 259.
11. Varelas, C. G.; Steiner, C. A. *J. Polym. Sci. B: Polym. Phys. Ed.,* manuscript in review.
12. Steiner, C. A.; Gelman, R. A. In *Cellulosic Utilization Research and Rewards in Cellulosics,* Inagaki, H.; Phillips, G. O., Eds.; Elsevier: London, 1989; p 132.
13. Dualeh, A. J.; Steiner, C. A. *Macromolecules* **1990,** *23,* 251.
14. Dualeh, A. J.; Steiner, C. A. *Macromolecules* **1991,** *24,* 112.
15. Steiner, C. A. *J. Appl. Polym. Sci.* **1991,** *42,* 1493.

16. Zugenmaier, P.; Aust, N. *Makromol. Chem., Rapid Commun.* **1990,** *11,* 95.

17. Kalyanasundaram, K.; Thomas, J. K. *J. Am. Chem. Soc.* **1977,** *99,* 2039.

18. Almgren, M.; Lofroth, J. E. *J. Colloid Interface Sci.* **1981,** *81,* 486.

19. Ferry, J. D. *Viscoelastic Properties of Polymers, 3rd Edition* Wiley: New York, 1980.

20. Rao, I. V.; Ruckenstein, E. *J. Colloid Interface Sci.* **1986,** *113,* 375.

21. Evans, D. F.; Wightman, P. H. *J. Colloid Interface Sci.* **1982,** *86,* 515.

22. Williams R. J.; Phillips, J. N.; Mysels, K. J. *Trans. Faraday Soc.* *51,* 728.

23. Kratohvil, J. P. *J. Colloid Interface Sci.* **1980,** *75,* 271.

RECEIVED September 30, 1991

Chapter 4

Characterization of Polyelectrolytes

D. Hunkeler[1], X. Y. Wu, and A. E. Hamielec

Institute for Polymer Production Technology, Department of Chemical Engineering, McMaster University, Hamilton, Ontario L8S 4L7, Canada

A series of polyacrylamide-co-sodium acrylates were prepared through alkaline hydrolysis of well fractionated polyacrylamides. The polymers were subsequently analyzed by light scattering and viscometry. The dilute solution properties such as the specific refractive index increment, intrinsic viscosity, and Mark-Houwink parameters were found to systematically depend on the ionic content of the copolymer. These relationships were used to develop reliable techniques for the molecular weight characterization of polyelectrolytes. A comparison of molecular weights of the polyelectrolytes with the parent nonionic polyacrylamides from which they were derived showed good agreement.

The molecular weight characterization of polyelectrolytes is relatively difficult, since variations in copolymer composition alter the electrostatic environment and hence the size and structure of the polymer chain. Hydrolyzed polyacrylamide is therefore often used for methods development since it has a random distribution of charged groups along the backbone, and the molecular weights of the parent polyacrylamide can be estimated accurately by conventional techniques.

Copolymers of acrylamide and sodium acrylate are used in aqueous solutions as drag reduction agents, flocculants and thickeners. These copolymers are also employed in tertiary oil recovery as thixotropic aqueous polymer emulsions. These solutions are pseudoplastics and exhibit typical polyelectrolyte behavior with respect to salinity. The polymers can be prepared by direct copolymerization or derivatization from polyacrylamides. When the ionogenic monomer is introduced through a free radical addition mechanism, long acrylate and amide sequences form, with the microstructure conforming to Bernouillian statistics (*1*,*2*). By

[1]Current address: Department of Chemical Engineering, Vanderbilt University, Nashville, TN 37235

0097–6156/92/0480–0053$07.75/0
© 1992 American Chemical Society

comparison, polymers produced through alkaline saponification of the amide side chain are atactic (*3*) and have a relatively uniform charge density distribution which maximizes the viscosity increase for a given molecular weight and improves the polymer's performance (*4*).

Typical of polyelectrolytes, the solution behavior of polyacrylamide-co-sodium acrylate has several peculiarities. For example, the intrinsic viscosity, radius of gyration, and second virial coefficient all show maximums at 50-70 percent acrylate levels (*5,6*). The maximum is due to a combination of electrostatic repulsion and intramolecular hydrogen bonding, the latter diminishing as the amide groups are hydrolyzed. The viscosity also reaches a limit at a degree of neutralization of approximately sixty percent. Above this level counterion shielding has been postulated to reduce the electrostatic repulsion (*7*).

The purpose of this research is to develop valid characterization methods for copolymers of acrylamide and sodium acrylate. Initially a series of polyacrylamides will be synthesized and fractionated to yield narrow molecular weight distributions. The fractions will first be measured in their nonionic form where light scattering and viscometry are more reliable and accurate than for ion containing polymers. The polyacrylamides will then be hydrolyzed to various degrees and analyzed in their ionic form. The measured weight average chain lengths will be used to evaluate the absolute accuracy of the polyelectrolyte characterization method. In anticipation that the results of these studies could be applied to GPC calibration, we employed aqueous Na_2SO_4 solution as a solvent for polymers in these measurements and chose the Mw range of PAM fractions from 1.4 $x10^4$ to 1.2 $x10^6$ daltons.

Experimental Methods and Procedures

Polymer Preparation. Polyacrylamides (PAMs) were synthesized by aqueous free radical polymerizations using potassium persulfate (BDH Chemicals, 99% purity) as an initiator and ethanol mercapton (BDH Chemicals) as a chain transfer agent. The polymers were heterodisperse in molecular weight, with polydispersities between 2.0 and 2.5 as determined by Size Exclusion Chromatography (SEC). SEC chromatograms were measured with a Varian 5000 liquid chromatograph, using Toya Soda columns (TSK 3000,5000,6000 PW) and an aqueous mobile phase (0.02 M Na_2SO_4 (BDH Chemicals), 0.1 wt% Sodium Azide (Aldrich) and 0.01% wt% Tergitol NPX (Union Carbide Corp.)). The chromatograms were recorded on a Varian CDS 401 Data Station. Molecular weight calibration was done using universal calibration with seven narrow polyethyleneoxide standards (Toya Soda Manufacturing Co., Ltd.) and two broad MWD PAM samples. Peak broadening correction was performed using standard methods (*8,9*).

The synthesis conditions did not lead to preliminary hydrolysis as was determined by the ^{13}C NMR spectra. The spectra were recorded for a 10 wt% D_2O solution at 125.76 MHz and 70.9^oC in the Fourier transform mode with inverse gate decoupling. The pulse width was 6.8 ms with an acquisition time of 0.557s.

Fractionation. Polymer fractionation can be performed by a variety of techniques based on differential solubility, sedimentation, diffusion and chromatographic exclusion. For the large scale fractionation of high molecular weight polyacrylamide, fractional precipitation is preferred since it provides good separation efficiency and high yields, without requiring excessively dilute solutions. Previous investigations on polyacrylamide fractionation have used several nonsolvents including methanol (*10-12*), isopropanol (*13*), acetone, dioxane and THF (*14*). Wu (*15,16*) found acetone to have the highest solvent power by the solvent-precipitation-fractionation (SPF) Method. Wu also observed the narrowest polydispersity for fractions separated with mixtures of water and acetone as a nonsolvent. However, at high molecular weights methanol performed better, providing a more uniform polymer distribution between the lean and rich phases. A summary of the fractionation conditions and procedures used in this research is given in Table I. This method is the first large scale procedure to be applied to polyacrylamide, although large scale apparatus have been used before (*17*). Figure 1 illustrates the magnitude in the reduction in polydispersity due to fractionation for fractions generated in this research.

Polymer Hydrolysis. The fractionated polyacrylamides were hydrolyzed in an aqueous solution containing 4 wt% polymer. The reaction was performed in a three neck round bottom Pyrex flask. The centre opening was tightly fit with a glass agitator equipped with two, one-inch teflon blades. The agitator was connected to an external motor (Type RZRI-66, Caframo). The two side necks housed a thermometer and served as a sample withdrawal port. The flask was immersed in a constant temperature bath, operating at $30 \pm 0.5^{o}C$. The aqueous polyacrylamide solutions were maintained in the bath for one hour prior to hydrolysis. Distilled deionized water with 0.5 M NaOH (BDH Chemicals) was used as the reaction medium. Experiments were carried out for six hours, with aliquots withdrawn periodically at approximately fifteen minute intervals.

Hydrolysis was performed on five nonionic fractions with nominal molecular weights of 30,000, 120,000, 400,000, and 1,000,000 daltons. At each molecular weight a range of polymer compositions from zero to fourty percent sodium acrylate was obtained. These polymers were precipitated in an excess of methanol (70-100 hrs). Methanol proved to be an ideal non-solvent since it solubilized the unreacted sodium hydroxide. For low molecular weights samples were centrifuged for 1/2 hour at 5000 rpm to recover the fine powder of hydrolyzed polyacrylamide (HPAM). Samples were dried under vacuum to constant weight, and prepared for analysis after being cooled to room temperature and removed from the oven. The unused portion of the sample was stored in a sealed glass desiccator over silica gel.

Copolymer Characterization. The composition of copolymers of acrylamide with sodium acrylate can be measured using a variety of chromatographic, spectrometric, elemental or titrative techniques. Early investigations into these copolymers measured the extent of hydrolysis by potentiometric (*7,18-23*) or conductiometric titrations (*24*). Elemental analysis (C-H-N) is also common (*25,26*). However, it is

Table I

Fractionation Conditions

Solvent	Distilled deionized water.
Non-solvent	Acetone, Methanol (BDH, Reagent grade)
Temperature	23 ± 2°C
Polymer concentration	1-8 wt% of initial solution[†]
Reservoir	Polyethylene: 20L, 50L
Agitation	1/8 HP heavy duty lab stirrer with torque limiting controller (Series H, G.K. Heller Corp. Floral Pk. N.Y.)
Non-solvent addition	Dropwise
Length of non-solvent addition	2-48 h[††]
Settling time	2 - 7 days
Method of phase separation	Supernatant layer was siphoned off and refractionated

[†] Polymer concentration was reduced for higher molecular weight samples to maintain low viscosities and high fractionation efficiencies.

[††] Larger amounts of non-solvent were needed to separate lower molecular weight polymers and refractionated samples.

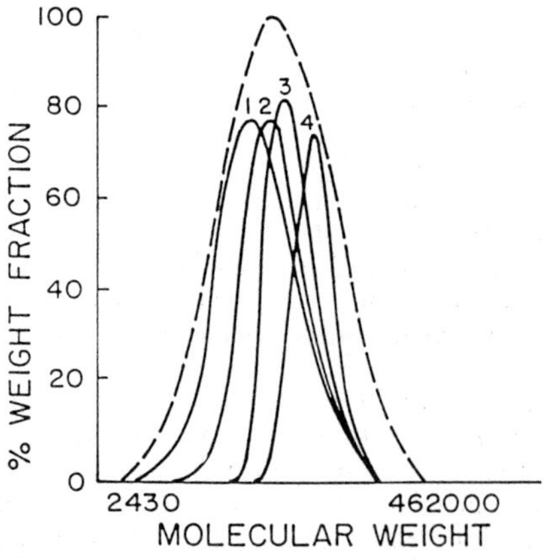

Figure 1: Unnormalized SEC chromatograms showing the reduction in polydispersity upon fractionation. Dashed line: original polymer, Solid lines: fractions 1-4.

sensitive to trace residuals of water which are difficult to completely remove. ^{13}C NMR techniques have been reported (*1-3*) and give excellent results for low to moderate molecular weight polymers, but require extensive data acquisition for chain lengths above approximately one thousand. Infrared methods have been developed (*5,27,28*) and while they provide quick estimates of the composition, their reliability and reproducibility are poor (*29*). Recently Maurer and Klemann (*30*) have used ion chromatography and found it provided excellent agreement with atomic absorption spectroscopy (for Na^+). Indeed, the elemental analysis of Na^+ directly is insensitive to residual water, but we have found it to underpredict the degree of ionization, likely because of incomplete neutralization. Elemental sodium methods are also susceptible to contamination by sodium ions, either from insufficient washing of NaOH or residual on the glassware. Therefore, although elemental methods have the best reproducibility, their absolute accuracy is insufficient.

In contrast with atomic absorption spectroscopy, the titration methods are insensitive to the carboxyl forms, either in H^+ form or in Na^+ form, and are also insensitive to the viscosity of solution. Hence, titrations are more reliable in analyzing the hydrolysis degree of HPAM.

In this work both potentiometric and conductiometric electrodes were placed in the titration reservoir. This resevoirwas equipped with a flow-through jacket to control temperature, and sealed such that an inert nitrogen atmosphere could be maintained over the sample. Entrance holes for the probes and titrant were included. Homogenized aqueous solutions of polyacrylamide-co-sodium acrylate were titrated with 1.0 M HCL (BDH Chemicals; reagent grade), and backtitrated with 0.1 M NaOH (BDH Chemicals; reagent grade). The potentiometric and conductiometric calculation gave compositions within one percent of each other. Duplicate titrations were performed on a representative number of samples to estimate the reproducibility.

Viscometry. The intrinsic viscosities of polymers in 0.2 Na_2SO_4 were obtained from the quadratic form as well as the conventional form of Huggins equation by the least squares technique (*31,32*). The efflux time was measured with a No.75 Cannon-Ubbelohde semi-micro dilution viscometer at 25 $\pm$ 0.05^oC.

Dialysis. A cellulose dialysis membrane (Spectra/Por 6) with a molecular weight cutoff of 1000 was purchased from Spectrum Medical Industries, Inc. (Los Angeles, CA). A small pore size was selected to prevent oligomers from diffusing into the dialysate. Prior to use the membranes were conditioned in the dialyzing buffer (0.2 M Na_2SO_4) for one hour, and rinsed with distilled deionized water.

Polymer solutions were prepared at a concentration suitable for light scattering (*33*). One hundred mL of these solutions were pipetted into the membrane which was sealed at one end with a dialysis tubing enclosure. After the second end of the membrane was secured, it was submersed in three litres of saline solution. This solution was housed in a four litre polyethylene container, isolated from the atmosphere.

After a predetermined time the vessel was opened and the membrane removed. Several concentrations of the polymer were prepared by diluting the dialyzed solution with the dialysate. The samples were immediately analyzed by light scattering or differential refractometry.

Light Scattering. The molecular weight of each fraction was measured using a Chromatix KMX-6 LALLS photometer, with a cell length of 15 mm and a field stop of 0.2. These parameters corresponded to an average scattering angle of 4.8°. A 0.45 μm cellulose-acetate-nitrate filter (Millipore) was used for the polymer solutions. A 0.22 μm filter of the same type was used to clarify the solvent. Distilled deionized water with 0.02 Na_2SO_4 (BDH Chemicals, analytical grade) was used as the solvent.

For characterization of dilute solution properties of polyacrylamide-co-sodium acrylate, Wu (16) has determined that polyelectrolyte interactions are suppressed for Na_2SO_4 concentrations of 0.2 mol/L. This salt concentration was consequently selected as a solvent for the light scattering characterization performed in this investigation.

The refractive index increment of the polymer in solution was determined using a Chromatix KMX-16 laser differential refractometer at 25°C and a wavelength of 632.8 nm. The dn/dc was found to be 0.1869 for acrylamide homopolymers.

Results and Discussion

I. Characterization of Polyacrylamide. Eight PAM fractions with polydispersity indices 1.2 - 2.0 were obtained by fractionation. The homopolymers were subsequently analyzed by viscometry and light scattering with the intrinsic viscosities and weight-average molecular weights summarized in Table II. The reduced specific viscosities, η_{sp}/c, when plotted versus concentrations gave straight lines with regression coefficients of 0.999 - 0.9999. The intrinsic viscosities from the quadratic equation were not significantly different from those from the conventional equation, with an average deviation of 0.46%.

The Mark-Houwink constants, K and a, were estimated with the Error-In-Variables method (*34,35*). The variances in molecular weight were evaluated from Hunkeler and Hamielec's data (*33*), while those in intrinsic viscosity were calculated from the present data using Chee's equation (*31*). The variance in concentration was assumed to be negligible. The following equation has been established for PAM in 0.2 M Na_2SO_4 at 25±0.05°C:

$$[\eta]= 2.43\times10^{-4}\ M_w^{0.69} \qquad (1)$$

The 95% confidence intervals for parameters K and a are:

$$K = 2.43\times10^{-4} \pm 0.36\times10^{-4}$$
$$a = 0.69 \pm 0.014$$

The double logarithm plot of $[\eta]$ vs M_w is given in Figure 2. The "a" value is consistent with that obtained by Kulicke (*36*) in 0.1 M Na_2SO_4 (0.7).

Table II

M_w, PDI and [η] of PAM fractions

Sample	F1	F2	F3	F4	F5	F6	F7	F8
M_w	1.244E6	9.90E5	4.01E5	2.01E5	9.90E4	3.60E4	2.69E4	1.39E4
PDI	1.8	2.0	1.8	1.7	1.6	1.5	1.3	1.2
[η] dl/g	3.804	3.555	1.733	1.096	.6502	.3220	.2741	.1754

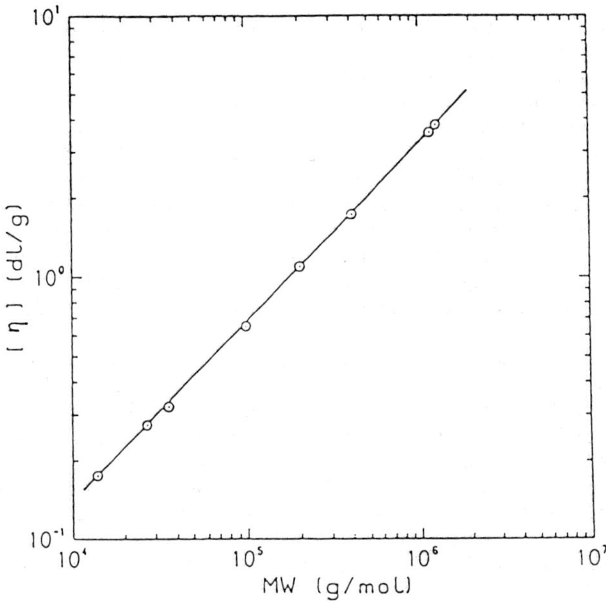

Figure 2: Plot of log [η] vs. log M_w for PAM in 0.2 M Na_2SO_4 at 25°C

II. Light Scattering Characterization of Polyacrylamide-Co-Sodium-Acrylate

Light Scattering of Polyelectrolytes. When performing light scattering on polyelectrolyte samples, the effect of ionogenic monomers is manifested in an apparent second virial coefficient and refractive index increment. Nonetheless, the light scattering equation derived for nonionic polymers (*36-38*) can be generalized for polyelectrolytes (*39*) provided the following criterion are met:

> 1) The local fluctuations in concentration, density and refractive index are electrically neutral (*40*).
> 2) The refractive index increment is obtained at constant electro-chemical potential of the counterions in the solution and the environment of the polymer coil (*41*).

The second criterion is satisfied by dialyzing the polymer and aqueous salt solutions against each other. At 'Donnan Equilibrium' the chemical potential of the counterion will be the same at all locations in the solution. Therefore, to correctly measure the molecular weights of polyelectrolytes we must first determine $dn/dc)_\mu$ as a function of the copolymer composition.

Determination of the Refractive Index Increment at Constant Chemical Potential. In order to determine the time required for Donnan equilibrium of the sodium ion in this particular solvent-membrane combination, several parallel experiments were performed. A polyacrylamide-co-sodium acrylate (PAM-NaAc) with thirty percent ionic content and a nominal molecular weight of 400,000 daltons was selected as a representative sample. A 0.15 wt% polymer solution was prepared and separated into six samples, which were dialyzed for 0, 24, 48, 72, 96 and 120 hours, respectively. After these periods the solutions were diluted and refractive index increments were determined against the dialysate. The equilibrium dialysis time was found to be 72 hours, although for subsequent measurements a dialysis time of 120 hours was used.

The refractive index increment at constant chemical potential was determined as a function of the extent of hydrolysis (Table III and Figure 3). The observed decrease in $dn/dc)_\mu$ with acrylate content, a manifestation of the negative selective sorption of sodium sulphate, has been reported previously (*5,22,28*). Only Kulkarni and Gundiah (*42*) have found a contradictory dependency. None of these authors have used the same solvent as in this work, and therefore direct comparison of the magnitude of dn/dc is not possible. Nonetheless, the validity of these measurements will be confirmed in the next section.

The equation:

$$dn/dc)_{\mu,\mathrm{PAM\text{-}NaAc}} = 0.1869\ (F_{\mathrm{NaAc}})^{-0.076} \qquad (2)$$

has been fit from this data so that the refractive index may be computed at any specific copolymer composition (F) between zero and thirty three percent sodium acrylate (NaAc).

Table III

Refractive index increments at constant chemical potential for various compositions of polyacrylamide-co-sodium acrylate

Acrylate content in copolymer (mol %)	$(\partial n/\partial c)_\mu$
0.0	0.1869
6.4	0.1624
9.6	0.1551
15.0	0.1503
33.0	0.1464

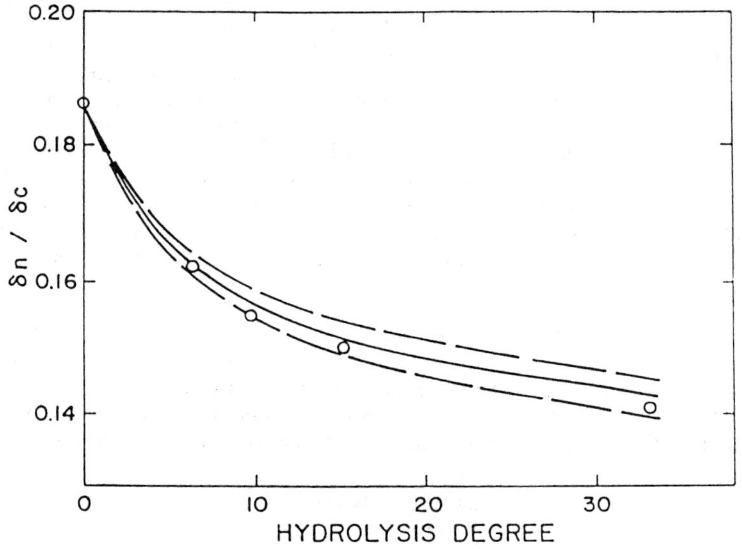

Figure 3: Refractive index increment at constant chemical potential as a function of the acrylate level in polyacrylamide-co-sodium acrylate. Solid line is the regressed equation; dashed lines are the 95% confidence limits.

Evaluation of the Molecular Weight Method for Polyacrylamide-co-sodium acrylate. Light scattering measurements were made for each of the five narrow polymer standards. A typical result, showing the Rayleigh factor as a function of composition, is given in Figure 4. From these linear plots the weight average molecular weights (M_w) were regressed. The molecular weight data were normalized with respect to composition, with the corresponding chain lengths summarized in Table IV. The polyelectrolyte chain lengths deviate on average by 7.68% from the original polyacrylamide homopolymer. Such an agreement is well within the random errors of aqueous light scattering (±10%)(*32*). We can conclude, therefore, that the measured molecular sizes of the ionic and nonionic polymers agree. It is worthwhile to note that without the correct refractive index increment, the estimation of the molecular weight of a thirty percent hydrolyzed polyacrylamide is 62% underpredicted! Therefore, the polyelectrolyte method, and the specific refractive index increments at constant chemical potential are accurate and reliable. Furthermore, it is sufficient to correct the optical constant for $dn/dc)_\mu$ in order to obtain accurate molecular weights of polyelectrolytes.

The following procedure is recommended for the molecular weight characterization of polyacrylamide-co-sodium acrylate:

A dilute solution of the polyelectrolyte should be prepared at an appropriate concentration for light scattering (*32*) in a solvent of high ionic strength, for example, 0.2 M Na_2SO_4. The polymer solution should subsequently be dialyzed against the saline solution for an equilibrium period in a membrane of small pore size. The dialysis will equilibrate the electrochemical potential of the counterion between the bulk solution and the domain of the polymer coil. Several dilutions of the base polymer solution should then be prepared by combining the dialyzed solution and the dialysate (These new polymer solutions will also be in electrochemical potential equilibrium with the dialysate since they are prepared from mixtures of two solutions with the same activity of Na^+ ions). The excess Rayleigh factors (R_θ) can then be measured by light scattering over the range of concentrations (c) prepared:

$$R_\theta = R_{\theta c} - R_{\theta,\mathrm{Dialysate}} \tag{3}$$

With an estimate of the copolymer composition (F), the specific refractive index increment at constant chemical potential, $(dc/dc)_\mu$, can be calculated and used to compute the optical constant (K_o):

$$K_o = \frac{2\pi^2 n^2 (dn/dc)^2_\mu}{\lambda^4\ Na} \tag{4}$$

where n is the refractive index increment of the solvent, Na is Avogadro's number, and λ is the wavelength of radiation. The weight average molecular weight (M_w) can subsequently be regressed from measurements of the excess Rayleigh factors at several polymer concentrations, using the classical light scattering equation valid for low observation angles:

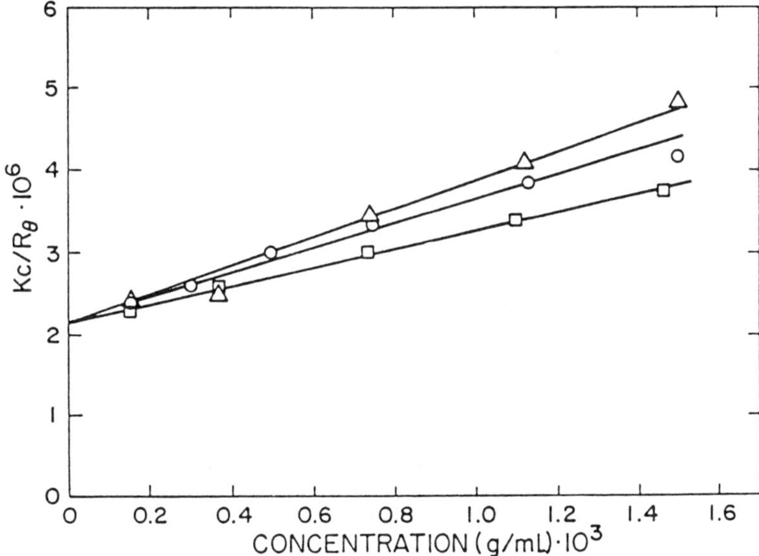

Figure 4: Light scattering plots showing the equivalence of molecular weight determination for polyelectrolytes and the nonionic polymer from which they were derived. (O): Polyacrylamide, (□): HPAM with 9.53% hydrolysis, (Δ): HPAM with 33.7% hydrolysis

Table IV

A comparison of the polyelectrolyte and nonionic molecular weight characterization procedures

Molecular weight of polyacrylamide measured by the nonionic method	Molecular weight measured on polyacrylamide-co-sodium acrylate by the polyelectrolyte method			
r_w (nonionic polyacrylamide)	r_w (10% hydrolysis)	r_w (30% hydrolysis)	Percent Deviation from Nonionic (10% hydrolysis)	(30% hydrolysis)
379	482	402	21.4	5.7
1,676	1,633	1,702	2.6	1.5
3,282	3,413	3,439	3.8	4.6
6,563	6,412	5,959	2.4	10.1
13,859	13,845	10,446	0.1	24.6
			Average deviation 7.68%	

$$\frac{K_0 c}{R_\theta} = \frac{1}{M_w} + 2\ A_2 c \qquad (5)$$

where A_2 is the experimentally measured second virial coefficient, an estimate of the interaction potential between two polymer coils in a given solvent.

III. Viscometric Characterization of Hydrolyzed Polyacrylamide. In this section the intrinsic viscosity of HPAM, as well as the Mark-Houwink parameters will be correlated with the copolymer composition. These relationships will then be used to calculate weight average molecular weights. A comparison of MW's determined from viscometric data with those measured by an absolute characterization procedure (light scattering) will be used to evaluate the accuracy of the new correlations.

Viscosity and the Square Root Law. As shown by the straight lines of η_{sp}/c vs concentration in Figure 5, HPAM in 0.2 M Na_2SO_4 behaves like nonionic polymer. However, the intrinsic viscosity of polyelectrolytes is uniquely dependent on the polymer composition and the salt concentration. Therefore, for a HPAM of a given molecular weight and a fixed salt level, the intrinsic viscosity will strictly vary as a function of the degree of hydrolysis (HD). This dependency has been verified experimentally (*5,7,25,43-47*); plots of $[\eta]$ vs HD are bell-shaped with a maximum $[\eta]$ at about HD = 40-50%. Kulkarni (*7*) has demonstrated that the data can be linearized by plotting $[\eta]$ vs $HD^{1/2}$. This square root law has also been found to be valid between 0 and 40% hydrolysis (Figure 6). It is described by the following emperical equation:

$$[\eta] = A{*}HD^{1/2} + B \qquad (6)$$

where A and B are the slope and intercept of a plot of $[\eta]$ vs $HD^{1/2}$, respectively. Although they are constant for a series of HPAM samples from the same parent PAM, the A and B will vary with the M_w of parent PAM as shown in Table V.

Obviously, both A and B are functions of the molecular weight of the parent PAM (Mw_0) and therefore can be related to the intrinsic viscosity of the parent PAM ($[\eta]_0$). The plots in Figure 7 and Figure 8 reveal fairly good linear relationships between A, B and $[\eta]_0$. The data calculated from Kulkarni's results are also plotted in the figures. The same trends have been observed.

The following equations have been regressed for the parameters A and B as a function of the intrinsic viscosity of the parent polyacrylamide:

$$A = f([\eta]_0) = 0.294\ [\eta]_0 - 0.0246 \qquad (7)$$
$$B = f([\eta]_0) = 0.266\ [\eta]_0 + 0.144 \qquad (8)$$

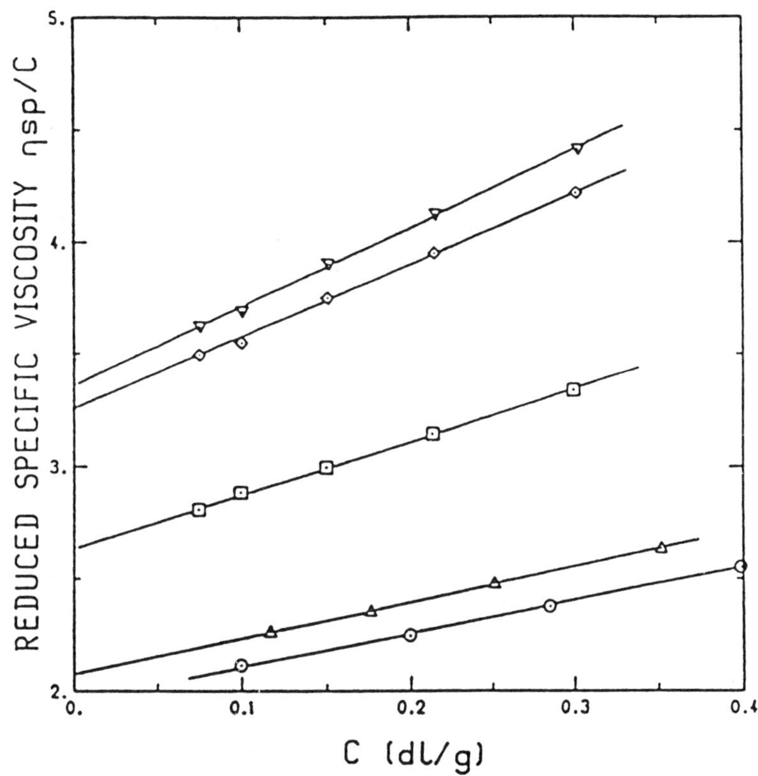

Figure 5: Reduced specific viscosity vs. concentration of polyacrylamide fraction F3HY in 0.2 M Na_2SO_4 at various degrees of hydrolysis (HD). (O): HD = 6.4%, (Δ) : HD = 9.5%, (□): HD = 14.9%, (◇): HD = 23.0%,(▽): HD = 26.0%

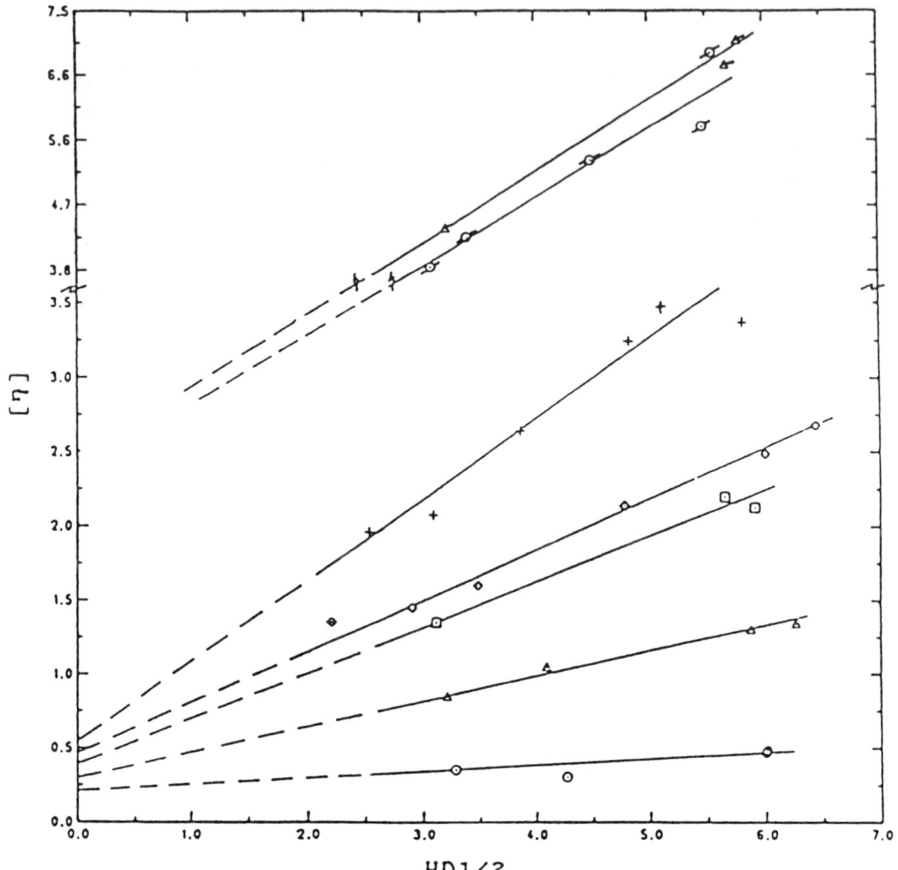

Figure 6: Dependence of intrinsic viscosity of HPAM on the square root of the hydrolysis degree: ⊙, F7HY; Δ, F5HY; ⊡, F4HY; ◇, HY2; +, F3HY; ⊙̸, HY5; and Δ̸, F2HY.

Table V

Parameters A and B in equation (6) for HPAM from the PAM with various molecular weights

Sample	$[\eta]_0$	A	B
F7HY	0.2741	0.0462 ± 0.011	0.201 ± 0.056
F5HY	0.6502	0.159 ± 0.027	0.363 ± 0.137
F4HY	1.096	0.300 ± 0.098	0.428 ± 0.495
HY2*	1.257	0.348 ± 0.026	0.428 ± 0.128
F3HY	1.733	0.507 ± 0.085	0.634 ± 0.375
F2HY	3.555	1.010 ± 0.173	1.109 ± 0.872

* M_{w0} = 246900; PDI = 1.8

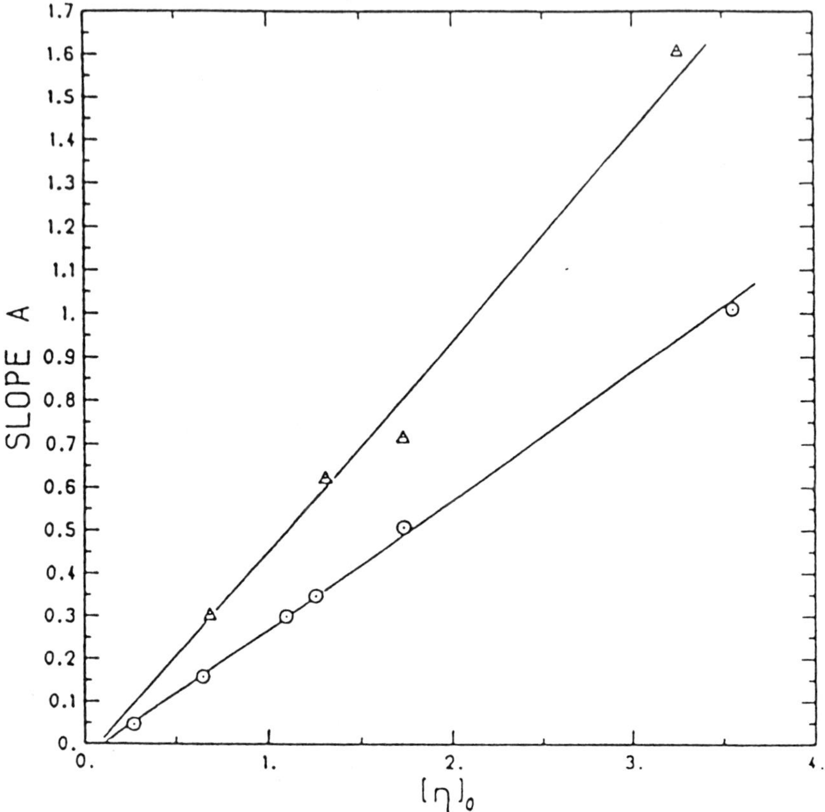

Figure 7: Slope of ($[\eta]$-$HD^{1/2}$ plot) for HPAM vs. $[\eta]$ of the parent PAM (⊙): this work, (Δ): Kulkarni's work

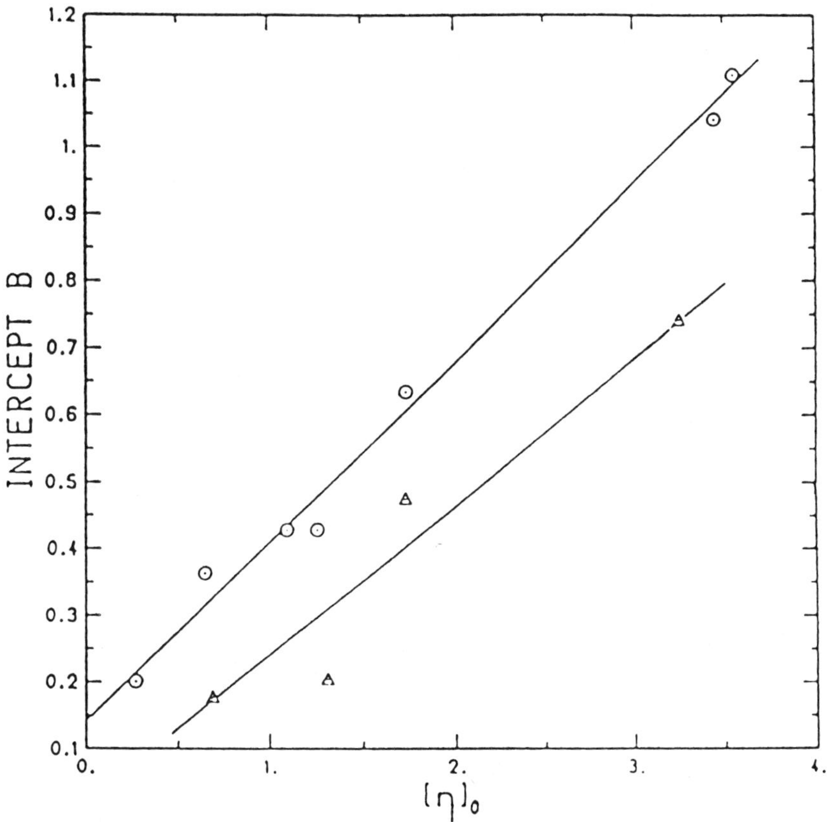

Figure 8: Intercept of ([η]-HD$^{1/2}$ plot) for HPAM vs. [η] of the parent PAM (⊙): this work, (Δ) Kulkarni's work

Hence, an empirical equation of the intrinsic viscosity, $[\eta]$, as a function of the intrinsic viscosity of the parent polyacrylamide, $[\eta]_0$, and the hydrolysis degree (HD) can be obtained for HPAM in 0.2 M Na_2SO_4 at 25.0°C by combining equations (6-8):

$$[\eta] = (0.266+0.294HD^{1/2})[\eta]_0+0.144-0.0246HD^{1/2} \quad (9)$$

From equation (9) one can readily estimate the intrinsic viscosity of HPAM of any composition in the range of 6 - 40% without a direct measurement provided the intrinsic viscosity of the parent polyacrylamide and copolymer composition are known.

Mark-Houwink Equations for HPAM

In the Mark-Houwink equation the parameters K and a are constant only if the polymer composition, solvent and temperature are unchanged. For HPAM under the given conditions, K and a are functions of the hydrolysis degree. Klein et al. (*46*) observed a maximum value of exponent a at about 40% HD and a minimum value of K at about 20% HD. McCarthy et al. (*43*) showed some changes in the values of K and a with HD but did not show definite trends.

Since the intrinsic viscosities of HPAM, especially those for high molecular weight samples, strongly depend on HD, it is necessary to choose HPAM samples with exactly the same HD to determine the parameters K and a. Because of the difficulties in preparing HPAM samples with exactly desired HD, interpolation from equations, such as the square root law, will be very useful.

Using the square root law (Equation 6) for various molecular weights, the intrinsic viscosities of HPAM at HD = 6, 10, 15, 20, 25, 30, 35, 40% were obtained. Assuming that all hydrolyzed acrylamide groups are in Na^+ form, the molecular weights of HPAM were calculated from the stoichiometric equation:

$$M_{ws} = M_{wPAM}/71.08(94.04x + 71.08(1 - x)) \quad (10)$$

where x is the mole fraction of hydrolyzed groups (x=HD/100). Correlating the intrinsic viscosities and the molecular weights, a set of Mark-Houwink constants has been determined and are listed in Table VI. Some examples of log-log plots of $[\eta]$ and Mw are given in Figure 9 for HPAM with various compositions.

Using polynomials to regress the data from Table VI, two empirical equations were obtained for K and a of HPAM. The polynomials were used since they provided a lower residual than other curves investigated, including exponentials. The equations, parameters and 95% confidence intervals are summarized below:

$$a = C_0 + C_1HD + C_2(HD^2)+ C_3(HD^3) \quad (11)$$

where $C_0 = 0.625 \pm 0.007$

$C_1 = 8.86x10^{-3} \pm 1.27x10^{-3}$

$C_2 = -2.405x10^{-4} \pm 0.617x10^{-4}$

$C_3 = 2.48x10^{-6} \pm 0.89x10^{-6}$

Table VI

Mark-Houwink constants K and a for HPAM with various HD

HD (mol%)	a	K (10^{-4})	Regression Coefficient
6	0.669	3.31	0.9982
10	0.694	2.85	0.9991
15	0.712	2.57	0.9995
20	0.725	2.41	0.9997
25	0.734	2.30	0.9997
30	0.742	2.22	0.9998
35	0.748	2.16	0.9997
40	0.753	2.12	0.9997

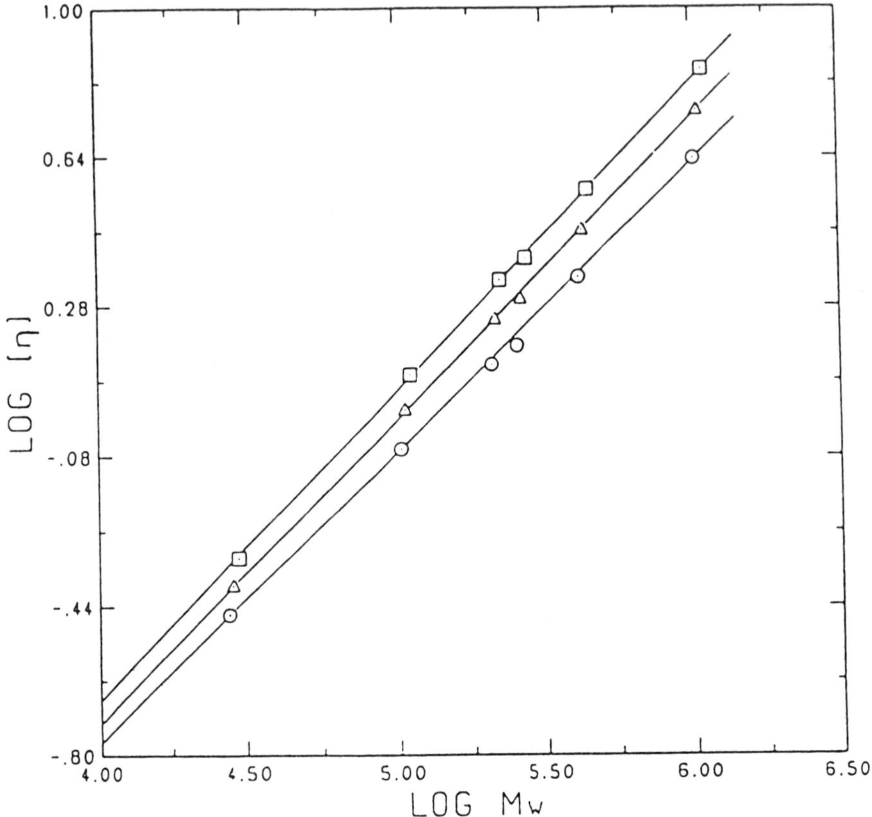

Figure 9: Plot of log [η] vs. LogM_w for HPAM at various hydrolysis degrees (Θ): HD = 10%, (Δ): HD = 20%,(□): HD = 35%

$$\log K = d_0 + d_1 HD + d_2(HD^2) + d_3(HD^3) \tag{12}$$

where
$d_0 = -3.36 \pm 0.024$
$d_1 = -2.39\times10^{-2} \pm 0.42\times10^{-2}$
$d_2 = 6.96\times10^{-4} \pm 2.05\times10^{-4}$
$d_3 = -7.37\times10^{-6} \pm 2.95\times10^{-6}$

The regressed equations (11 and 12) show a good fit with the Mark-Houwink parameters (Figure 10). Therefore, one can accurately calculate the values of K and a at any polymer composition of interest over the range 6 - 40% acrylate from these polynomials. The reliability of these Mark-Houwink parameters will be demonstrated in the following section of the paper.

Molecular Weights of Hydrolyzed Polyacrylamide. The molecular weights of HPAM samples were calculated using the Mark-Houwink equations (as M_{wv}) established in this work as well as from the stoichiometric equation (as M_{ws}) based on light scattering measurements. In the viscometric calculations the values of K and a were obtained from the polynomials (equations 11,12) with the $[\eta]$'s and HD's obtained experimentally. The two kinds of molecular weights together with those measured by light scattering are listed in Table VII.

An average error in molecular weight determination of 4.7% is relatively low as compared with light scattering, osmometry and GPC. The agreement between the viscometric method and stoichiometric method is good even for the samples of high polydispersities (e.g. HY5). This suggests that the polynomial equations for the Mark-Houwink parameters and equation (5) which relates the intrinsic viscosity to the copolymer composition are reliable. For broader samples (Mw/Mn > 2.5) a polymolecularity correction might be necessary to reduce the error (*43,48-50*).

Recommended method for the Calculation of Molecular Weights from Viscometric Data. The molecular weight of a hydrolyzed polyacrylamide can be based on either the viscosity of the copolymer, if it is available, or on the viscosity of the parent polyacrylamide from which the copolymer was derived. The two possible cases are summarized below:

1. If $[\eta]_{HPAM}$ is known, the molecular weight can be calculated from the Mark-Houwink equation (equation 1) provided the hydrolysis degree has been determined. In this case the parameters K and A are calculated at the exact HD from the polynomials given by equations (11) and (12).

2. If $[\eta]_{PAM}$ is known the molecular weight of the copolymer can be calculated directly from equations 1 and 10. The latter requires a measurement of the hydrolysis degree.

Mark-Houwink Constants. As shown in Table VI and Figure 10, the value of K decreases exponentially with an increase in the hydrolysis degree. This is similar to Klein's data (*46*). As a measure of the

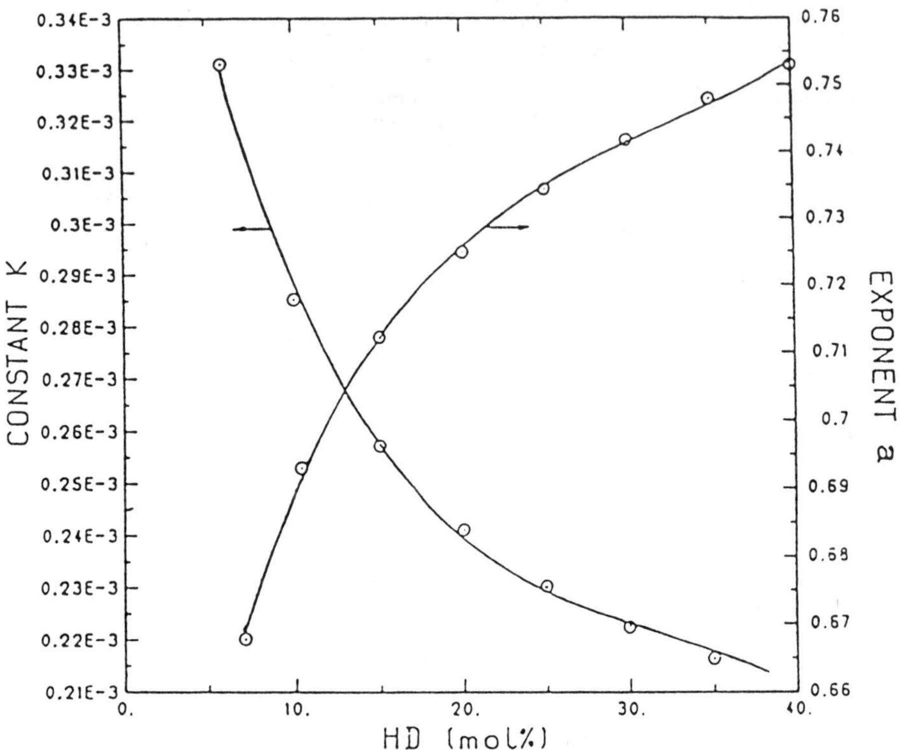

Figure 10: Mark-Houwink constants K and a for HPAM at various hydrolysis degrees (——); from correlated polynomials, (⊙): data point

Table VII

Molecular weight of Polyacrylamide-co-sodium acrylate from: the Mark-Houwink equations (M_{wv}), the stoichiometric equation (equation 7) (M_{ws}) and from light scattering ($M_{w\,LS}$)

HD (mol%)	[η] (dl/g)	M_{wLS}	M_{wv}	M_{ws}	ERR (%)[a]
		F7HY			
10.77	0.3524	35800	28300	27800	1.53
36.00	0.4692	31700	28800	30000	−4.19
36.08	0.4872	/	30200	30000	0.64
		F5HY			
10.30	0.8468	119600	101500	102300	−0.80
16.70	1.049	/	112300	104300	7.61
34.42	1.303	/	114700	110000	4.30
39.20	1.342	135500	112900	111500	1.22
		F4HY			
9.68	1.353	247000	205000	207300	−1.08
31.90	2.200	/	238600	221700	7.62
34.90	2.127	268000	219900	223700	−1.67
		HY2			
4.86	1.355	/	267400	250800	6.62
8.42	1.451	/	241100	253600	−4.92
12.10	1.601	/	235800	256600	−8.10
22.80	2.137	/	262200	265100	−1.11
36.00	2.486	/	267300	275600	−3.01
41.50	2.677	/	273100	280000	−2.46
		F3HY			
6.40	1.961	/	421400	409300	2.97
9.53	2.075	468000	383800	413300	−7.15
14.90	2.640	/	431200	420300	2.59
23.20	3.242	/	459800	431100	6.68
26.00	3.470	/	479700	434700	10.35
33.10	3.372	467000	417200	443900	−6.00
		HY5[b]			
9.46	3.979	/	925500	1061500	−12.81
11.50	4.228	/	969500	1068300	−9.25
20.15	5.329	/	974700	1097000	−11.16
36.60	6.893	/	1126800	1132500	−0.51
		F2HY			
10.30	4.354	1014000	1076500	1022900	5.24
32.20	6.709	/	1064500	1093000	−2.60
33.40	7.067	822000	1122000	1096800	2.30
		average absolute error (%)			4.70

a Err (%) = $(M_{wv} - M_{ws}) / M_{wS} \times 100\%$

b HY5 is from unfractionated PAM with PDI 2.5

flexibility of the polymer chains, K is proportional to the viscosity constant ϕ (*50,51*).

$$K \sim \phi \tag{13}$$

where $$\phi = \phi_0 (1-2.63\varepsilon + 2.86\varepsilon^2) \tag{14}$$

for nondraining polymer coils. Since ϕ will become smaller when the polymer chains become more rigid, it is reasonable for K to decrease when the concentration of charged groups increases.

The values of the exponent a obtained in this work are in the range, 0.67 to 0.76. This is in agreement with the theoretical predictions (*50*) for unbranched, non-solvent-draining coils with excluded volumes and implies that 0.2 M Na_2SO_4 is not a good solvent for polyacrylamide or poly(acrylamide-co-sodium acrylate). Therefore the polymer chain has a random coil conformation in this solvent.

The constant K and exponent a are also not independent parameters since from the theoretical derivation, "a" is a function of "ε":

$$a = 0.5 (1 + 3\varepsilon) \tag{15}$$

Hence, K and a have a certain relationship as summarized by Elias (*50*) for various coil-like polymers:

$$\log K = C_1 - C_2 a \tag{16}$$

where C_1 and C_2 are positive constants. Our data show a similar relationship between K and a for the copolymer with various compositions as follows and in Figure 11:

$$\log K = -1.946 - 2.302a \tag{17}$$

(In our case constant C_1 is negative). The agreement again implies a coil-like conformation of poly(acrylamide-co-sodium acrylate) in the salt solution.

Conclusions

The first large scale fractionation method for polyacrylamides is reported. This procedure has been optimized to prepare a series of narrowly distributed polymer standards with molecular weights between 10,000 and 1,000,000. The homopolymers were subsequently hydrolyzed to between 5 and 40% sodium acrylate content through a saponification reaction.

It has been found that the classical light scattering theory, derived for nonionic scattering bodies, can be applied to polyelectrolytes provided the specific refractive index increment (dn/dc) is evaluated at constant electrochemical potential (μ). Equilibrium is obtained by dialyzing the polymer solution against the dialysate (0.2 M Na_2SO_4 for polyacrylamide-co-sodium acrylate) for a predetermined equilibrium period, specific to the pore size of the membrane used. From such measurements at various copolymer compositions (F), the following empirical relationship is established:

$$dn/dc)_{\mu,PAM\text{-}NaAc} = dn/dc)_{PAM} (F_{NaAc})^a \tag{18}$$

where a was found to be -0.076.

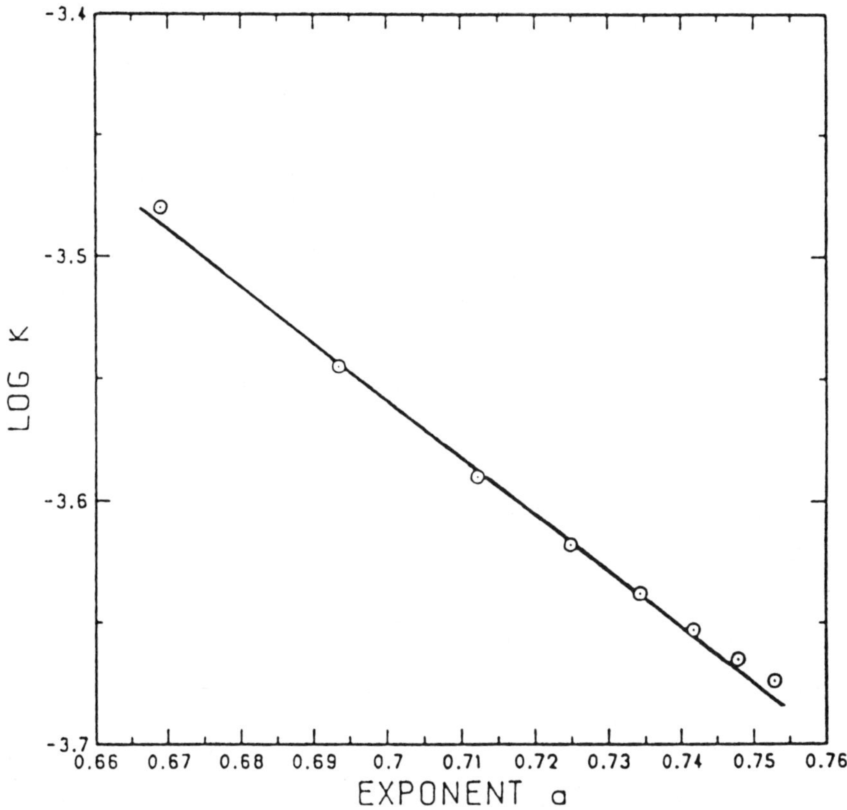

Figure 11: logK vs. exponent a of HPAM at various hydrolysis degrees (HD) from top left to bottom right, the data points are at: HD = 6, 10, 15, 20, 25, 30, 35, 40%

The validity of the specific refractive index increment at constant chemical potential is established through light scattering measurements of the weight average chain length of the hydrolyzed polyacrylamides. These chain lengths were then compared with the chain lengths of the parent nonionic polymers from which they were derived. Excellent agreement is observed, implying the polyelectrolyte characterization method developed, specifically the $dn/dc)_{\mu}$'s, are accurate and reliable.

Mark-Houwink constants for polyacrylamide and poly(acrylamide-co-sodium acrylate) in 0.2 M Na_2SO_4 were also measured. The dependence of K and a on the copolymer compositions was found for the range of acrylate content 6 - 40 mol%. A relationship between intrinsic viscosity and acrylate content in the form of the square root law was also found. A method was therefore developed from which the molecular weight of polyacrylamide-co-sodium acrylate can be estimated without recourse to direct light scattering measurements. Knowledge of the hydrolysis degree and intrinsic viscosity of the parent polymer are sufficient. This method has been evaluated against experimental data, and good agreement is observed.

Literature Cited

1. Troung,N.D.;Galin,J.C.;Francois,J.;Pham,Q.T.*Polymer*.**1986**,*27*,459.
2. Candou,F.;Zekhini,Z.;Heatley,F.*Macromolecules*.**1986**,*19*,1895.
3. Troung,N.D.;Galin,J.C.;Francois,J.;Pham,Q.T.*Polymer*.**1986**,*27*,467.
4. Ellwanger,T.E.;Jaeger,D.A.;Barden,E.*Polymer Bulletin*.**1980**,*3*,369.
5. Kulicke,W.-M.;Horl,H.H.*Colloid and Polymer Sci*.**1985**,*263*,530.
6. Candou,F.;Zekhini,Z.;Heatley,F.;Franta,E.*Colloid and Polymer Sci*. **1986**,*264*,676.
7. Muller,G.;Laine,J.P.;Fenyo,J.C.*J.Appl.Polym.Sci.Polym.Chem.Ed*. **1979**,*17*,569.
8. Hamielec,A.E.;Ray,W.H.*J.Appl.Poly.Sci*.**1969**,*13*,1317.
9. Balke,S.T.;Hamielec,A.E.*J.Appl.Poly.Sci*.**1969**,*13*,1381.
10. Shulz,A.R.*J.Am.Chem.Soc*.**1954**,*76*,3422.
11. Baysal,B.*J.Polym.Sci*.**1963**,*1*,257.
12. Baysal,B.*J.Polym.Sci.C*.**1963**,*4*,935.
13. Venkatarao,K.;Santappa,M.*J.Polym.Sci.A-1*.**1970**,*8*,1785.
14. Ramazanov,K.R.;Klenin,S.I.;Klenin,V.I.;Norichkova,L.M.*Vysokomol. soyed*.**1984**,*A26*,2052.
15. Wu,X.Y.;Hunkeler,D.;Pelton,R.H.;Hamielec,A.E.*Polym.Mater.Sci.Eng*. **1988**,*58*,792.
16. Wu,X.Y.;Hunkeler,D.;Pelton,R.H.;Hamielec,A.E.;Woods,D.R.,*J.Appl. Polym.Sci*.,**1991**,*42*,2081.
17. Kamide,K.;Miyazaki,Y.;Yamaguchi,K.*Makromol.Chem*.**1973**,*173*, 175.
18. Kurenkov,V.F.;Verizhnikova,A.S.;Myagchenkov,V.A.*Doklady Akademii Nuak SSR*.**1984**,*278*,1173.
19. Nagase,K.;Sakaguchi,K.*J.Polym.Sci.Pt.A*.**1965**,*3*,2475.
20. Fenyo,J.C.;Laine,J.P.;Muller,G.*J.PolymSci.Chem*.**1979**,*17*,193.
21. Muller,G.;Fenyo,J.C.;Selegny,E.*J.Appl.Polym.Sci*.**1980**,*25*,627.

22.Gunari,A.A.;Gundiah,S.*Makromol.Chem.***1981**,*182*,1.
23.Schwartz,T.;Francois,J.*Makromol.Chem.***1981**,*182*,2757.
24.Moens,J.;Smets,G.*J.Polym.Sci.***1957**,*23*,931.
25.Klein,J.;Heitzmann,R.*Makromol.Chem.***1978**,*179*,1895.
26.Kulicke,W.-M.;Kniewske,R.*Makromol.Chem.***1981**,*182*,2277.
27.Shaglayeva,N.S.;Brodskaya,E.I.;Rzhepka,A.V.;Lopyrev,V.A.; Voronkov,M.G.*Vysokomol.soyed.***1979**,*A21*,950.
28.Kulicke,W.-M.;Siesler,H.W.*J.Polym.Sci.Polym.Phys.***1982**,*20*,557.
29.Hansen,L.D.;Eatough,D.J.*Thermochimica Acta.***1987**,*111*,57.
30.Maurer,J.J.;Klemann,L.P.*J.Liquid Chrom.***1987**,*10*,83.
31.Chee,K.K.*J.Appl.Polym.Sci.***1985**,*30*,2607.
32.Fanood,M.H.R.;George,M.H.*Polymer.***1987**,*28*,2241.
33.Hunkeler,D.;Hamielec,A.E.*J.Appl.Polym.Sci.***1988**,*35*,1603.
34.Sutton,T.L.;MacGregor,J.F.;*CanadianJ.Chem.Eng.***1977**,*55*,609.
35.Reilly,P.M.;Patino-Leal,H.*Technometrics.***1981**,*23(3)*,221.
36.Kulicke,W.-M.;Bose,N.*Polymer Bulletin.***1982**,*7*,205.
37.Debye,P.*J.Appl.Phys.***1944**,*15*,338.
38.Debye,P.*J.Phys.Coll.Chem.***1947**,*51*,18.
39.Ewart,R.H.;Roe,C.P.;Debye,P.;McCarthy,J.R.*J.Chem.Phys.***1946**,*14*, 687.
40.Alexandrowicz,Z.*J.Polym.Sci.***1959**,*40*,91.
41.Richards,E.G., *An Introduction to the Physical Properties of Large Molecules in Solution*, Cambridge University Press: New York,NY, 1980.
42.Nagasawa,M.;Takahashi,A. in *Light Scattering from Polymer Solutions*;Huglin,M.B.,Ed.;Academic Press: London,U.K. 1972.
43.McCarthy,K.J.;Burkhardt,C.W.;Parazak,D.P.*J.Appl.Polym.Sci.***1987**, *33*,1699.
44.Kulkarni,R.A.;Gundiah,S.*Makromol.Chem.***1984**,*185*,957.
45.McCarthy,K.J.;Burkhardt,C.W.;Parazak,D.P.*J.Appl.Polym.Sci.***1987**, *33*,1683.
46.Klein,J.;Conrad,K.-D.*Makromol.Chem.***1978**,*179*,1635.
47.Kowblansky,M.;Zema,P.*Macromolecules*,**1981**,*14*,1451.
48.Brandup,J.;Immergut,E.H.*Polymer Handbook*,2nd Ed., John Wiley & Sons:New York, NY,1975.
49.Kulicke,W.-M.;Kniewske.R.;Klein,*J.Prog.Poly.Sci.***1982**,403.
50.Elias,H.-G.*Macromolecules.*1,2nd Ed;Plenum Press: New York, NY,1984;358-365.
51.Vollmert,B.*Polymer Chemistry*;Springer-Verlag: New York,NY, 1973;513.

RECEIVED September 6, 1991

Chapter 5

Polyethylene Film Acrylic Acid Grafted by Electron-Beam Preirradiation Method

Properties of Grafted Materials

J. Harada[1,2], R. T. Chern[1], and V. T. Stannett[1]

[1]Department of Chemical Engineering, North Carolina State University, Raleigh, NC 27695–7905
[2]Tsukuba Research Laboratories, Mitsubishi Paper Mills, Ltd., 46 Wadai, Tsukuba City, Ibaraki 300–42, Japan

The properties of acrylic acid grafted high density polyethylene films prepared by the electron beam preirradiation method have been studied. At high extents of grafting, the products are essentially poly(acrylic acid) hydrogels which are crosslinked by polyethylene. The hydrogels have a rubberlike elasticity. Although the tensile strength decreased with the extent of grafting, the highest grafted material which contained only 0.6 wt% polyethylene (on a dry basis) still had a tensile strength of 4.0 MPa at an elongation of about 790% after the sample was equilibrated with water. The water absorption capacity was also affected by the preirradiation dose, with the 8 Mrad irradiated sample exhibiting the highest water uptake.

Grafting of hydrophilic monomers throughout the whole thickness of non-porous hydrophobic polymer films is known to lead to hydrophilic membranes. These hydrophilic membranes have been considered as biomedical materials because of their excellent biocompatibility, high water permeability, and characteristics desirable for enzyme immobilization and controlled release of drugs (1). The polarized gels have also been studied for use as mechano-chemical materials (2, 3), including pH sensors, polymer actuators, and separation membranes.

The grafting reaction, including the use of high energy radiation, of acrylic acid or methacrylic acid onto polyethylene has been studied by many researchers because of its simplicity, high grafting ratio, and many application possibilities. These applications include water absorbents (4-7), battery separators (8-16), ion exchange and ion trap materials (17-21), selective separation membranes (19, 22, 23), anti-static materials, deodorant materials, enzyme-immobilization substrates (24, 25), metal coatings (26), and so on.

In this paper, we will present the properties of acrylic acid grafted polyethylene films, prepared by the electron beam preirradiation method at a comparatively high reaction temperature, high irradiation doses, and under redox reagent-free conditions (27, 28).

0097–6156/92/0480–0080$06.00/0
© 1992 American Chemical Society

Experimental

High density polyethylene (HDPE) film (d = 0.963 g/cm^3, 95 mm thick, crystallinity=70%, Mitsubishi Chemical Product) was used. Acrylic acid (Aldrich, containing 200 ppm hydroquinone monomethylether as a stabilizer) was used without purification. Potassium hydroxide and sodium hydroxide were A.C.S. reagent grade (Aldrich).

An electron curtain type electron beam accelerator (175Kv, Energy Sciences Inc.) was used. The preirradiation was conducted under nitrogen atmosphere and room temperature. The maximum irradiation dosage for a single pass was 16 Mrad. For doses larger than 16 Mrad, the sample was exposed several times from both sides of the sample. According to the dose-depth profile at 175Kv (29), the irradiation dosage was expected to be almost the same (>95%) throughout the sample thickness.

The gel fraction of irradiated polyethylene was determined by weighing the sample after 48 hours of extraction in hot toluene.

Monomer solutions were aqueous and were bubbled with nitrogen in the Erlenmeyer flask for at least 30 minutes to degas the solution before the grafting reaction. The irradiated polyethylene samples were dropped into the reaction flask which was kept in an isothermal (75° C) water bath. Nitrogen bubbling was continued during the grafting reaction. After reaction, the grafted samples were taken out and washed by running hot water for one day, followed by vacuum drying for 48 hours at 55° C. By this washing method, the grafted polymer weight reached almost the same value (within 2 wt. %) as the complete extraction results (2 days water reflux and 2 days methyl alcohol extraction). Grafting ratio (%) was calculated according to the following equation,

100* (weight after grafting - original weight) / original weight

Strictly speaking, the "grafting ratio" is actually the fractional unextractable poly(acrylic acid).

A microtome was used to slice the grafted film and a scanning electron microscope (JEOL, JXA-840) was used to observe the surfaces and cross sections of the grafted films. The grafted samples were neutralized by a 1% potassium hydroxide solution to enhance the contrast between the grafted layer and the ungrafted layer.

A Monsanto Tensometer 10 was used to measure the tensile strength of grafted samples. The sample width and length were 2 cm and 5 cm, respectively. The elongation rate was 300 mn/min.

Standard pH buffer solutions (Fisher) were used to check the pH dependence of swelling of the grafted polyethylene samples. The degree of swelling was defined as 100 X (length after swelling / original length).

Results and Discussion

Asymptotic Grafting Ratio and Crosslinking Table I shows the asymptotic grafting ratio of polyethylene samples irradiated to different doses. The asymptotic grafting ratio can be fitted to the following equation:

$$\text{Asymptotic grafting ratio} = K * [D]^b$$

where K and b are constants, [D] is the irradiation dose (Mrad). The b value for HDPE is about 0.5 between 1 Mrad to 16 Mrad. Above 32 Mrad the asymptotic grafting ratio was much higher than that calculated from this equation (27). It is not clear why this was so. Presumably, a combination of reduced termination and chain transfer reactions or grafting to existing grafted poly(acrylic acid) has led to these results.

Table I. Asymptotic Grafting Ratio of Irradiated Polyethylene Samples

Irradiation Dose (Mrad)	*Asymptotic Grafting Ratio (%)*	*Size Swelling L/Lo (%)**	*Gel Fraction*** (%)*
0	0	100	0
1	428	192	0
2	659	250	0
4	993	310	0
8	1240	335	0
16	1709	370	43
32	3645	432	66
48	8685	590	71
64	26086	1000	78
96	40653	1170	84

* Lateral dimension of the sample at the end of the grafting reaction.
** Gel fraction of the irradiated but ungrafted HDPE.

Above 16 Mrad, effective crosslinking happened and an insoluble fraction of polyethylene remained after extraction with toluene for 48 hrs. The unextractable gel fraction increased with preirradiation dose. Crosslinking was found to affect the grafting reaction rate, penetration of grafting layer and the mechanical properties of grafted materials; these results will be presented below.

Water Absorption of Samples in Acid Form Acrylic acid grafting of polyethylene has been shown by us to proceed from the surface to the center of the film (27). Figure 1 shows the equilibrium water uptake of the un-neutralized samples as a function of the grafting ratio, expressed in terms of g water / 100 g poly(acrylic acid) grafted. Water uptake increased initially with grafting ratio and asymptoted to a constant value after 20% grafting. Clearly, the grafted layer can absorb a significant amount of water even if the grafting ratio is very small (<5%). This water absorption property is quite different from that of poly(acrylic acid) grafted polyethylene produced by the mutual irradiation method. In the case of mutual irradiation, there was little water uptake below 15% grafting as reported by Lawler and Charlesby (10).

Although the grafted layer produced by the preirradiation method could absorb water, the water absorbing ability was restricted by the polyethylene chains. As shown in Figure 2, the water uptake per grafted poly(acrylic acid) stayed about the same up to 60 wt% poly(acrylic acid) in the polymer. It then increased dramatically with the poly(acrylic acid) content and reached a maximum at a poly(acrylic acid) content around 90% (grafting ratio ca. 900%). At this peak point, water uptake reached ca. 70 wt% of the wet sample. The dramatic increase in water uptake above 60 wt% poly(acrylic acid) may be attributed to the fact that the grafted layer merged at the center of the film sample when the poly(acrylic acid) content reached about 60% (grafting ratio 150%). Below this level, the grafted material has a three-layer structure, including two grafted layers and one center ungrafted polyethylene layer. The ungrafted polyethylene layer might have restricted swelling of the sample and thus reduced the equilibrium water uptake. Finally, the decrease in water uptake with

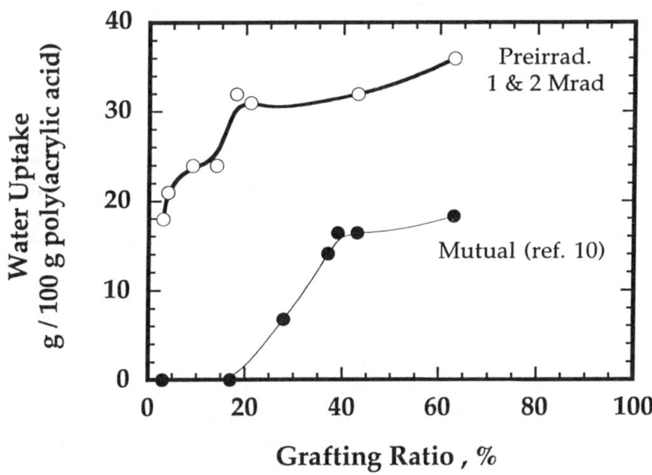

Figure 1. Equilibrium water uptake of poly(acrylic acid) grafted HDPE. A comparison between pre- and mutual-irradiation methods.

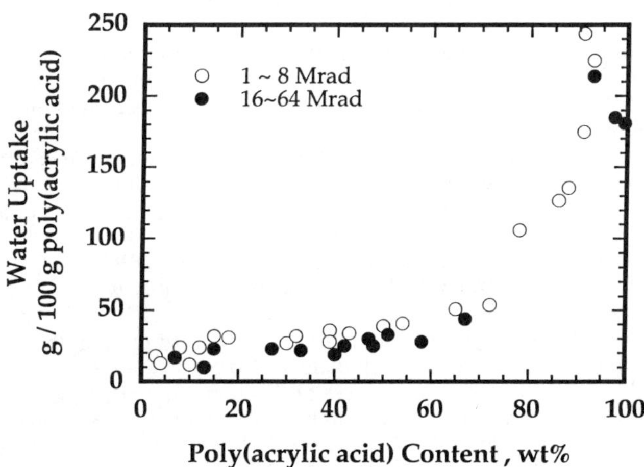

Figure 2. Equilibrium water uptake vs poly(acrylic acid) content. The influence of irradiation dose.

grafting above 90% of poly(acrylic acid) content (Figure 2) is probably due to crosslinking of polyethylene; high grafting ratios (P-AA content >90%) were obtained only for samples irradiated to doses greater than 16 Mrad which was enough to crosslink the polyethylene.

A hydrogel, according to definition, should have the ability to swell in water and retain a significant fraction (e.g. >20 wt%) of water within its structure but will not dissolve in water (30). The poly(acrylic acid) grafted HDPE with grafting ratio over 150% satisfies this criterion.

Swelling of Neutralized Samples in Water Figure 3 shows the swelling, expressed in terms of (swollen sample weight / dry sample weight), of samples both before and after being neutralized in a 1.0 w/v% sodium hydroxide solution for 20 hours. These samples have been grafted to their respective asymptotic grafting ratios. Before neutralization, the extent of swelling is essentially independent of the preirradiation dose; the 1 Mrad sample is the only exception which seems to have retained too much of the polyethylene structure to absorb much water. After neutralization, the difference in the swelling was enhanced; the swelling peaked at 8 Mrad irradiation. The increase in swelling between 1 Mrad and 8 Mrad is due to an increase in the grafting ratio, the decrease above 8 Mrad can be attributed to the crosslinking effects addressed above. The curve designated "Neutralization" in Figure 3 denotes the amount of swelling when the samples were equilibrated with the 1.0% NaOH solution. After neutralization, the samples were soaked in deionized water for 1 hr. and 4 hrs, respectively. The maximum water uptake is about 35 times the dry weight of the grafted sample, much greater than what was observed before neutralization, reflecting the effects of osmotic swelling.

The appearance of the grafted samples in the dry state also strongly depends on the preirradiation dose. For example, the 8 Mrad sample with 800% grafting ratio was milky white while the 64 Mrad sample with 800% grafting ratio was completely transparent; the equilibrium water contents of the two samples were roughly the same. No definitive explanation can be given at the present time for this observation.

Tensile Strength Figure 4 shows the tensile stress-strain curves, measured at 30 mm/min, for the irradiated but ungrafted HDPE samples. At this speed, the un-irradiated sample was torn along the elongation direction into pieces without breaking at roughly 100% strain. The maximum tensile strength did not vary much with irradiation between 1 to 64 Mrad. However, a mere 2 Mrad irradiation is enough to increase the elongation at break from ~100% to 950%. All the samples showed an extended plateau in the stress-strain curve. The tensile strength increases mildly and the elongation at break decreases with the irradiation dose.

Figure 5 shows the stress-strain curves for wet grafted HDPE samples preirradiated at 2 Mrad; the samples had been washed thoroughly with hot water after grafting. At this preirradiation dose, the grafted layers in the HDPE film met at the center of the film at ca. 150% grafting ratio. Below this grafting ratio, the grafted materials have a three-layer structure which include two grafted layers and one ungrafted central layer. The stress-strain curve of this three-layer material appears to be sensitive to the extent of grafting. For example, the sample with 30% grafting ratio failed in a brittle manner at a strain smaller than 50%, but the stress-strain curve of the 70% sample exhibited a plateau before break. After the grafted layers merged (>180% in this case), rubberlike elasticity was observed; the strain at break increased and the maximum tensile stress decreased with increasing grafting ratio, mainly resulting from becoming thicker because the force at break was not much changed. However, the maximum tensile strength of samples with grafting ratios above 3000% was at about the same level (4.2 ± 0.6 MPa). These values are much higher than

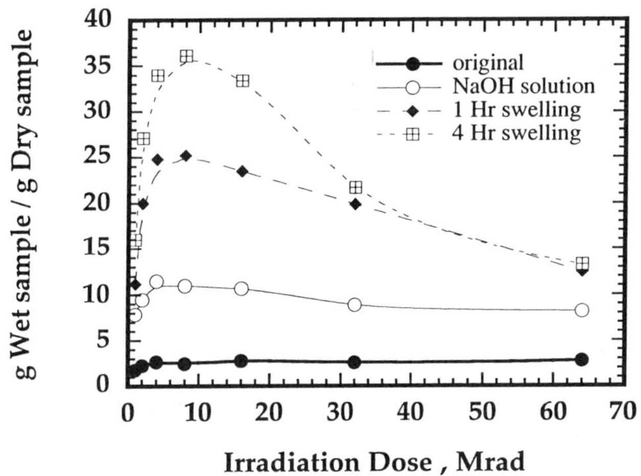

Figure 3. Effects of neutralization on the extent of swelling. Samples grafted at 70°C for 120 minutes in 50% acrylic acid aqueous solutions. Grafting ratios are 352%, 635%, 753% 1051%, 1389%, 3979% and 21810% for samples preirradiated to 1, 2, 4, 8, 16, 32 and 64 Mrad, respectively.

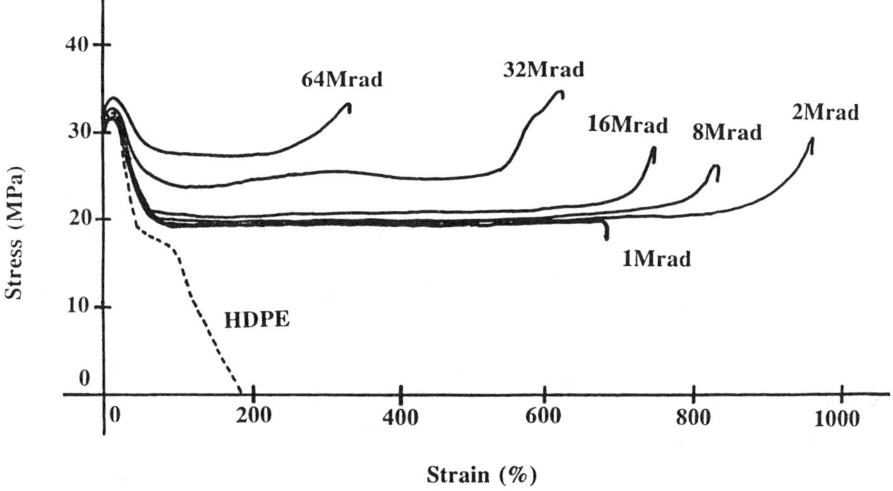

Figure 4. Stress-strain relationship for irradiated but ungrafted HDPE.

those of ordinary hydrogels, e.g. poly (2-HEMA) has a maximum tensile strength of 0.6 MPa and a 250% elongation at break (31). Some typical mechanical properties of the poly(acrylic acid) grafted HDPE samples are shown in Table II.

Table II. Mechanical Properties of Grafted HDPE

Irradiation Dose	64Mrad	32Mrad	8Mrad	1Mrad
Grafting ratio	17640%	2477%	619%	383%
Reaction Time	120min .	120min.	60min.	120min.
*Water Content**	59%	58%	51%	42%
Max. Strength	4.0MPa	5.9MPa	11.1MPa	8.2MPa
Max. Strain	790%	362%	647%	550%
Elastic limit	<150%	-	-	-

*Water content is expressed in (g H_2O/g gel)

Figure 6 shows the relationship between the maximum tensile strength and the grafting ratio for a variety of preirradiation doses. Obviously, the tensile strength decreases with increasing grafting ratio, and increases with increasing preirradiation dose.

The Young's modulus also depends on both grafting ratio and preirradiation dose. However, the effect of grafting ratio dominates. The modulus decreased rapidly with grafting up to ca. 2000% grafting ratio, from 60.1MPa (grafting ratio 182%) to 4.2MPa (1803%), then it gradually decreased to 1.3MPa (20261%).

pH Sensitivity It is well known that the pKa value of poly(acrylic acid) is about 4. When the carboxyl group is dissociated, the polymer chains are extended by the repulsion between the negative charges. Below pH 4, most of the carboxyl groups are not dissociated and the polymer remains unextended. Figure 7 shows the quasi-equilibrium swelling of a 64 Mrad preirradiated sample with a grafting ratio of 20,000% at various pH values. The sample that has been equilibrated at pH 4 was designated to be the reference ($L_o/L_o = 1$). In the first swelling cycle, the measurements were made by soaking the samples in buffered solutions for 20 hours except for pH=7 and 6 which were given 40 and 100 hours, respectively. Swelling between pH 4 and 6 was very minor after soaking for 20 hours (<5% change). Line 2 in Figure 7 shows the deswelling of the samples after they were kept at pH = 10 for 15 hours. The rate of deswelling at pH ≥4 was very slow; e.g. the swelling ratio was 145% after 4 hours and 138% after 80 hours at pH=4. The extent of swelling remained around 150% until the pH was dropped to below 4.

Clearly, there is a pronounced hysteresis. Line 3 shows the second swelling of the hydrogel which was kept at pH = 1 for additional 15 hours after the deswelling cycle described above. The rate of reswelling hydrogel at pH >2 was rapid. Moreover, there was a much smaller hysteresis between the quasi-equilibrium values of the deswelling and the reswelling cycles (lines 2 and 3). (Because of slight differences in the swelling history of the samples, the degree of swelling at the two extremes of the curves in Fig. 7 do not coincide).

Figure 8 shows the pH sensitivity of the swelling-deswelling properties of a sample with a lower grafting ratio. The sample started to swell at pH = 7 but the rate of swelling was very slow. The quasi-equilibrium swelling reached only 135% at pH = 10 after 30 hours. Similar to the data in Figure 7, there is a large hysteresis between the swelling and the deswelling cycle, but a smaller hysteresis between the deswelling and reswelling cycles.

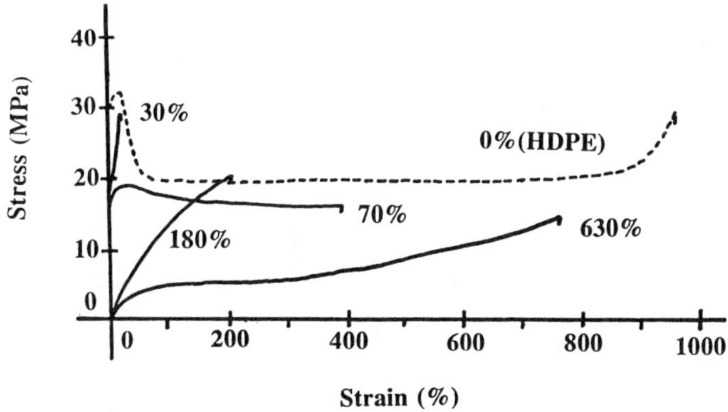

Figure 5. Stress-strain relationship for grafted HDPE.

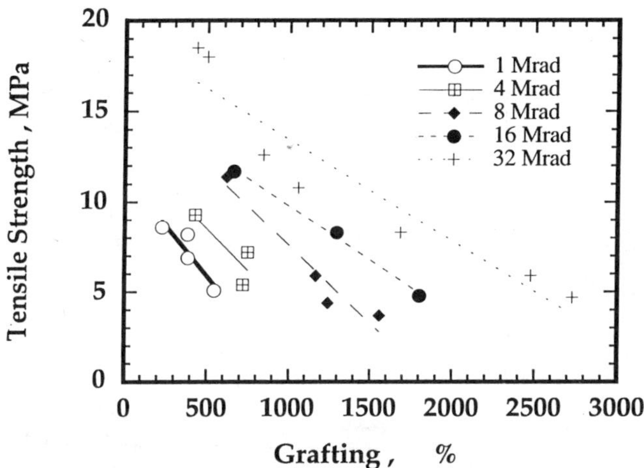

Figure 6. Relationship between tensile strength and grafting ratio for samples irradiated to various doses.

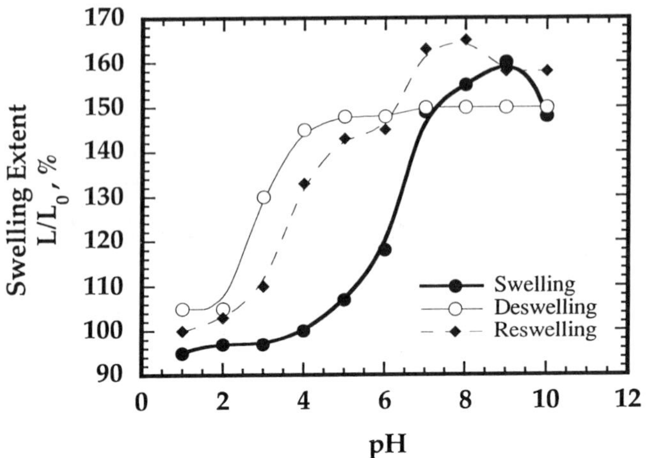

Figure 7. pH dependence of the quasi-equilibrium swelling of highly grafted HDPE. Grafting ratio = 20,000%.

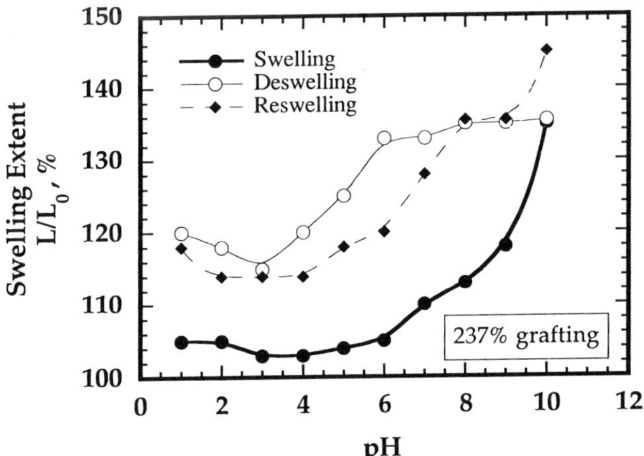

Figure 8. pH dependence of the quasi-equilibrium swelling of relatively lightly grafted HDPE. Grafting ratio = 237%.

The large hysteresis between the first swelling cycle and the deswelling cycle may be attributed to changes in the poly(acrylic acid) structure. Before neutralization, the grafted poly(acrylic acid) chains might be tightly coiled because the grafting reaction was carried out inside polyethylene. This highly coiled structure may significantly retard swelling at a moderate pH. After neutralization, all the carboxyl groups have been dissociated, and the ionic repulsion greatly expanded the originally tight coils of poly(acrylic acid). Upon deswelling, the expanded chains contracted to a much looser coil structure. During the subsequent reswelling, the conformation of the grafted poly(acrylic acid) never experienced the same degree of reorganization, therefore a much smaller hysteresis was observed.

Conclusions

Poly(acrylic acid) grafted HDPE gel was obtained easily by the electron beam preirradiation method. The gel had a high tensile strength (>4.0 MPa) even if it contained more than 50 wt% of water. The preirradiation dose strongly affected the tensile strength of the grafted HDPE. Higher preirradiation dose yielded higher tensile strength at the same grafting ratio. The amount of water uptake by the grafted sample before neutralization of the acid by NaOH was insensitive to the radiation dose or the grafting ratio as long as the grafting was above 200%. After neutralization, the sample preirradiated to 8 Mrad absorbed about twice as much water as samples of higher doses. Crosslinking of polyethylene at higher doses might have contributed to this reduced water absorption capacity. The un-neutralized hydrogel exhibited swelling-deswelling hysteresis when equilibrated in buffered solutions. The hysteresis probably originated in the transformation of the conformation of the poly(acrylic acid) chains from a tightly coiled state to a loosely coiled state after the expansion-contraction cycle caused by the pH changes.

This paper contains a great deal of interesting new data but with only tentative explanations. Some additional data and a more detailed discussion will be presented in Part III of this series.

Acknowledgment

We thank Energy Sciences, Inc. for providing access to electron beam accelerator. The microscopy analyses provided by the Analytical group of Mitsubishi Paper Mills, Ltd. Central Research Laboratories are also highly appreciated.

Literature Cited

(1) Niemoller A; Ellinghorst,G. *Angew. Makromol. Chem.*, **1987**, 148, 1.
(2) Suzuki M.; Ushida, T.; Fujishige, S. *Biorheology*, **1986**, 23, 274.
(3) Osada, Y.;Hasebe, M. *Chem. Lett.*, **1986**, 1285.
(4) Japan Patent 53-46199, **1978**.
(5) Vitta, S. B.; Stahel, E. P.; Stannett, V. T. *J. Macromol. Sci.*, A, **1985**, 22, 579.
(6) Vitta, S. B.; Stahel, E. P.; Stannett, V. T. *J. Appl. Polym. Sci.*, **1986**, 32, 5799
(7) Hegazy, E.-S. A.; Mokhtar, S. M.; Osman, M. B. S.; Motafa, A. I.-K. B. *Radiat. Phys. Chem.*, **1990**, 36, 365.
(8) Charlesby, A.; Fydelor, P. J. *Radiat. Phys. Chem.*, **1972**, 4 107.
(9) Guimon, C. *Radiat. Phys. Chem.*, **1979**, 14, 841.
(10) Lawler, J. P.; Charlesby, A. *Radiat. Phys. Chem.*, **1980**, 15, 595.
(11) Sidorova, L. P.; Aliev, A. D.; Zlobin, V. B.; Aliev, R. E.; Chalkh, A. E.; Kabanov, V. Ya. *Radiat. Phys. Chem.*, **1986**, 28 407.

(12) Grushevskaya, L. N.; Aliev, R. E.; Kabanov, V.Ya. *Radiat. Phys. Chem.*, **1990**, 36, 475 .
(13) Ishigaki, I.; Sugo, T.; Senoo, K.; Takayama, T.; Machi, S.; Okamoto, J. ; Okada, T. *Radiat. Phys. Chem.*, **1981**, 18, 899.
(14) Ishigaki, I.;Sugo, T.; Senoo, K.; Okada, T.; Okamoto, J.; Senoo, K. *J. Appl. Polym. Sci.*, **1982**, 27, 1033.
(15) Ishigaki, I.; Sugo, T.; Takayama, T.; Okada, T.; Okamoto, J.; Machi, S. *J. Appl. Polym. Sci.*, **1982**, 27, 1043.
(16) Tanso, S.; Yoshida, S.;Senoo, K. *Yuasa-Jiho* (Japanese), **1985**, 59, 25.
(17) Omichi, H.; Okamota, J. *J. Appl. Polym. Sci.*, **1985**, 30, 1277.
(18) Okamoto, J.; Sugo, T.; Katakai, A.;Omichi, H. *J. Appl. Polym. Sci.*, **1985**, 30, 2967.
(19) Furusaki, S.; Okamoto, J.; Sugo, T.; Saito, K. *Chem. Eng.*, **1987**, 521.
(20) Saito, K.; Yamada, S.; Furusaki, S.; Sugo, T.; Okamoto, J. *J. Membr. Sci.*, **1987**, 34, 107.
(21) Omichi, H.; Chundury, D.; Stannett, V. T. *J. Appl. Polym. Sci.*, **1986**, 32, 4827.
(22) Okamoto, J. *Membrane* (Japanese), **1989**, 24, 277.
(23) Pimg, Z. H.; Nguyen, Q. T.; Clement and, R.; Neel, J. *J. Membr. Sci.*, **1990**, 48, 297.
(24) Hoffman, A. S. *IAEA-TECDOC-486*, Viena, Austria, **1988**, 25.
(25) Kaetsu, I. *IAEA-TECDOC-486*, Viena, Austria, **1988**, 153.
(26) Guimon, C. *Radiat. Phys. Chem.*, **1986**, 14, 841.
(27) Harada, J.; Chern, R. T.; Stannett, V. T. *RadTech '90 in North America*, **1990**, 1, 493.
(28) Harada, J.; Chern, R. T.; Stannett, V. T. In *Radiation Effects on Polymer*; ACS Symp. Series, in press **1991**.
(29) Nablo, S. V.; Frutiger, W. A. *Radiat. Phys. Chem.*, **1981**, 18, 1023.
(30) Ratner, D. B.; Hoffman, A. S. In *Synthetic Hydrogels for Biomedical Applications* , Andrade J. D., Ed.; ACS Symp. Ser. 31, **1976**, 1.
(31) Stoy, V. A. *J. Biomat. Appl.*, **1989**, 3, 552.

RECEIVED August 26, 1991

Chapter 6

Relationship Between Poly(acrylic acid) Gel Structure and Synthesis

Yu-Ling Yin[1], Robert K. Prud'homme[1], and Fredrick Stanley[2]

[1]Department of Chemical Engineering, Princeton University, Princeton, NJ 08544
[2]Dow Chemical Company, Midland Division, Building 768, Midland, MI 48674

Water swellable gels are used widely in consumer product applications where their ability to imbibe up to 1000 times their own weight in water make them uniquely suited as absorbents. Their ability to swell arises from polyelectrolyte interactions and molecular structure of the gel network. A brief review of theories that relate the molecular structure of gels to their properties is given in the first part of this chapter. The main properties of interest are swelling volumes and elastic moduli. In the second part of the chapter we present a study of poly(acrylic acid) (PAA) gels crosslinked with 4 or 6-functional crosslinkers (bisacrylamide and TMPTA). The goal of the study is to relate the chemistry used to form the gel to the molecular structure, and finally to the end-use performance of the gel. When the degree of the neutralization of the monomer is zero, a linear relation exists between shear modulus and both crosslinker concentration and functionality. This is in accord with classical network theories. Increasing the neutralization of the monomer increases electrostatic interactions in the gel which increases swelling, but it also impedes efficient formation of the gel network structure. At high degrees of neutralization the crosslinkers are not incorporated uniformly and the extractable sol-fraction increases sharply while the modulus decreases. A correlation is presented that relates swelling to both the degree of neutralization and the crosslinker concentration.

Crosslinked polymer networks were studied initially because of their importance in vulcanized (i.e. crosslinked) rubber. These studies provide the basis for the classical theory of rubber elasticity. The relationship between molecular network architecture and bulk properties is relatively well understood for these systems, although debates remain about fine points of the theory and experimental problems in making model networks with which to test the theory. Recently, there has been considerable interest in water-swellable polyelectrolyte gels (1-5). The commercial interest has come from applications of these gels as absorbents in diapers and personal products, and from the use of these gels in controlled drug release. The development of gels for these

0097–6156/92/0480–0091$06.75/0
© 1992 American Chemical Society

applications has come without benefit of theories to relate molecular structure to swelling capacity. The reasons that these gels are difficult to model are three-fold: 1) polyelectrolyte effects are dominant and satisfactory theories for the properties of polyelectrolyte chains have been developed only recently, 2) water is a particularly difficult solvent to treat theoretically-- hydrogen bonding, and specific polar interactions with itself and the polymer must be incorporated, and 3) it is difficult to make model networks with the free-radical chemistry that is required for the synthesis of these gels.

At the same time that there has been commercial interest, the chemical physics community has also begun work on these systems because they display interesting critical phenomena and phase transitions (6-11). For polymer gels the time scales for critical fluctuations are long and can be more easily studied than fluctuations in low molecular weight fluids.

In this chapter we first present a brief review of models that relate molecular structure of gels to their bulk properties. There are two primary "bulk property" measurements that are used to study networks: measurements of the shear modulus, and measurements of swelling. The modulus measurements are more straightforward because they do not explicitly require a model of the solution thermodynamics. Swelling measurements are simple to perform, but to interpret the results, a model is required both for the mechanical properties (i.e. modulus in dilatation) and for the water/polymer solution thermodynamics. The three components needed to model swelling are: 1) the elastic free energy, 2) the free energy of mixing of the polymer and solvent, and 3) the electrostatic free energy. We present experimental results on the relationship between properties and synthesis of poly(acrylic acid) (PAA) gels crosslinked by bisacrylamide (Bis) and trimethylolpropanetriacrylate (TMPTA) -- a sparingly soluble crosslinker. The primary variables of interest are the degree of neutralization of acrylic acid during the synthesis, the concentration of crosslinker, and the functionality of the crosslinker. Many of the experimental results link this work to previous, well-established results for other gel systems -- such as the dependence of swelling on the functionality of the crosslinker (for gels formed from non-ionized monomer) that is predicted by classical network theory. However, there are striking deviations from known theories -- as in the case of swelling of gels formed from partially ionized monomers. Some qualitatively explanations are given to account for the observed experimental phenomena.

Theories

Network Theories. Theories of rubber elasticity which neglect interaction between network strands are called "phantom" network theories. The strands comprising the network are assumed to behave as random coils, the displacements of the mean junction positions are assumed to be affine, and the free energy change of the network is assumed to be the simple sum of contributions from changes in the distribution of configurations of individual strands. For these theories, the relation between the shear modulus, G, and network structure is given by the following equation (12,13)

$$G=(\nu - \mu)kT \tag{1}$$

where ν and μ are the concentrations of elastically active strands and junctions, respectively. A junction is elastically active if three or more of its arms are independently attached to the network. A strand is elastically active if it is attached at both ends to an active junction. If all strands are attached and there is only one type of junction with functionality f, then

$$\mu=2\nu/f \tag{2}$$

Thus, for a perfect tetrafunctional network

$$G=\nu kT/2 \tag{3}$$

In the phantom network, junctions can fluctuate about their mean position due to Brownian motion. This gives rise to the μ term in the modulus expression. The higher the junction functionality, f, the less will be the fluctuations. If all junction fluctuation is suppressed, then

$$G=\nu kT \tag{4}$$

which is the result obtained in Flory's early work and can be obtained directly if one assumes affine deformation of the network junctions. To allow for intermediate behavior, Dossin and Graessley have used (14)

$$G=(\nu-h\mu)kT \tag{5}$$

where h is an empirical parameter between 0 and 1.

Measurements on rubber networks have shown that the shear modulus can be substantially larger than νkT. The major source of these deviations is argued to be chain-chain interactions termed "entanglements". Entanglements arise because the chains are, in fact, not phantom-like and cannot pass through each other. At small strains, these entanglements can raise the modulus considerably. It has been found that when a high molecular weight un-crosslinked polymer is tested under dynamic conditions, the shear modulus has an essentially constant value, G°_{N} over a wide range of frequency. This suggests that a temporary entangled network is present with a lifetime longer than the experimental time scale. G°_{N} is the "plateau" modulus. If a portion of this network becomes permanently fixed or trapped by the addition of cross-links, then the modulus should be greater than a phantom network having the same structure. Graessley (14) and Langley (15) have suggested that these trapped entanglements can be simply added to the modulus given by Eq.1:

$$G=(\nu-h\mu)RT+G^{\circ}_{N}T_{e} \tag{6}$$

where T_e is the fraction of the temporary intanglements arising from topological interactions in an uncrosslinked polymer system, that are permanently trapped by covalent crosslinks. Macosko and co-workers (15) have shown that ν and T_e are sensitive to the extent of polymerization reaction. The entanglement contribution to the plateau modulus G°_{N} decreases with concentration according to

$$G^{\circ}_{N}\sim c_p^2\, G^{\circ}_{Nb} \tag{7}$$

where c_p is the polymer concentration on a volume fraction basis and G°_{Nb} is the plateau modulus of the undiluted (ie. bulk) sample. Therefore, it is expected that the contributions from entanglements should be relatively small for swollen polyelectrolyte gels which have polymer concentrations from 25 % to 0.025 %.

It has been noted that the random coil model is unsatisfactory at high extension and that it is totally unacceptable for displacement lengths approaching full extension of the chains. For polyelectrolyte networks the charges on the polymer backbones repel each other, which leads to local chain stiffening and long-range excluded volume effects. These effects cause the network to swell tremendously. Thus in the treatment of swelling of polyelectrolytes, non-Gaussian elasticity should be taken into account in some cases. At present, there are two ways to cope with the problem, one is by using the inverse Langevin function (16), another is by using the wormlike chain model (17).

Prediction of Swelling. As mentioned above, the three factors that contribute to the swelling behavior of a polyelectrolyte gel are the elastic free energy, ΔF_c, the free energy of mixing, ΔF_m,, and the electrostatic free energy, ΔF_e. At equilibrium, the total free energy of the system equals zero:

$$\Delta F = \Delta F_c + \Delta F_m + \Delta F_e = 0 \tag{8}$$

How to model the three terms in the above expression is the main issue in developing a theory of the swelling of polyelectrolytes. In the case of low ionic strength solutions, Flory obtained a simple expression for the swelling equilibrium of a polyelectrolyte by considering only the osmotic effect and elastic free energy (16),

$$q^{\frac{2}{3}} = \left(\frac{V_0}{\nu_e}\right)\frac{i}{z v_u} \tag{9}$$

where q is swelling ratio, i.e. the swollen volume of the gel relative to its initial volume, V_0 is the molar volume of the solvent, ν_e is the number density of strands, i is the fractional degree of neutralization per monomer unit, v_u is the volume of a monomer unit, and z is the valence of ionizable group. Flory's theory predicts that the swelling ratio increases with the degree of neutralization and decreases with crosslink density, monotonically. In addition to Flory's model, several other more complicated models have been proposed-- Katchalsky (18,19), Hasa (20,21), and Konak (22) -- which either treat the electrostatic interactions in a different way or consider a non-Gaussian chain extension correction. Recently, Prausnitz and co-workers (23,24) considered specific interactions, such as hydrogen bonding, in hydrogel systems. They have been able to predict the changes of swelling with crosslink density quite well in some gel systems in which electrostatic interactions are not dominant. All models predict that swelling ratio decreases with crosslink density. But some of these models have contradictory predictions on how the degree of neutralization affects the swelling behavior. Katchalsky's model, like Flory's, predicts that the swelling ratio increases monotonically with the degree of neutralization; Hasa's model, however, predicts that when the degree of neutralization reaches a critical value, the swelling ratio will not change further due to counter-ion condensation; Konak's model predicts an even more dramatic result: a maximum will exist on the curve of the swelling ratio versus degree of neutralization. More work is needed, both experimentally and theoretically, to clarify the regions where each theory is appropriate.

Experiments

Synthesis. Materials. Acrylic acid (AA) with an inhibitor content of 200 ppm was obtained from Polysciences. Inc.. The monomer was vacuum distilled to remove the inhibitor and subsequently stored in a refrigerator. The melting point of poly(acrylic acid) is 13 C, so the monomer is a crystalline solid during storage. It must be thawed without heating to avoid runaway exothermic polymerization of the unstabilized monomer. Monomer was also obtained from Hoechest Celanese Corporation. It is reported that the commercial monomer has lower stabilizer levels than the monomers from laboratory supply companies. It was used without vacuum distillation. The following substances are obtained from the indicated sources and use as supplied:

Trimethylolpropanetriacrylate (TMPTA)
(Aldrich Chemical Company, Inc.)
Vinol- 523 polyvinyl alcohol
(Air Products Inc.)
L-Ascorbic acid (sodium salt 99%)
(Aldrich Chemical Company, Inc.)
Sodium persulfate (98%)
(Aldrich Chemical Company, Inc.)
TM-80 (Diethlenetriaminepentaacetic acid pentasodium salt (40% in water)
(Kodak Inc.)
Bisacrylamide (Bis)
(Bethesda Research Laboratories Inc.)

Synthesis procedure(25). A procedure was developed to make gel slabs and disks. In a 135 mm I-Chem bottle, the following substances are added in order: 20.83 g acrylic acid , TMPTA, 0.5ml Vinol-523 (1 g Vinol-523 in 278 g water), 1 drop of TM-80, 50% NaOH solution and 53.76 g DI water. House vacuum is used to de-gas the solution for 1 hour. Then, 200 ml ascorbic acid (0.1550 g in 10ml DI water) and 400 ml sodium persulfate solution (0.7750 g in 10ml DI water) are added to the mixed solution and the solution is degased for more 5 minutes. Using a pipet, 5ml of solution are taken from the bottom of the bottle and poured between the parallel glass plate mold. When TMPTA is used as the crosslinker, Vinol-523 (a polyvinyl alcohol) is used as dispersing agent. The free radicals are generated using sodium persulfate as initiator and TM-80 and ascorbic acid as co-initiators. The mold is placed into a water bath maintained at a temperature of 81°C. After 1 hour, the mold is taken out of the bath. The slab gel is removed from the mold and wraped with two layers of plastic film and placed in a plastic bag. The sample is stored in the refrigerator until use. Since the thickness of the sample is only 1 mm, the temperature can be better controlled during these polymerizations than in the bulk polymerizations.

Samples. The compositions of a typical series of samples is given in Table 1, and Table 2, where the "appearance" and "pH" refer to that of monomer solutions before polymerization.

Table 1. Variation of Crosslinking Density
(Slab gels,DN=70%,TMPTA)

	AA (g)	TMPTA (g)	NaOH (50%,g)	DI (g)	V-523 (ml)	TM-80 (drop)
1*	20.83	0.0175	16.17	53.76	0.50	1
2	20.83	0.0261	16.17	53.76	0.50	1
3	20.83	0.0344	16.17	53.76	0.50	1
4*	20.83	0.0525	16.17	53.76	0.50	1
5	20.83	0.0688	16.17	53.76	0.50	1
6*	20.83	0.1050	16.17	53.76	0.50	1
7*	20.83	0.1575	16.17	53.76	0.50	1
8*	20.83	0.2100	16.17	53.76	0.50	1

* The corresponding samples with same compositions(mole) but with Bis as crosslinker have been made.

Extractables Analysis. To assay for monomer and polymer not covalently linked to the gel network, the following procedure was developed. About 0.2 g of sample is put into 100 ml of 0.1538M NaCl solution and shaken once a day for one week. then, 40 ml sample of the salt solution with extracted acrylic acid is titrated by first raising the pH to 7.7 with 0.02 N NaOH solution and then titrating to pH 4.2-4.3 using a standard HCl solution.

A 40 ml sample of 0.9% NaCl solution is titrated from pH 7.7 to pH 4.2-4.3 as a control, and the volume of the titrate required is called V_b.

The following equation is used to calculate the percent extractables (E%):

$$E\% = \frac{(V_t - V_b) \times N}{1000} \times \frac{(\bar{M} \times 2)}{W_0} \times \frac{100}{40} \times \frac{1}{c_p} \tag{10}$$

where, N = normality of the standard HCl solution
V_t = volume of titrant (ml)
V_b = volume of titrant used in blank titration (ml)
W_0 = weight of the gel sample used (g)
c_p = the polymer content in the gel sample (wt %)
$\bar{M}$ = the "equivalent" molecular weight of monomer, which depends on the degree of neutralization

Swelling Experiments. The swelling of gel samples was measured using the following procedure: A circular gel with a diameter of 14-16 mm, and thickness of 1.0-1.5 is cut from a slab gel. The circular gel is put into a 135 ml glass bottle with sealed cap, and 100 g of 0.1538 M NaCl solution (or DI water) is poured into the

Table 2. Variation of Both Crosslinking Density and Degree of Neutralization(Slab Gels)*

	TMPTA (mole%)	DN(%)**	Appearence	pH
1-1	0.02	0	clear	1.75-1.76
1-2	0.04	0	clear	1.75-1.76
1-3	0.06	0	clear	1.75-1.76
1-4	0.12	0	clear	1.75-1.76
1-5	0.18	0	clear	1.75-1.76
1-6	0.24	0	clear	1.75-1.76
2-1	0.02	20	clear	3.58-3.60
2-2	0.04	20	clear	3.58-3.60
2-3	0.06	20	clear	3.58-3.60
2-4	0.12	20	slightly cloudy	3.58-3.60
2-5	0.18	20	cloudy	3.58-3.60
2-6	0.24	20	cloudy	3.58-3.60
3-1	0.02	40	clear	4.00-4.10
3-2	0.04	40	clear	4.00-4.10
3-3	0.06	40	slightly cloudy	4.00-4.10
3-4	0.12	40	cloudy	4.00-4.10
3-5	0.18	40	cloudy	4.00-4.10
3-6	0.24	40	cloudy	4.00-4.10
4-1	0.02	60	clear	4.42-4.48
4-2	0.04	60	slightly cloudy	4.42-4.48
4-3	0.06	60	cloudy	4.42-4.48
4-4	0.12	60	cloudy	4.42-4.48
4-5	0.18	60	cloudy	4.42-4.48
4-6	0.24	60	cloudy	4.42-4.48
5-1	0.02	80	slightly cloudy	4.88-4.97
5-2	0.04	80	cloudy	4.88-4.97
5-3	0.06	80	cloudy	4.88-4.97
5-4	0.12	80	cloudy	4.88-4.97
5-5	0.18	80	cloudy	4.88-4.97
5-6	0.24	80	cloudy	4.88-4.97

* Total molarity of monomer is 3.2 M.
** Degree of neutralization.

bottle. The bottle is sealed to prevent the uptake of carbon dioxide which might cause shifts in pH. The swelling polymer is allowed to stand for two weeks at room temperature with one change of the solvent after the first week. After the completion of swelling, the swollen polymer is removed from the bottle, free water is removed from the disk, and the disk is weighed. The swelling ratio is given by:

$$\frac{V}{V_0} = \frac{W}{W_0 c_P} \tag{11}$$

where W and W_0 are the weight of swollen and unswollen gel, respectively, and c_p is the polymer content in gel as polymerized. This equation neglects the extraction of monomer not covalently attached to the network.

Modulus Measurement. The storage and loss moduli, G' and G", respectively, of the un-swollen slab gels are measured using a Rheometrics System IV rheometer with a parallel plate geometry at room temperature. Rate sweeps were performed to determine the frequency dependence of the elastic modulus, and strain sweeps were performed to determine the region of linear viscoelastic response.

Results and Discussion

Extractables. The results of measurements of the amount of extractable acid are shown in Table 3. The extractables decrease with increasing crosslink density and increase with increasing degree of neutralization of the monomer. The former is easy to understand since with the increase of crosslinking density, the polymer chains will have more chance to be incorporated into the network and the number of "single" polymer chains will decrease. The effect of monomer neutralization has not been reported previously. It is probably due to the ionized monomer being electrostatically repelled from the gel network, and therefore not being efficiently incorporated.

Table 3. Extractables Analysis

Sample	Degree of Neutralization DN (%)	Crosslinker Concentration (mole %)	M(g /mol)*	E (Sol fraction) (wt %)
4-1	60.00	0.02	85.20	11.37
4-2	60.00	0.04	85.20	7.02
4-3	60.00	0.06	85.20	8.65
4-4	60.00	0.12	85.20	6.70
4-5	60.00	0.18	85.20	5.39
4-6	60.00	0.24	85.20	4.68
1-3	00.00	0.06	72.00	5.13
2-3	20.00	0.06	76.40	5.32
3-3	40.00	0.06	80.80	6.14
4-3	60.00	0.06	85.20	8.65
5-3	80.00	0.06	89.60	13.35

* The equivalent molecular weight of monomer, which depends on the degree of neutralization.

Comparison of Crosslinkers. In commercial applications, TMPTA is used to form polyacrylate gels because it is is biologically safe, whereas bis-acrylamide (Bis) is a neurotoxin. From a fundamental point of view the two crosslinkers are interesting for two reasons: 1) tetra-functional and hexa-functional gels can be formed using the two crosslinkers and these can be used to test classical network theories, and 2) because TMPTA has only a limited solubility in neutralized monomer, we can study the affect on gel structure of continuously supplying crosslinker during the course of the reaction versus having all the crosslinker available initially. The comparison of the gels made by the two crosslinkers help us to understand the differences in gel structure produced by different in crosslinkers. The structure of TMPTA is:

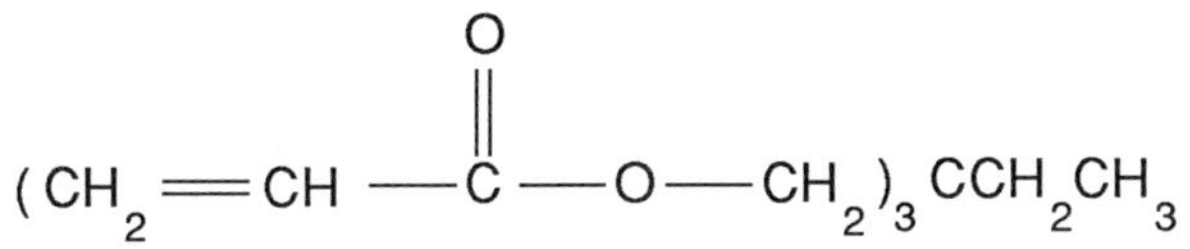

$$(CH_2{=}CH{-}\overset{\overset{\displaystyle O}{\|}}{C}{-}O{-}CH_2)_3CCH_2CH_3$$

TMPTA is soluble in the acrylic acid solution but is sparingly soluble when acrylic acid solution is partially neutralized. Bis-acrylamide is soluble in the acrylic acid monomer solution whether the monomer is in the acid or neutralized form. Its structure is:

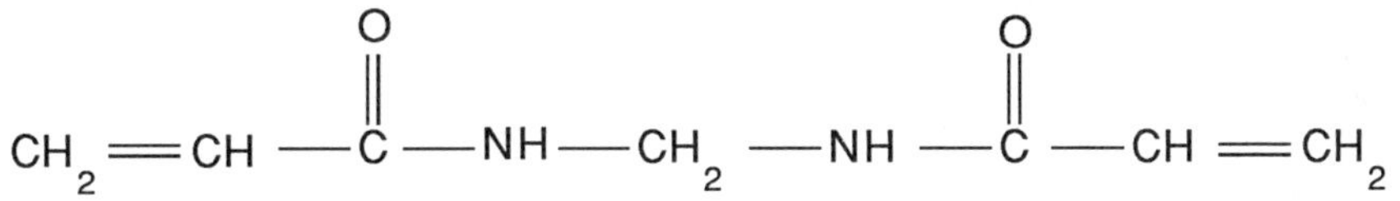

$$CH_2{=}CH{-}\overset{\overset{\displaystyle O}{\|}}{C}{-}NH{-}CH_2{-}NH{-}\overset{\overset{\displaystyle O}{\|}}{C}{-}CH{=}CH_2$$

We first compare the swelling behavior of the samples with the degree of neutralization equal to zero (DN=0, i.e. the monomer is in the acid form) because in this case both crosslinkers are soluble. Figure 1 gives the results of swelling in salt solution for gels with different crosslinkers. The horizontal axis is the concentration in mole percentage of the crosslinker with respect to the monomer . Noting that the functionality of the TMPTA is 6 and that of Bis is 4, if we multiply the mole percentage of TMPTA by 3/2 and Bis by 1 and replot the the swelling ratio versus the "normalized" crosslink density, Fig. 2 is obtained. The figure shows that the swelling ratio data from the gels with different crosslinkers coincide almost perfectly. When normalizing for the functionality of the crosslinker, the same result is obtained if the samples are swollen in DI water as shown in Figs. 3 and 4. Those results suggest that TMPTA is as effective as Bis as long as it is soluble in the solution and the difference in the functionality has been considered. This shifting, according to Eq. 4, suggests that junction fluctuations are not significant. It might be argued that the highly swollen state of the gel and high molecular weight between crosslinks make junction fluctuations small relative to mean junction spacing. It seems also reasonable from the results to assume that two crosslinkers may have the similar reactivity toward the monomer, and gels made using the two crosslinkers have similar microstructure, since the gels made using two crosslinkers have the same swelling properties. Here "similar" means the length distribution of the polymer chains is similar. When the acrylic acid solution is partially neutralized, the picture is quite different. Figure 5 gives the comparison of the swelling behavior of the gels with different crosslinkers (DN=70%) in salt solution. The most obvious difference is that the swelling ratios of the samples with TMPTA crosslinker are greater than those of the samples with Bis crosslinker when the swelling ratios are plotted versus the

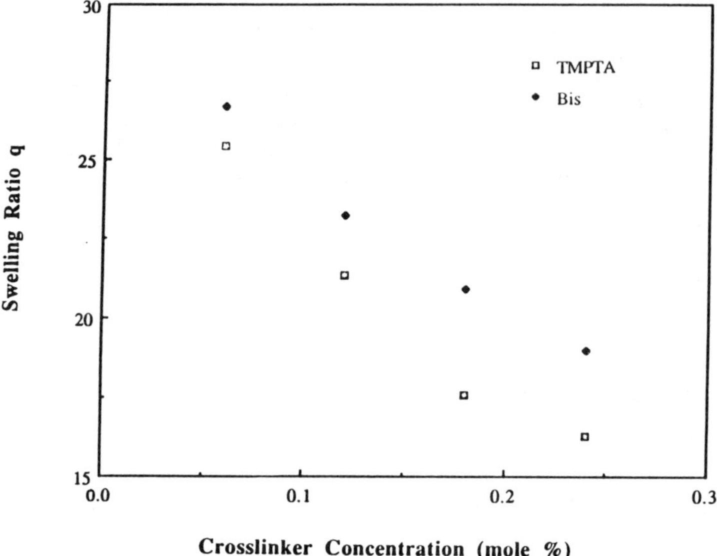

Figure 1. Swelling ratio versus crosslinker concentration for gels made from 4-functional (bisacrylamide) and 6-functional (TMPTA) crosslinkers with DN (degree of neutralization) = 0 %. Gels are polymerized at 3.2 M monomer concentration and swollen in 0.1538 M (0.9 % wt.) NaCl solution.

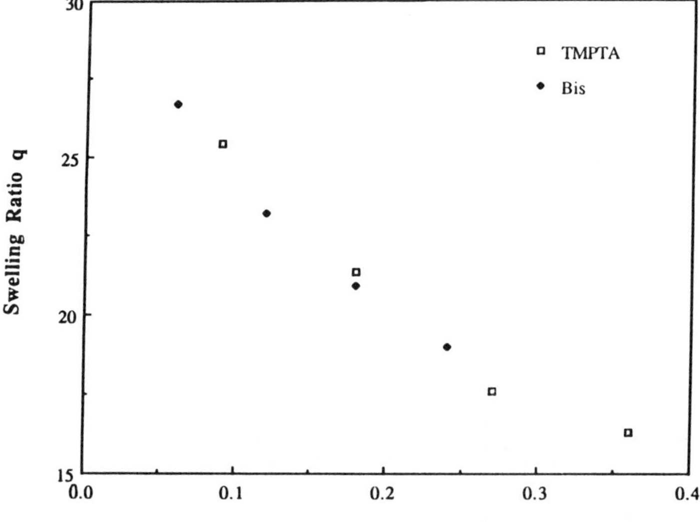

Figure 2. Swelling ratio versus normalized crosslinker concentration. The data from Fig.1 are reploted versus the crosslinker concentration multiplied by a coefficient that is proportional to the functionality of the crosslinker.

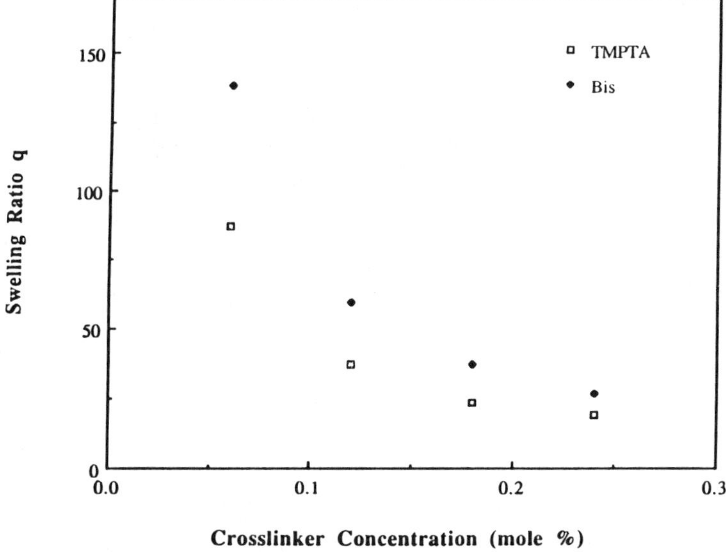

Figure 3. Swelling ratio versus crosslinker concentration for gels made from 4-functional (bisacrylamide) and 6-functional (TMPTA) crosslinkers with DN = 0 %. Gels are polymerized at 3.2 M monomer concentration and swollen in DI water.

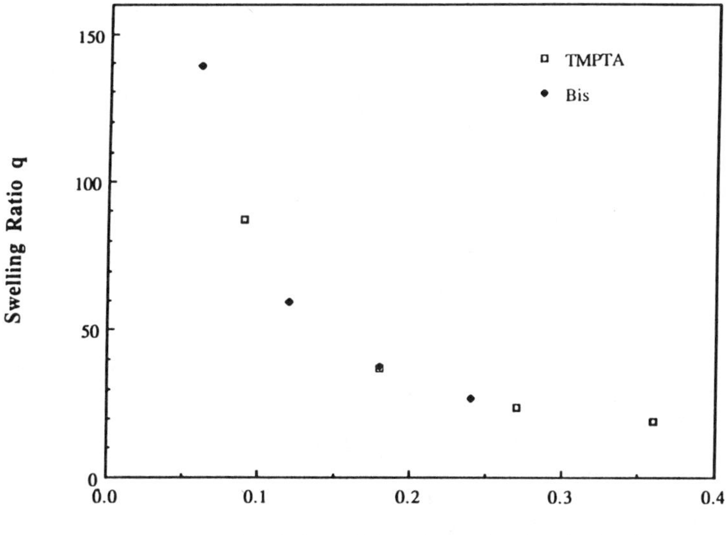

Figure 4. Swelling ratio versus normalized crosslinker concentration. The data from Fig.2 are reploted versus the crosslinker concentration multiplied by a coefficient that is proportional to the functionality of the crosslinker.

crosslinker concentrations. If we replot the same experimental data by using the "normalized" crosslink density, the difference become even more obvious as shown in Fig. 6. Apparently, the sparingly soluble TMPTA is not as "efficient" as soluble crosslinkers. We use the following procedure to find the efficiency of the crosslinker from Fig. 6 when compared at equal functionality. First we draw horizontal line on the diagram. From the intersection we draw lines down to the x-axis and read the values of the crosslink density for the two samples. Next we assume that Bis is uniformly incorporated into the network. Since the swelling ratio is the same along the horizontal line, the effectiveness of TMPTA at forming network strands can be deduced from the ratio of normalized TMPTA concentration to the normalized Bis concentration. This defines the "efficiency" of TMPTA relative to Bis. Some efficiency data thus obtained for the samples with DN=70% are given in Table 4.

Table 4 The calculation of the Efficiency of TMPTA

TMPTA Conc. Added (mole%)	Equivalent Conc. of Bis (mole %)	TMPTA Conc. Apparent (mole %)	Efficiency of TMPTA Crosslinking (%)
0.02	0.018	0.012	60
0.06	0.05	0.033	55
0.12	0.08	0.053	44
0.18	0.13	0.086	47
0.24	0.18	0.012	50

The TMPTA crosslinker efficiency is approximately 50% over a wide range of crosslinker concentrations. How this efficiency varies with the degree of neutralization of the monomer and why adding crosslinker slowly during the course of the reaction increases swelling is a subject of our continued research.

Swelling Behavior. The change in swelling ratio of the gel polymerized from 70% neutralized monomer (with TMPTA as crosslinker) in salt solution versus crosslinker concentration is shown in Fig. 7. The swelling ratio monotonically decreases with an increase of crosslink ratio. When the crosslink ratio is below 0.06, the swelling ratio increases dramatically. The swelling behavior of samples with other degrees of neutralization in salt solution is shown in Fig. 8. All curves show similar trends, but the degree of neutralization changes the slopes of the curves: the higher the degree of the neutralization, the steeper is the curve. When the samples are swollen in DI water rather than salt solution large difference in the swelling behavior exists between the samples with DN less than 20% and those with DN higher than 20% as shown in Fig. 9. For example, the swelling ratio of the sample with DN=0 and crosslink density equal to 0.06 is about 400 while the sample with DN=20% and the same crosslink density is 1600. However for the sample with DN higher than 20%, the shapes of the curves are very similar and the absolute difference in swelling ratio is relatively small.

When the amount of the crosslinker is fixed, the change in the swelling behavior with degree of neutralization (DN) in salt solution is shown in Fig. 10. The figure shows that the swelling ratio increases monotonically with the degree of the neutralization. The slope of the curves decrease with increasing crosslink density.

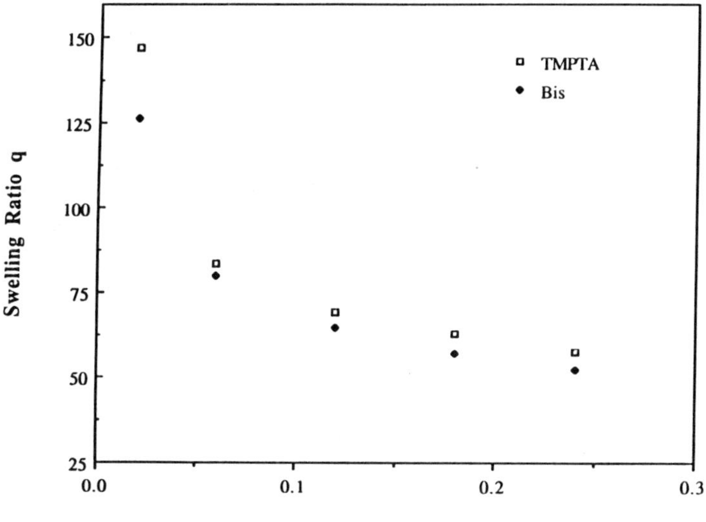

Figure 5. Swelling ratio versus crosslinker concentration for gels made from 4-functional (bisacrylamide) and 6-functional (TMPTA) crosslinkers with DN (degree of neutralization) = 70 % Gels are polymerized at 3.2 M monomer concentration and swollen in 0.1538 M (0.9 % wt.) NaCl solution.

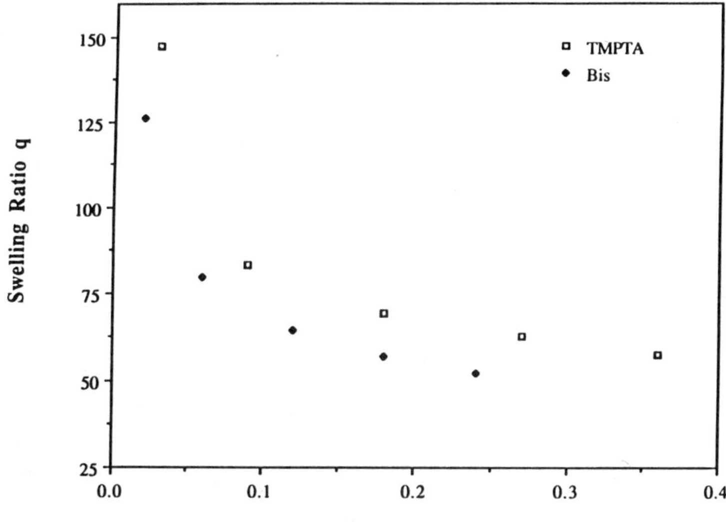

Figure 6. Swelling ratio versus normalized crosslinker concentration. The data from Fig. 5 are reploted versus the crosslinker concentration multiplied by a coefficient that is proportional to the functionality of the crosslinker used.

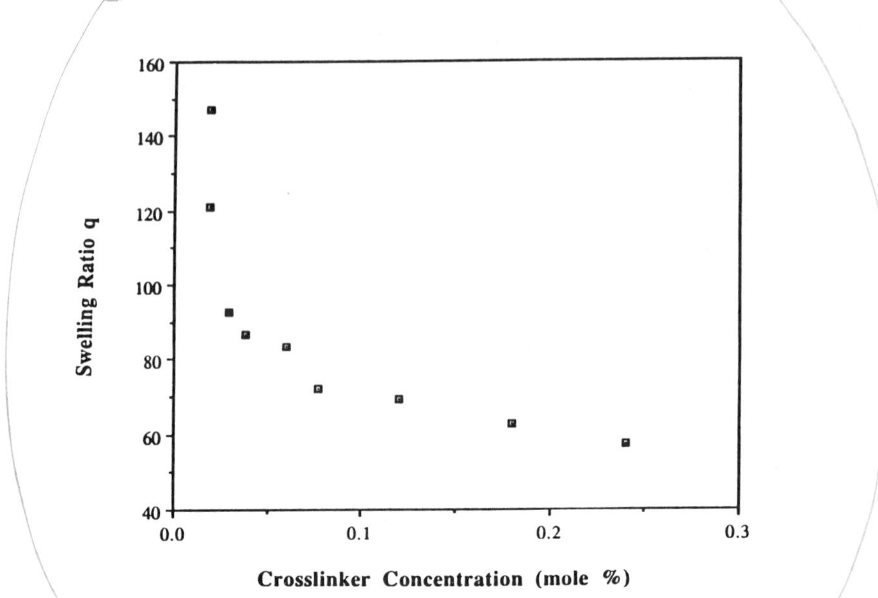

Figure 7. Swelling ratio versus crosslinker concentration for the gels with DN=70 % in 0.1538 M (0.9 % wt.) NaCl solution.Gelsare polymerized at 3.2 M monomer concentration and swollen in 0.1538 M (0.9 % wt.) NaCl solution.

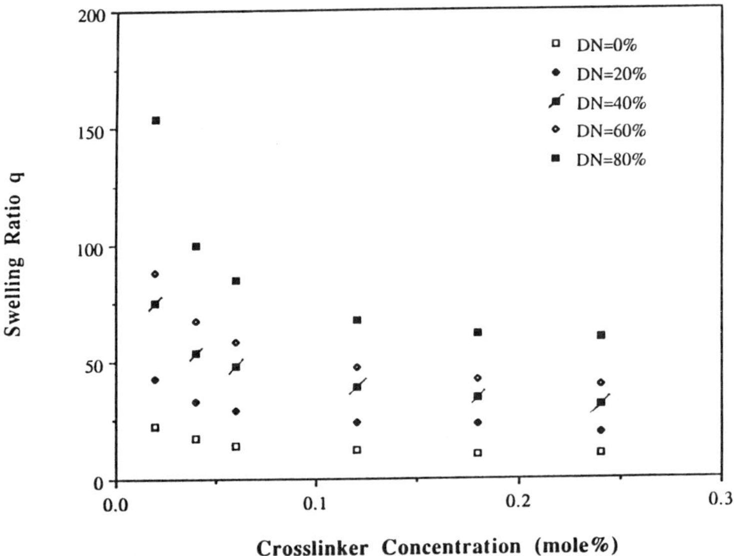

Figure 8. Swelling ratio versus crosslinker concentration for the gels with different degrees of neutralization. Gels are polymerizd at 3.2 M monomer concentration and swollen in 0.1538 M (0.9 % wt.) NaCl solution.

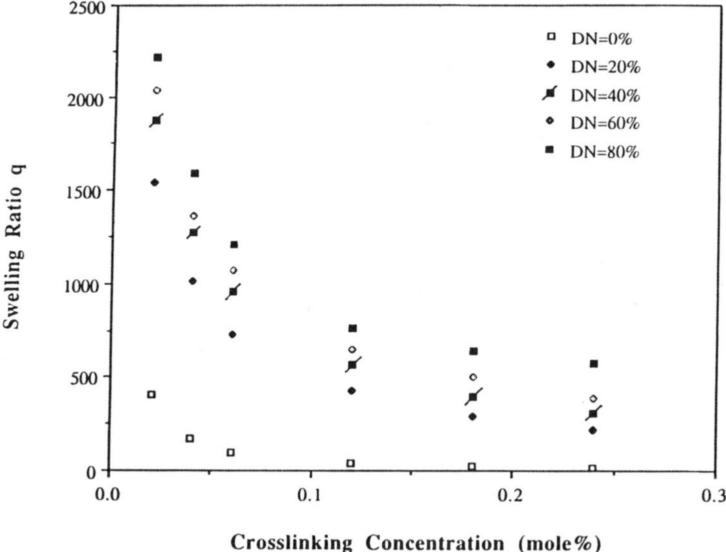

Figure 9. Swelling ratio versus crosslinker concentration for the gels with different degrees of neutralization. Gels are polymerizd at 3.2 M monomer concentration and swollen in DI water.

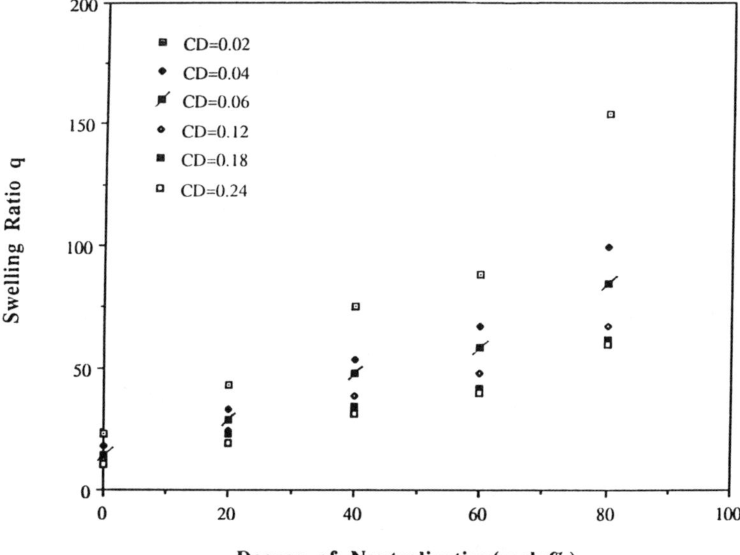

Figure 10. Swelling ratio versus degree of neutralization for the gels with different crosslinker concentrations. Gels are polymerizd at 3.2 M monomer concentration and swollen in 0.1538 M (0.9 % wt.) NaCl solution.

Table 2 shows the crosslinker becomes less soluble with increases in the degree of neutralization, and the "efficiency" of the crosslinker decreases, also. The low efficiency of crosslinks leads to a swelling ratio higher than expected. However, according to Manning's counterion condensation theory (26,27), the free ions will condense on the polymer chains when the degree of neutralization exceeds a critical value. In other words, increasing the degree of neutralization will not increase the number of free ions when the degree of neutralization exceeds a critical value. Counter-ion condensation makes the swelling ratio less than expected. These two factors act in opposite directions. For all samples with DN = 60%, the swelling ratio of the samples is lower than would be expected from an interpolation between the swelling ratios of the samples with DN=40% and DN=80%.

The swelling behavior of the samples in DI water is different from that in the salt solution as shown in Fig. 11. A large increase in swelling occurs in the region between DN=0% and DN =20%. After that, the curves "flatten". The relative insensitivity in the swelling ratio for degrees of neutralization above 20% in DI water is probably due to the chains reaching their limits of extensibility. In this regime the gels have swelled over 1000 times their original volume. In salt solution the swelling is much less and continues to increase as the electrostatic interactions are increased with increasing degrees of neutralization.

To obtain some correlations that would be useful to guide synthesis work, we replot the swelling data in Fig. 7 versus the reciprocal of the crosslinker concentration. The data fall on a series of straight lines as shown in Fig. 12; the swelling is inversely proportional to the crosslinker concentration and this relation can be expressed by the following equation:

$$q\,(DN,x) = q_o\,(DN) + K(DN)/\,x \tag{12}$$

where x is crosslinker concentration, q_o is the intercept and K is the slope of the straight lines in Fig. 7, both of which are functions of the degree of neutralization. Plotting the q_o and K versus degree of neutralization as shown in Figs. 13 and 14, it is found that q_o and K can be fitted by:

$$q_o\,(DN) = 9.656+0.496\ DN \tag{13}$$

and,

$$K(DN) = 0.292\ e^{\,0.024\ DN} \tag{14}$$

The intercept q_o has the interpretation of being the swelling ratio at infinite crosslinker concentration for a fixed degree of neutralization. The slope K is a measure of the sensitivity of the swelling to crosslinker concentration. Therefore, for the poly(acrylic acid) gel crosslinked by TMPTA, the dependence of the swelling ratio on the degree of neutralization and crosslinker concentration can be correlated by the following expression:

$$q(DN,x) = 9.656 + 0.496\ DN + (0.292\ e^{\,0.024\ DN})/\,x \tag{15}$$

Modulus. Fig. 15 shows how the storage modulus G' and loss modulus G" change with measuring frequencies for two samples: both G' and G" increase with increasing frequency. When the degree of neutralization is zero, a linear relation exists between G' and the crosslink ratio as shown in Fig. 16. A theoretical modulus calculated from

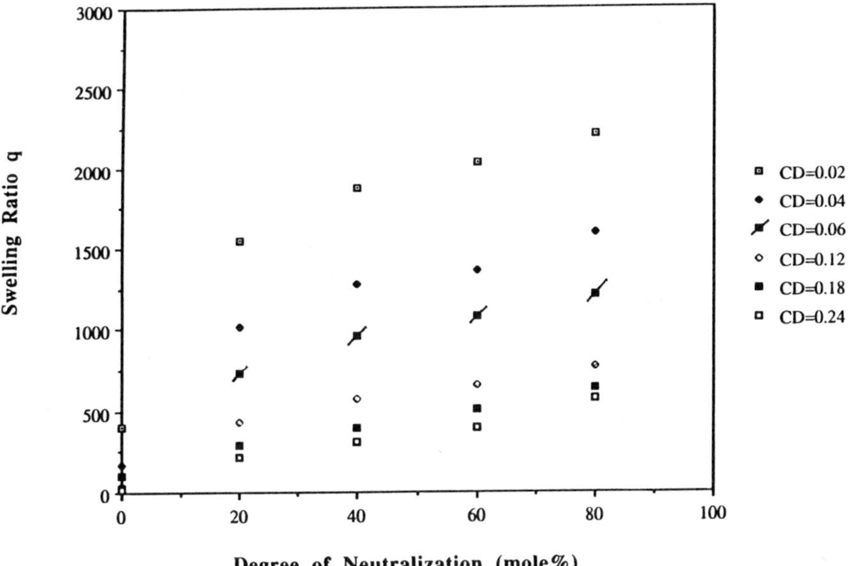

Figure 11. Swelling ratio versus degree of neutralization for the gels with different crosslinker concentrations. Gels are polymerizd at 3.2 M monomer concentration and swollen in DI water.

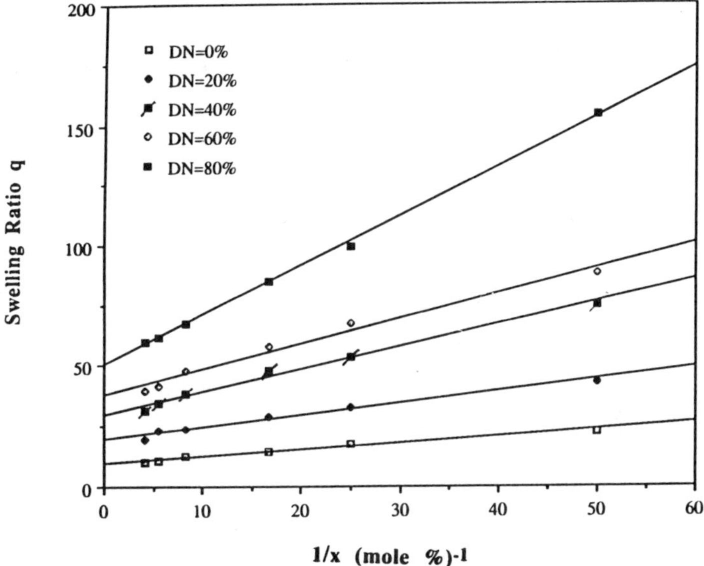

Figure 12. Swelling ratio versus (crosslinker concentration)$^{-1}$ for the gels with different degrees of neutralization. Gels are polymerizd at 3.2 M monomer concentration and swollen in 0.1538 M (0.9 % wt.) NaCl solution.

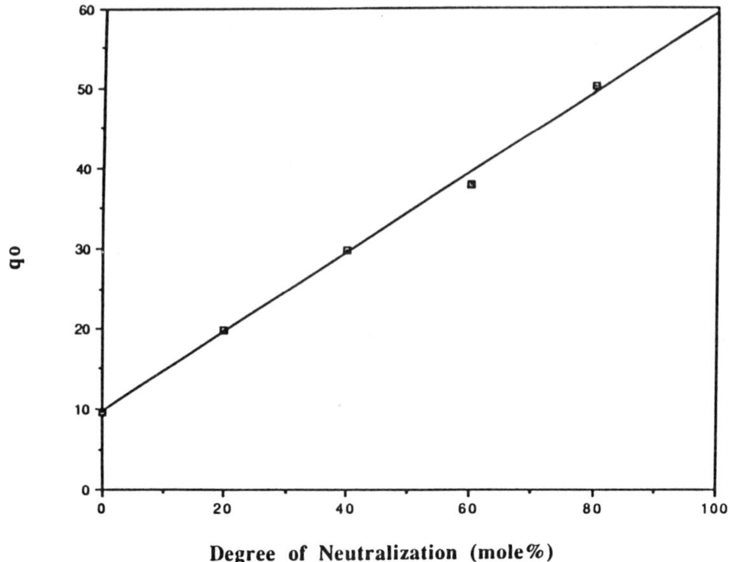

Figure 13. Slope obtained from Fig.12 versus degree of neutralization.

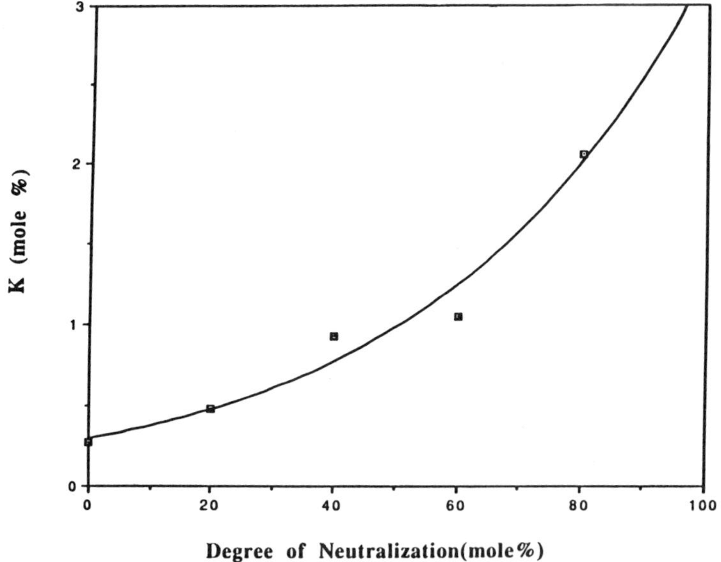

Figure 14. Intercept obtained from Fig.12 versus degree of neutralization.

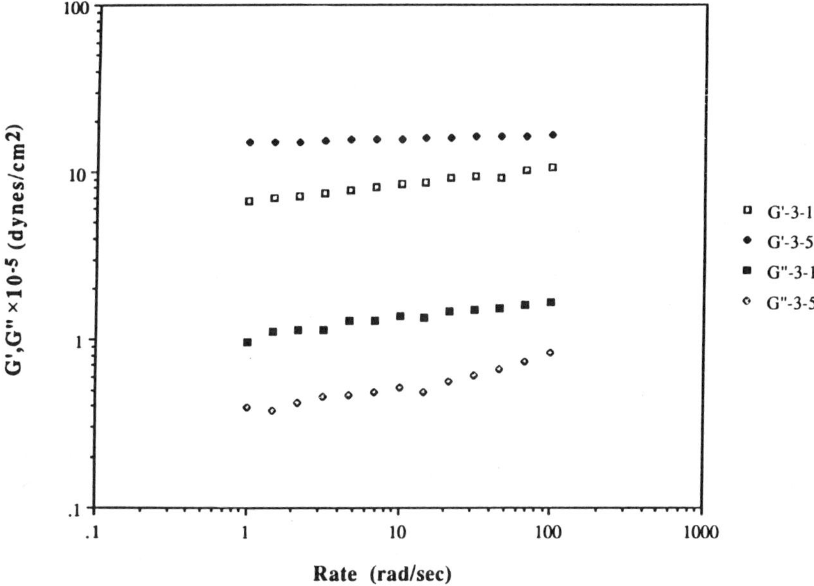

Figure 15. Frequency dependence of the modulus for samples 3-1 and 3-5.

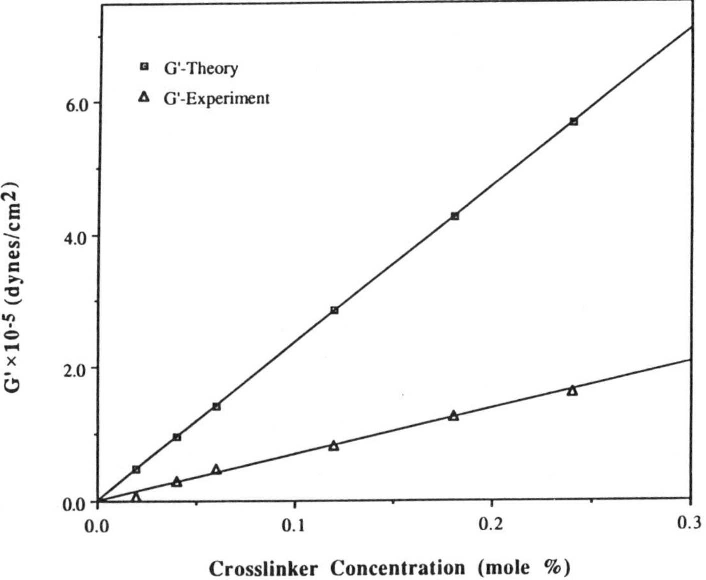

Figure 16. Comparison of the storage modulus G' with the theoretical modulus predicted by Eq.1 for the gels with DN=0. Gels are polymerizd at 3.2 M monomer concentration and measured at rate = 1 rad/sec.

Eq. 9 is also given in the diagram. All experimental data are lower than the theoretical value, suggesting that incompleteness may exist in the network structure. The relationship between G' and crosslinker concentration is shown in Fig. 17 for gels with different degrees of neutralization. The linearity gradually disappears as the degree of neutralization increases; the higher the degree of neutralization, the lower the modulus and the less effective the crosslinker is. At low frequencies the storage modulus, G', equals the static shear modulus, G, which is the modulus predicted by the polymer kinetic theories. The upturn in G' at high frequencies (Rate = 100 rad/s) seen for low crosslink ratios in Fig. 17 represents recoverable elastic interactions between polymer chains that do not relax over the time-scale of a single oscillation. More highly ionized gels show higher values of G'. The upturn in G' becomes less pronounced at low frequency (Rate = 1 rad/s) as shown in Fig. 18. This same behavior would be seen even for uncrosslinked high molecular weight poly(acrylic acid) polymers in solution. The upturn in G' at low crosslink ratios is also reflected in an upturn in the loss modulus, G", shown in Fig. 19. The energy dissipation in the crosslinked network increases as the network becomes less perfectly connected with more dangling polymer strands. The effect of degree of neutralization on gel structure is seen in Fig. 20. For the gel formed with DN=0 the modulus increases with increasing crosslink density in a linear fashion as shown in Fig. 16. Increasing the degree of neutralization decreases the slope of the modulus/crosslink-ratio curve; the crosslinker is less and less efficient at producing elastically active strands as the neutralization of the monomer increases. Finally, at DN= 80% we see that the modulus is independent of the crosslink concentration. Presumably, strong electrostatic repulsions limit the number of elastically effective strands that are formed, rather than the number of added crosslink sites. Extra crosslinker forms dangling chains that increase the level of the loss modulus, G", but do not contribute to G'.

Summary

Acrylic acid gels formed from uncharged (i.e. un-neutralized) monomers follow the rules of classic network theory; the swelling ratio depends on the functionality of the crosslinker in the manner predicted by theory. The swelling of acrylic acid gels in DI water is larger than the swelling in salt solution, but the same scaling with crosslinker functionality is observed in either case.

The degree of neutralization of the monomer prior to polymerization makes a significant difference to the structure of the gel as demonstrated by differences in modulus and swelling. Increased neutralization of the monomer increases electrostatic interactions and increases the swelling, decreases the elastic modulus, and increases the amount of extractable material in the gel. A correlation relating swelling ratio to crosslink ratio and degree of neutralization has been obtained. The swelling ratio varies inversely with crosslink ratio, and the swelliing ratio at high crosslink densities is linearly proportional to the degree of neutralization. Molecular modeling of chain growth with electrostatic interactions is being pursued to understand the basis for the relatively simple dependences seen in the correlation. The elastic modulus measurements show a decreasing efficiency at creating gel networks at higher degrees of neutralization. The networks formed have higher levels of dangling ends and other imperfections that lead to increases in the loss modulus, G".

The performance of absorbent gels depends upon both the swelling ratio and the "strength" of the network. It is desirable to have as large a swelling ratio as possible, while still having enough mechanical strength in the gel network to prevent collapse of the gel under load. Understanding the interaction of modulus and swelling

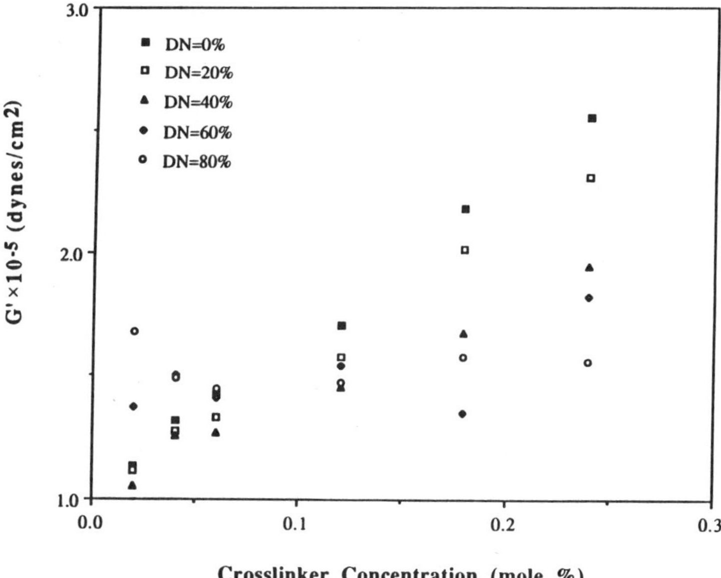

Figure 17. Storage modulus G' versus crosslinker concentration for the gels with different degrees of neutralization. Gels are polymerizd at 3.2 M monomer concentration and measured at rate = 100 rad/sec.

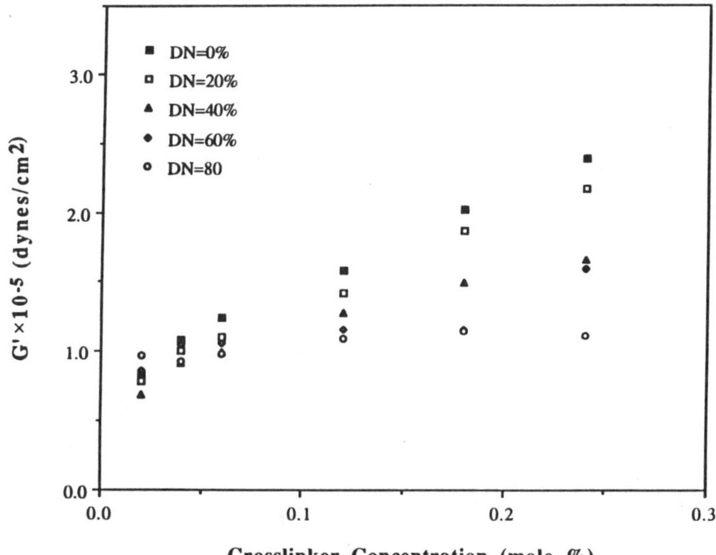

Figure 18. Storage modulus G' versus crosslinker concentration for the gels with different degrees of neutralization. Gels are polymerizd at 3.2 M monomer concentration and measured at rate = 1 rad/sec.

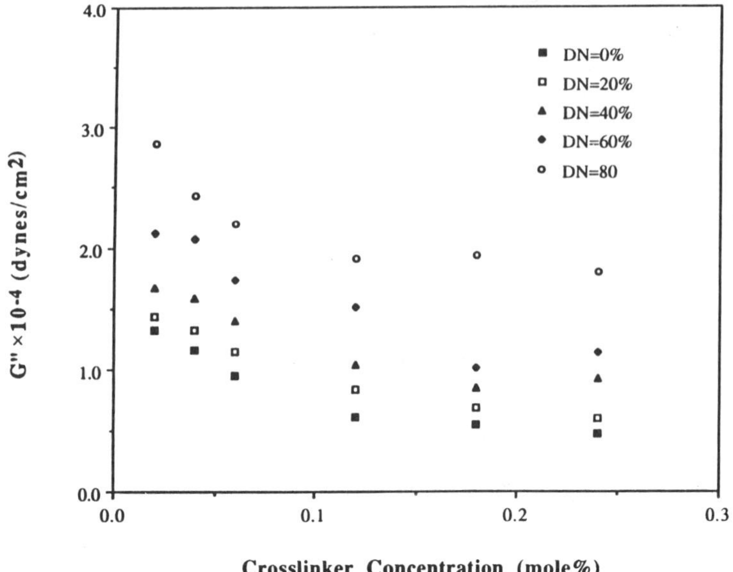

Figure 19. Loss modulus G" versus crosslinker concentration for the gels with different degrees of neutralization. Gels are polymerizd at 3.2 M monomer concentration and measured at rate = 100 rad/sec.

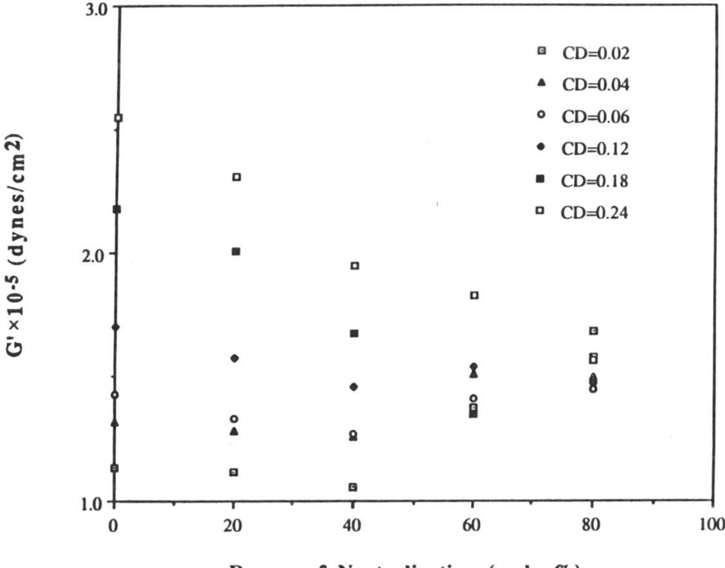

Figure 20. Storage modulus G' vs. degrees of neutralization for the gels with different crosslinker concentrations. Gels are polymerizd at 3.2 M monomer concentration and measured at rate = 100 rad/sec.

ratio, and their dependence on chemical composition during polymerization, allows the rational development of optimum gel compositions.

Literature Cited

1. Tanaka, T. *Phys. Review Lett.* **1980**, 45, 1636.
2. Gelman, R. A.; Nossal, R. *Macromolecules* **1979**, 12, 311.
3. Oppermann, W.; Rose, S.; Rehage, G. *British Polymer J.* **1985**, 17, 175.
4. Ilavsky, M. *Polymer* **1981**, 22, 1687.
5. Ilavsky, M.; Hrouz, J.; Havlicek, I. *Polymer* **1985**, 26, 1514.
6. Nossal, R.; Jolly, M. *J. Appl. Phys.* **1982**, 53, 5518.
7. Tanaka, T. et. al. *Nature* **1987**, 325, 796.
8. Tanaka, T. et. al. *Macromolecules* **1984**, 17, 2916.
9. Hirokawa, Y.; Tanaka, T. *J. Chem. Phys.* **1984**, 81, 6379.
10. Tanaka, T.; Hocker, L. O.; Bendek, G. B. *J. Chem. Phys.* **1973,** 59, 5151.
11. Tanaka, T. *Scientific American* **1981**, 244(1), 124.
12. Graessley, W. W. *Macromolecules* **1975**, 8, 186.
13. Gottlieb, M.; Macosko, C. W. et.al. *Macromolecules* **1981**, 14, 1039.
14. Dossin, L. M.; Graessley, W. W. *Macromolecules* **1979**, 12, 123.
15. Langley, N. K. *Macromolecules* **1968**, 1, 348.
16. Flory, P. J. *The Principles of Polymer Chemistry* **1953,** Connell University Press, Ithaca and London.
17. Yamakawa, H. *Modern Theory of Polymer Solutions* **1971,** Harper & Row.,New York.
18. Katchalsky, A.; Lifson, S. *J. Polym. Sci.* **1953**, 5, 409.
19. Katchalsky, A.; Micheali, I. *J. Polym. Sci.* 1955, **9**, 69.
20. Hasa, J.; Ilavsiky, M.; Dusek, K. *J. Polym. Sci.* **1975**, 13, 253.
21. Hasa, J.; Ilavsky, M. *J. Polym. Sci.* **1975**, 13, 263.
22. Konak, C.; Bansil, R. *Polymer* **1989**, 30, 677.
23. Prange, M. M; Hooper, H. H; Prausnitz, J. M. *AICHE Journal* **1989**, 35, 803.
24. Hooper, H. H.; Baker, J. P. ; Blanch, H. W. ; Prausnitz, J. M. *Macromolecules* **1990**, 23, 1096.
25. Siddle, J. H.; Jonson, T. C. *US Pat., 4,833,222*, **1989.**
26. Manning, G. *J. Chem. Phys.* **1969**, 51, 924.
27. Manning, G. *J. Chem. Phys.* **1969**, 51, 934.

RECEIVED August 20, 1991

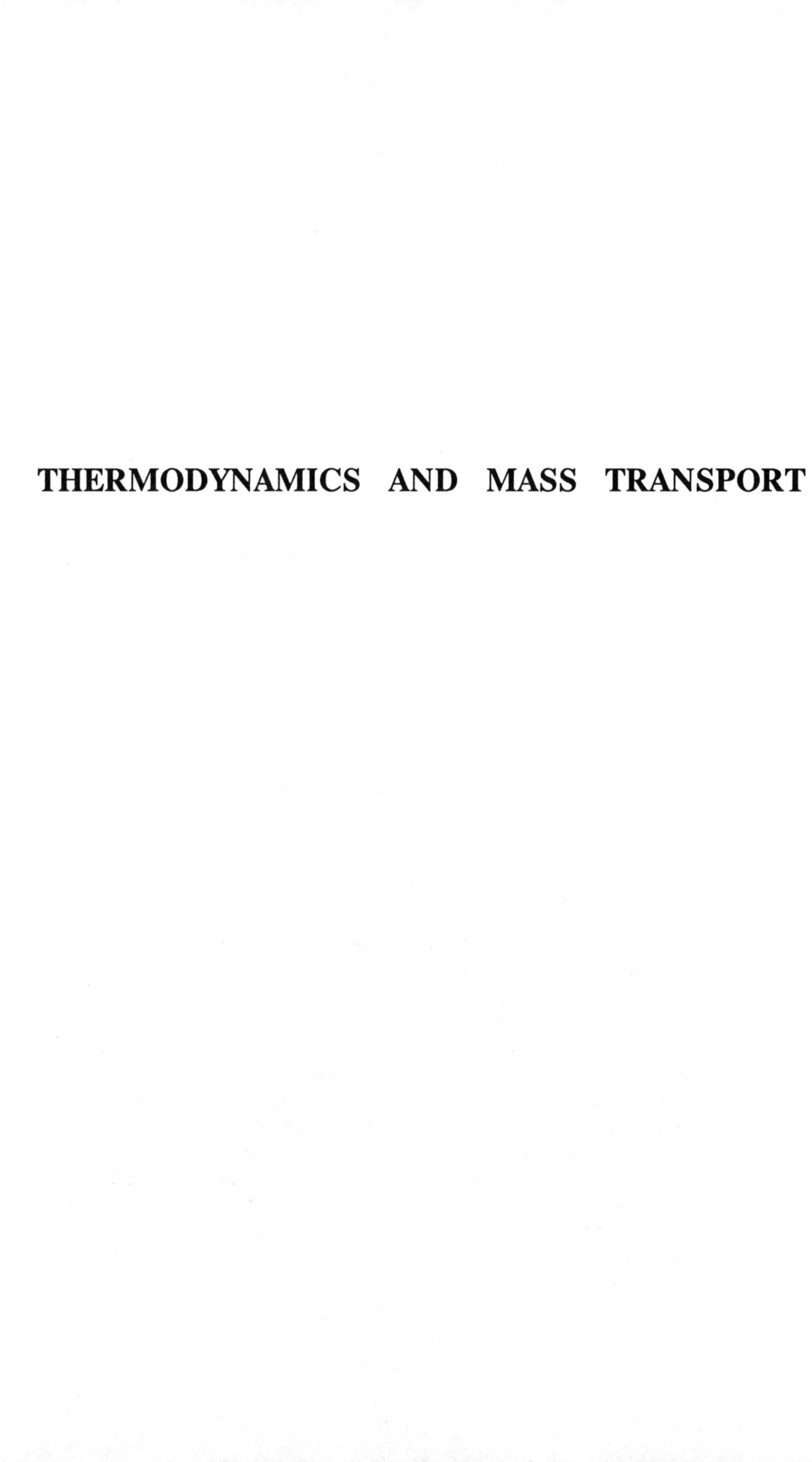

THERMODYNAMICS AND MASS TRANSPORT

Chapter 7

Hyaluronic Acid Gels

Chirag B. Shah and Stanley M. Barnett

Department of Chemical Engineering, University of Rhode Island, Kingston, RI 02881

In recent years numerous polymer hydrogel systems have been investigated for a wide variety of applications. The selection of a polymer system is very important for the development of pharmaceutical, medical or cosmetics applications. Important applications include controlled release systems and prosthetic and biomedical devices. The knowledge of interaction of the chosen system with different surrounding conditions is necessary due to the fact that polymer gel properties are greatly affected by pH, ionic strength, temperature, electric field and solvent system (*1*). As a controlled release system to deliver desired material over a long period of time, a polymer should have appropriate barrier and transport properties since it is necessary to control the permeability or diffusion through the system for desired release rates. Other important physical properties are mechanical strength and flexibility. To be able to use gels for separation of biomolecules, it is necessary to control exclusion properties which might be dependent on size or the charge on the molecule (*2*). In biomedical applications the polymer system must be nontoxic and bio- compatible. Biodegradable material should not yield toxic byproducts. Hence an understanding of the various parameters affecting the performance is critical.

Hyaluronic acid is a biopolymer. It is found in the connective tissue of vertebrates, the synovial fluid of joints and the vitreous humor of the eye (*3*). It is mainly extracted from the rooster comb. Other sources include the human umbilical cord and shark skin. It can be synthesized by microbial fermentation using streptoccoci (*4*). It is identical in pure form irrespective of the source, and hence it does not elicit foreign reactions in the human body, regardless of the source (*5*). Hyaluronic acid is biocompatible and biodegradable. Highly purified hyaluronic acid is non-inflammatory, non-pyrogenic, non-cytotoxic and non-immunogenic (*6*). It can be formulated in a variety of ways to enhance the natural functional activities in the body. Various available forms include the hyaluronate gels, microparticles, membranes and gel composites (*6*). Commercially, hyaluronic acid (sodium hyaluronate) is known as hyaluronan.

0097–6156/92/0480–0116$06.00/0
© 1992 American Chemical Society

The biological functions of hyaluronic acid (*7*) such as protection and lubrication and separation of cells, regulation of transport of molecules and cell metabolites, maintenance of the structural integrity of connective tissues and fluid retention in intercellular matrix, have created significant interest among researchers. As a result a variety of biomedical applications have been developed. Hyaluronic acid has been successfully used in ophthalmic surgery (Healonid; Pharmacia, IAL;Fidia S.P.A), in treatment of arthritis and in visco surgery (*3*). As a biomaterial it can be applied as surface coatings for artificial organs to prevent them from attack by the body's immune system. (*5*). Hyaluronic acid films (Connectivina, Gauze,Fidia) are useful in wound healing as they allow oxygen to pass through while keeping bacteria away (*5*). Its moisturization and water retention capabilities have given hyaluronic acid important properties for a variety of cosmetic formulations and cosmetic delivery systems (*8*). Balazs et al. at Biomatrix Inc. are developing contact lenses having 95% water using hyaluronic acid gels (*5*). The potential application of hyaluronic acid as a drug delivery system and as a delivery vehicle for growth factors has been under development using hylan gel derivatives (*9*). Shah et al. (*10*) have shown the utility of hyaluronic acid for use as a swelling controlled release system. Larson et al. (*6*) have studied the release of antibiotic gentamicin from hylan gels. The study of diffusion and release rate of drug through films of ester derivatives of hyaluronic acid has shown considerable promise in applications where rapid drug release is desired (*11*). The sodium salt of hyaluronic acid enhance the miotic effect of pilocarpine in rabbits and in humans as confirmed by Gurny et al.(*12*). The bioavailability of ophthalmic drugs such as pilocarpine and tropicamide have found to increase by preparations based on the hyaluronic acid. Based on these observations Saettone et al.(*13*) have reported the excellent bioadhesive properties of hyaluronic acid salts and its derivatives. The complexes formed by interaction with chitosan gels also have been studied for drug release (*14*). The major objective of this article is to provide a perspective on the extent to which crosslinked gels interact with variations in the surroundings. This knowledge will be beneficial in the optimization of the design of ultimate properties. Initially we will evaluate the structure of hyaluronic acid considering molecular weight and its distribution, conformation and stereochemistry in the repeat units, functional groups and charge density. The solution properties of hyaluronic acid will be reviewed considering the effect of various physicochemical parameters. Then swelling properties of the crosslinked gels and the implications will be discussed.

Structure of Hyaluronic acid

Hyaluronic acid is an ubiquitous component of the intercellular matrix of connective tissues. The molecular weight of hyaluronic acid is usually between 1 to 6 million, and it is always polydisperse. Hyaluronic acid has a glycosaminoglycan structure (Figure 1) with a repeat disaccharide unit of N-acetyl glucosamine and glucuronic acid linked together by alternate β (1--4) and β (1--3) glucosidic bonds forming a long linear unbranched polysaccharide chain (*15*). It is a natural polyelectrolyte with carboxyl and acetoamido functional groups. The acidic pendent groups are positioned on both sides of the chain which allow for maximum water binding or holding capacity (*15*). The carboxyl groups acquire ionic form at higher pH. Blumberg et al. (*16*) have confirmed by electrometric titration that 80% of the acidic groups are ionized at acetate buffer pH of 5.

Several successful modifications of the hyaluronate molecule have been made. The esterification of carboxylic acid groups with alcohol produces ester derivatives having different physical properties (*17*). The ionic capacity depends on the degree of esterification. The increasing degree of esterification makes the molecule insoluble in water as opposed to the high water solubility of the hyaluronate molecule. Hyaluronic acid can form interpolymer complexes with charged cationic polymers due to its polyanionic nature. The water solubility of the resulting complex depends on the charge density of the cation. These anionic-cationic blends could have hybrid properties of both the hyaluronic acid and cationic polymer. Band et al. (*14*) have investigated compatibility of several cationic cellulose polymers with hyaluronate. Hyaluronic acid is also susceptible to enzyme hyaluronidase which hydrolyzes the N-acetyl glucosaminic bonds and splits the molecule (*19*).

Hyaluronic Acid in Solution

Hyaluronic acid mainly exists as sodium hyaluronate under physiological conditions, forming a hydrated polymeric network with remarkable biophysical properties. Hyaluronate molecules behave as random coil molecules in the dilute solution at neutral pH and physiological strength. Each molecule has its own solvent surrounding domain. The chain-chain interaction is less. This condition is highly dependent on the various parameters. An increase in the concentration of the polymer in the solution forces the individual chains to interpenetrate and form a highly entangled network (*8*). At the incipient point of the entanglement, a sharp discontinuity occurs in various hydrodynamic functions such as viscosity and elasticity. The higher the molecular weight of the polymer, the lower the concentration required to form the matrix. To predict the formation of a matrix, it is necessary to calculate the coil overlap parameter which is defined as the product of the concentration in g/cc and (n), the limiting viscosity number in cc/g. A typical value of this parameter is reported to be 2.0 to 2.5 (*8*, *19*) as opposed to the value of 10 for the onset of concentration dependence for most synthetic polymers. This value is found to be independent of pH, ionic strength and molecular weight of the polymer. Upon formation of the network, the molecular chain attains an extra degree of stiffening which could have been due to the formation of hydrogen bonding between two residues. The nature of the polymer chain-chain interaction now controls the viscoelastic properties. The changes in the pH or ionic strength varies the degree of ionization of the carboxyl groups, which in turn influence the electrostatic repulsions between chains (*19*). Due to this dynamic interaction, conformation of the molecular chains changes. This natural viscoelastic matrix has an extremely high shear dependent viscosity, frequency dependent elasticity and high pseudo plasticity.

Gibbs et al. (*20*) have obtained dynamic viscoelastic properties of hyaluronate over a range of conditions of pH and concentration. These researchers concluded that the hyaluronate molecule is very stiff at a pH of 2.5, which they attributed to the critical balance between the repulsive and the attractive forces operating between the molecular chains. The interpenetrating hyaluronic acid chains could literally hold the solvent water like a molecular sponge (*3*). Hydration characteristics are also greatly affected by the solution pH and ionic strength. Davies et al. (*19*) have investigated the influence of solution ionic strength, temperature and counter ion type on the hydration of sodium hyaluronate. A decrease in hydration was found with an increase in ionic strength as compared with its value in salt-free aqueous solution, within the temperature range 20-50 °C.

Laurent (*21*) and Ogston (*22*) have independently studied the transport properties of hyaluronic acid solutions. These researchers found that the entangled network restricts the movement of diffusing molecules by a molecular sieve action according to size and the charge of the molecule. Laurent (*23*) has further demonstrated potential application of hyaluronic acid (gels) in gel filtration. The exclusion properties of the hyaluronic acid gels were found to be similar to those of hyaluronic acid solutions.

Thus, pH, ionic strength and temperature greatly affect the solution properties of hyaluronic acid. It can be envisioned that these factors will also significantly influence the swelling of crosslinked hyaluronic acid. Depending on the concentration of hyaluronic acid in the solution, an entangled network will form. The crosslinking of this network in solution will restrain the degree of swelling due to extra elastic force. Although considerable research has been done on the solution behavior of hyaluronic acid, the information on the swelling behavior of crosslinked gels is not well studied. This data will be very valuable for the potential applications such as controlled release of drugs or cosmetics or as an extraction solvent.

Gel Preparation

Hyaluronic acid, in sufficiently high concentrations, exists as a three dimensional network of macromolecular chains due to entanglement of individual molecules. This entangled network exhibits some of the properties of gels. To create water insoluble and highly swellable gels, it is necessary to react the base polymer with a crosslinking agent. A few crosslinking procedures are available.

In the method developed by Laurent (*24*), NaOH solution is added to a known quantity of hyaluronic acid. The resulting solution is allowed to swell to a thick suspension overnight in a refrigerator, then 1,2,3,4 di-epoxybutane is added as a crosslinking agent. After stirring the mixture must be kept at 50°C for 2 hrs. Acetic acid is then added to neutralize pH, and the formed gel is washed repeatedly with water.

In the second procedure developed by Balazs et al. (*25*), the gel is obtained by using divinyl sulfone as a crosslinking agent which reacts with the OH groups of the polysaccharide chain and forms an insoluble gel network through sulfonyl-bis ethyl crosslinks. In this method hyaluronic acid is added to NaOH solution. After dissolving the polymer in the alkali solution, divinyl sulfone is added in known quantity to crosslink the solution. The reaction is carried out at 20°C or lower, and within 10-15 minutes a strong gel forms. The rigidity of the resultant gel depends on the amount of crosslinking agent. It is possible to mold the gel as a film, cylindrical rod or spherical particle. Due to the viscosity of the solution, it is necessary to stir the solution to get an even crosslinking reaction. Similarly, crosslinking processes with other agents such as formaldehyde, dimethylourea, dimethylolethyleneurea, ethylene oxide, a polyaziridine and a polyisocynate have also been developed (*26*). The crosslinked gels of hyaluronic acid are given the generic name 'hylan gels' suggesting that hyaluronate chains are associated with each other by covalent bonds and the crosslinking reaction do not alter carboxylic acid and N-acetyl groups. The biocompatibility of crosslinked gels is also found to be similar to that of the water soluble hyaluronic acid (*27*).

The characteristic of the resultant gel mainly depends on the amount of crosslinking agent, molecular weight of the polymer and polymer concentration in reaction mixture. It is observed that a higher molecular weight polymer at a lower concentration and with less crosslinking agent produces a gel with the best ultimate swelling capacity (*25*).

Swelling behavior of hyaluronic acid gels

The equilibrium swelling of hyaluronic acid gels under different pH conditions is shown in Figure 2. The gel does not swell at very low pH, and the major changes occur over the pH range of 2 to 4. The degree of swelling increases continuously at higher pH; this pH dependent swelling behavior is due to ionizable carboxylic groups. At lower pH the gel remains neutralized while the degree of ionization increases with the increase in pH, increasing the hydrophilicity and in turn the equilibrium water fraction. The fixed negatively charged carboxylic groups on the polymer chain also set up an electrostatic repulsion, and the polymer chain expands. Hence the water molecule can penetrate the gel, and the degree of swelling increases. However, this electrostatic repulsion is reduced by the inevitable presence of other ions. The crosslinking between molecules plays a major role in limiting the ultimate swelling. After some free expansion upon water absorption, depending on the amount, the crosslinks provide contraction forces against the expansive osmotic forces. Hence equilibrium swelling is the resultant of two forces. The pH dependent swelling is found to be completely reversible. This reversible nature is shown (Figure 3) for periodic swelling and deswelling cycles in deionized (Millipore Q™ water system, 18 Megohm-cm resistivity) water and .1 N HCl, respectively. In the case presented here, the swelling after the first cycle decreases and remains the same during the rest of the cycles. One possible reason for the initial change could be leaching of some unwashed salts or impurities in the polymer.

This pH sensitivity establishes the hyaluronic acid gel system as one of the few available candidates for pH sensitive swelling controlled release systems. Examples of other systems include hydrophobic polymeric copolymer hydrogels (*28*) and poly(hydroxy ethyl methylacrylate-co-methacrylic acid) hydrogels (*29*).

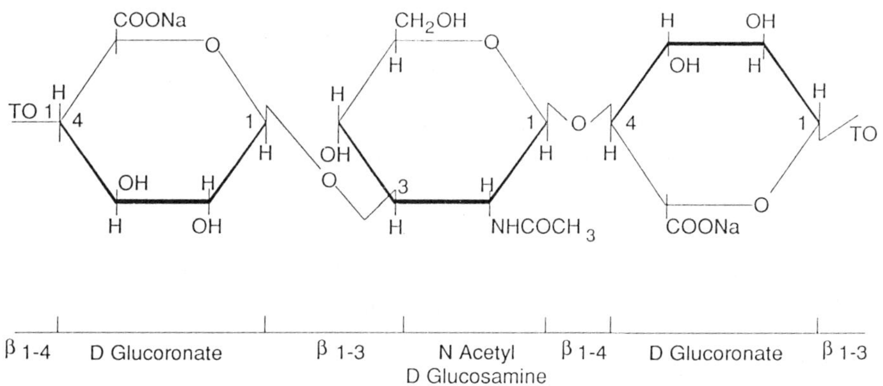

Figure 1. Structure of hyaluronic acid showing D-Glucuronate and N-acetyl glucosamine bonded by alternate β(1-3) and β(1-4) glucosidic bonds.

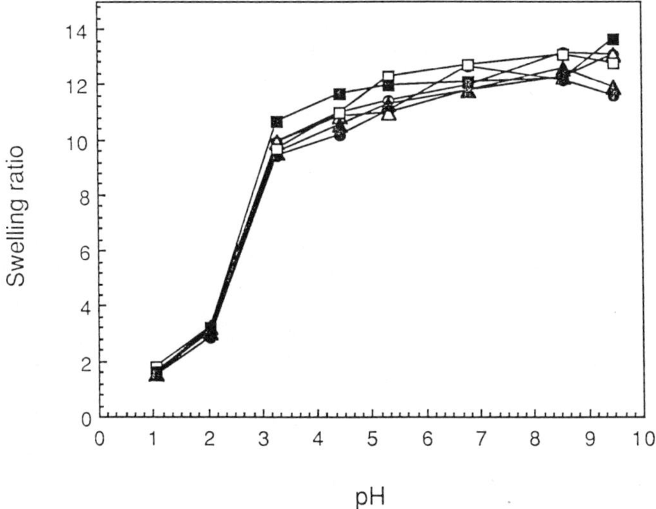

Figure 2. pH sensitive swelling of hyaluronic acid gel (at 1.27 crosslinking ratio).

The hydrophilicity of hyaluronic acid gels could best be used to protect acid labile drugs from the low pH environment of the stomach and for site specific drug release based on the regional pH differences. The variation in the pH of the surrounding solution will lead to modification in the drug release mechanisms. Shah et al. (*10*) have shown the pH induced changes in the release characteristics of hyaluronic acid gels using FD&C blue dye as a model system. The release rate was found to be lowered upon decrease in the pH of the medium, yet the results were not conclusive. Currently the effect of various parameters such as drug concentration, conditions of loading, and the pH dependent drug solubility are being analyzed using phenylpropanolamine hydrochloride (acidic) and sodium salicylate (basic) drugs. The reversibility of swelling and collapsing is significant for use of the gels as an extractant for chemical separations. Applications of gels in chemical separation are reported by Vartak (*30*) and Cussler et al. (*2*). Cussler et al. have used polyacrylamide derivatives and Vartak et al. have used polyacrylate gels in their study. Since hyaluronic acid gels can swell greatly while acting as molecular sieves they could be an efficient medium for extraction.

The extent of equilibrium swelling decreases considerably in the presence of salt, as shown in Figure 4. The ionic strength of the solution neutralizes some of the ionized groups causing the deswelling of the initial highly swollen condition or allowing a gel to swell to a lesser degree. Along with concentration the nature of the salt also affects the swelling degree as shown in Figure 5. This figure shows that in $CaCl_2$ solutions, swelling is appreciably reduced at low concentrations compared to the similar reduction obtained in NaCl solutions, but only at higher concentrations. This behavior could be due to the fact that only half of the cations are needed to neutralize the ionic groups of the gel molecules.

The observed salt effect of a buffer solution is explained in Figure 6. The degree of swelling in various pH buffer solutions is significantly less compared to the swelling of the gel, having the equivalent crosslinking, in deionized water. However, the extent of swelling is comparable to the swelling level observed in the presence of the salt in the water. This observation clearly demonstrates the effect of buffer salts on gel-water interaction. The equilibrium swelling level of the gel samples, swollen to various degrees in different solutions as shown in Figure 6 by a, b, c, d, increases upon immersion in deionized water to the condition shown by e in Figure 6.

The swelling levels of the gel in equilibrium with deionized water are shown in Table I as a function of the amount of the crosslinking agent (*31*). The crosslinking ratio is defined as the ratio of the amount of polymer to the amount of the crosslinking agent. The swelling ratio is defined as the ratio of swollen gel weight to initial gel weight. Due to the inhomogeneity observed in crosslinking for different gel samples, the swelling ratio shown is the mean value of at least 20 gel samples; the range of values obtained for each sample is also shown. As expected the swelling ratio increases with the increase in the crosslinking ratio. However, after some critical ratio (about 2.5) the mechanical strength of the gel is drastically reduced.

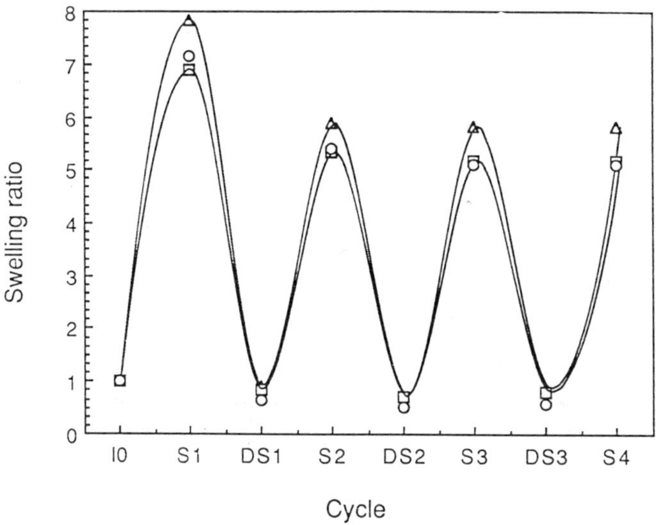

Figure 3. Reversible nature of the pH dependent swelling of hyaluronic acid gel. This data is shown for the gel of crosslinking ratio of 1.0

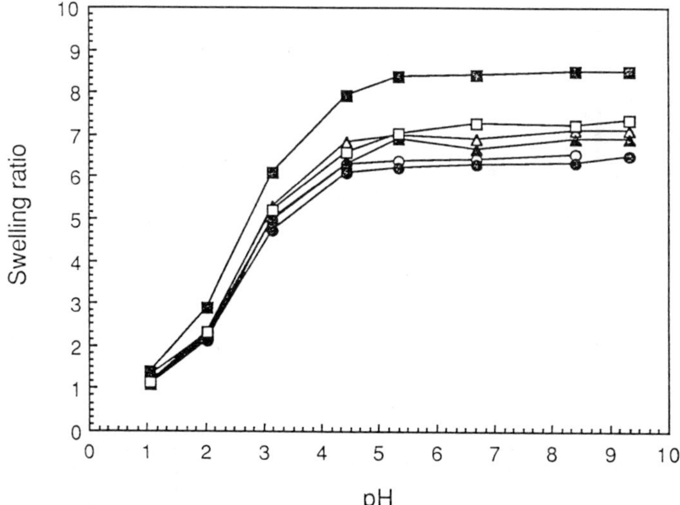

Figure 4. Salt induced depression in pH dependent swelling of hyaluronic acid gel (at 1.27 crosslinking ratio).

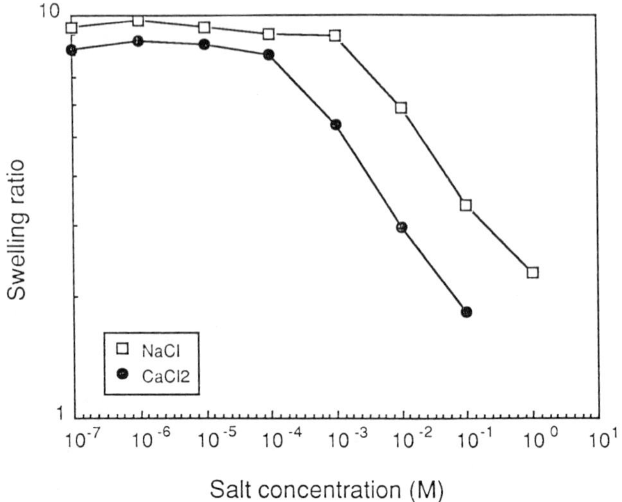

Figure 5. Swelling of hyaluronic acid gel in different salts solutions at different concentrations (at 1.27 crosslinking ratio).

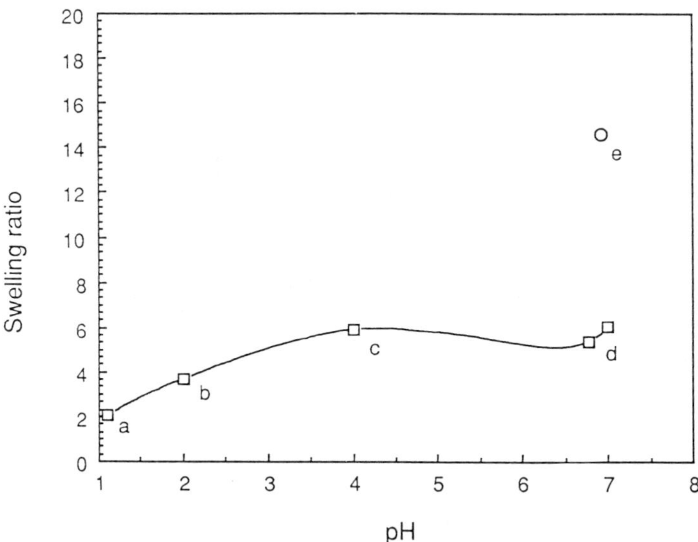

Figure 6. Swelling of hyaluronic acid gels in different buffer solutions. a,b,c,&d indicate the equilibrium swelling in buffer solution of respective pH. The conditions of point e are achieved upon immersion in deionized water (crosslinking ratio of gel = 1.27)

Table I. Equilibrium swelling of hyaluronic acid gel at 25°C, in water

Crosslinking Ratio	Swelling Ratio
1	8 (6-9)
1.27	14 (12-18)
2.1	28 (25-30)
Higher[a]	120

[a]This crosslinking ratio is approximately 5.

In dynamic swelling the cylindrical samples show a phenomenon which is described as surface driven instabilities by Tanaka (*32*) and Peppas (*33*). The swelling of virgin gels (used immediately after their removal from the mold) is anisotropic. The initial cylindrical shape instantly changes to a dumbbell shape with a distinct outside swollen layer and inside unswollen core. The swelling at the two edges is significantly more than that in the middle segment, and the surface of the gel is filled with numerous line segments which form a regular pattern. Upon more swelling this pattern grows and ultimately disappears. Then the dumbbell shape changes back to cylindrical and the succeeding swelling is isotropic. These observations are illustrated in Figure 7. If the gel is dried, this type of behavior can be observed very easily irrespective of the prior swelling history of the gel. However, it is not observed if the initial gel is lightly swollen. This phenomenon is qualitatively explained as the local stretching of the gel due to osmotic pressure changes at the surface (*32*). Initially when the osmotic pressure is high, the outer surface is forced to move inside due to the mechanical restraint offered by the inside unswollen core. Upon reduction in the osmotic pressure after some water absorption, the forces are resolved by the unidirectional swelling of the gel.

The initial drying of the gel affects the ultimate swelling capacity as shown in Table II. The gel when dried without any prior swelling, swells only 2-3 times. The highly swollen gel upon drying does not reswell to the original swollen condition (only 50% of the original swelling). A similar observation was made by Laurent et al. (*24*). This drying was done at 35° C for 24 hours. It is possible that upon drying the pores are closed completely bringing about an irreversible change in the structure. Initial swelling might be preventing complete pore elimination. Hence these pores reopen when the gel is rewetted. Another possibility is the formation of new junctions between molecules. The junctions formed during drying will then act as crosslinks in the structure.

Table II. Effect of drying on the swelling of gel

Condition	No. of samples	Mean swelling ratio	Std Deviation
Directly dried	10	1.2943	.2411
Directly swollen	24	15	1.949
First swollen, then dried and reswollen	24	6.5	.6538

The temperature sensitivity of the gel is illustrated in Figure 8. The swelling of the gel continuously decreases upon an increase in temperature. This behavior is similar to the behavior of the hyaluronate molecule in the solution. The interaction of the gel with aqueous solutions of acetone or n-propyl alcohol is illustrated in Figure 9. In Figure 9a we can see that the hyaluronic acid gel does not swell at all above 65% acetone concentration. The gel with the higher degree of crosslinking shows reduced swelling, but the behavior is the same. Similarly in Figure 9b the identical behavior of hyaluronic acid gels with n-propyl alcohol can be observed. This behavior for both solvents is found to be irreversible. The gel swollen at a higher solvent concentration does not swell significantly when immersed in the water. This result suggests that at higher concentrations of solvent, some irreversible changes in the structure are induced in the gel.

Conclusions

Crosslinked gels of hyaluronic acid reversibly swell about 15 times in water at a crosslinking ratio of 1.27. The equilibrium swelling level increases with the decrease in the degree of crosslinking. However, at higher swelling this gel loses mechanical integrity. The swelling level for the equivalent degree of crosslinking is higher than that shown by other synthetic polymers. The problem of the low mechanical strength should be resolved by increased crosslinking, grafting or coating. The swelling is significantly dependent on the pH and the ionic strength of the solution. The gel is highly swollen at higher pH and at lower ionic strength. The absorbed water fraction of the gel decreases with the increase in temperature. The similar effects are shown by hyaluronate molecules in the solution at various pH, ionic strength and temperature (*19-20*). During dynamic swelling this gel also shows the surface instabilities observed for the synthetic polymers. However, the interaction with organic solvents at higher concentrations is found to be different than that observed for most synthetic and some natural polymers. More work is needed to clarify these solvent induced irreversible swelling observations. Hyaluronic acid gels can be of use as pH and ionic strength dependent drug delivery systems.

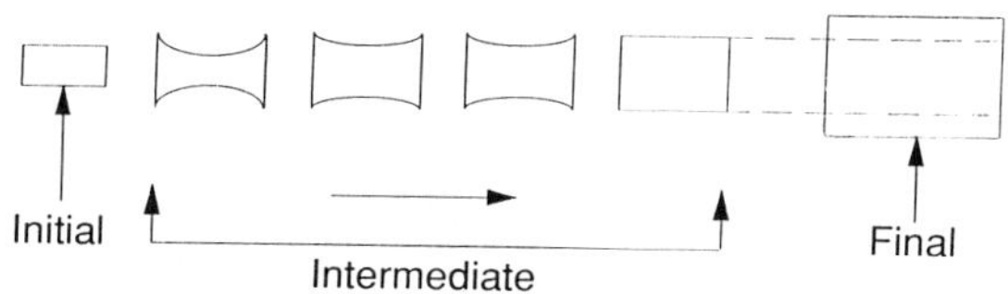

Figure 7. Alterations in the surface and shape of the hyaluronic acid gel samples during swelling.

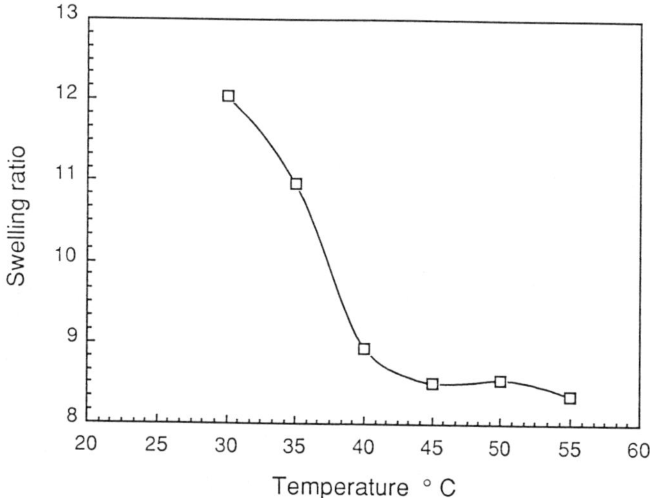

Figure 8. Swelling of hyaluronic acid gels at different temperatures (24 hr equilibration at each temperature).

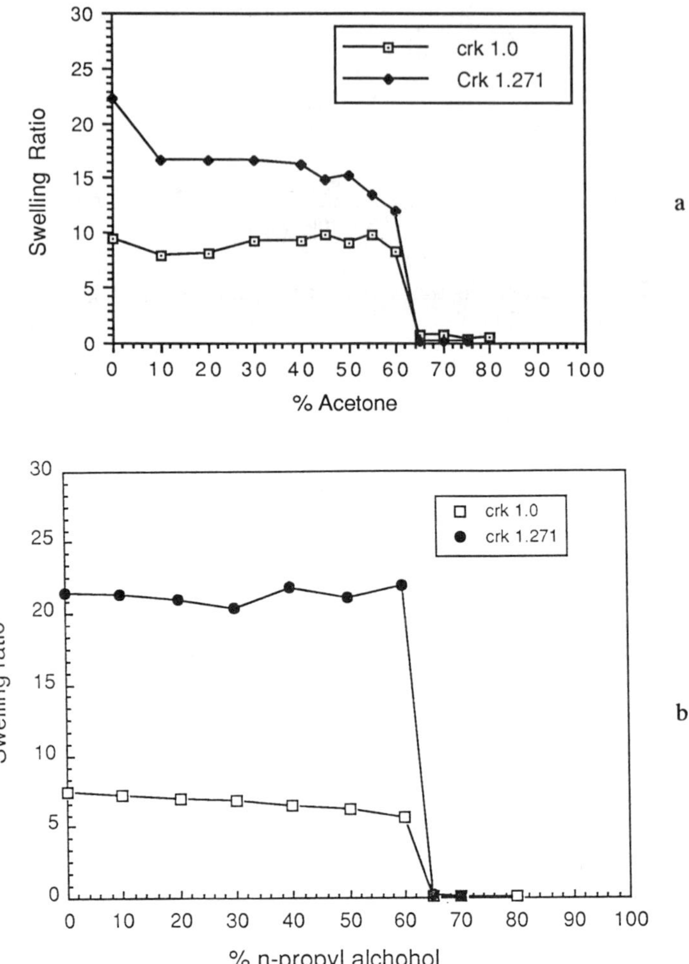

Figure 9. Swelling of hyaluronic acid gels in a). acetone-water and b). n-propyl alcohol-water solutions. Swelling of gels of crosslinking ratio of 1.0 and 1.27 is shown in both a and b.

Literature Cited

1). Tanaka, T. *Science*, **1980**, p.125.
2). Cussler, E.L.; Strokar, M.R.; Varberg J.E. *AIChE J*, **1984**, *30*, 4, 578.
3). Balazs, E.A.; Band, P. *Cosmetics and Toiletries*, **1984**, *99*, 65.
4). Sugahara, K.; Schwartz, N.B.; Dorfman, A. *J.Biol.Chem.*, **1979**, *254*, 14, 6252.
5). Mind & Body, *Science Digest*, **1989**, *90*, 97.
6). Larson, N.; Leshchiner, E.A.; Parent, E.G.; Balazs, E.A. *ACS Polymeric Materials Science & Engineering Fall meeting Proceedings*, **1990**, *63*, p.341.
7). Technical catalog, *Genzyme Corporation*, Boston, **1990**.
8). Band, P. *D & CI*, **1985**, *October*, p. 54.
9). Report, *Genetic Engineering News*, April **1990**, p. 11.
10) Shah, C.B.; Barnett, S.M. *AICHE Summer National Meeting*, San Diego **1990**, Paper 11D.
11). Hunt, J.A.; Joshi, H.N.; Stella, V.J.; Topp, E.M. *J.Cont.Release*, **1990**, *12*, 159.
12). Camber, O.; Edman, P.; Gurny, R. *Curr. Eye. Res.*, **1987**, *6*, 779.
13). Saettone, M.F.; Chetoni, P.; Torracca, M.T.; Burgalssi, S.; Giannaccini, B. *Int J. Pharm*, **1989**, *51*, 203.
14). Band, P.; Brode, G.L.; Goddard E.D.; Barbone, A.G.; Leshchiner, E.; Harris, W.C.; Pavlichto, J.P.; Partin, E.M. III, Leung, P.S. *ACS Polymeric Materials Science & Engineering Fall Meeting Proceedings*, **1990**, *63*, p.692.
15). Burchard, W. *British Polym.J.*,**1985**, *17*,(2),154.
16). Blumberg, B.S.; Oster, G.; Meyer, K. *J.Clin.Invest.*,**1955**, *34*, 1454.
17). Della Valla, F.; Romeo, A. *Eur. Pat. Appli*, **1989**, EP 216453, 129.
18). Rapport, M.M.; Linker, A.; Meyer, K. *J.Biol.Chem*, **1951**, *192*, 4283.
19). Davies, A.; Gormally, J.; Jones, E.W.; Wedlock, D.J.; Phillips, G.O. *Biochem J.*,**1983**, *213*, 363.
20). Gibbs, D.A.; Merril, E.W.; Smith, K.A.; Balazs, E.A. *Biopolymers*, **1968**, *6*, 777.
21). Laurent, T.C.; Bjork, I.; Pietruszkiewicz, A.; Person, H. *Biochem. Biophys. Acta*, **1963**, *78*, 351.
22). Ogston, A.G.; Sherman, T.F. *J. Physiol.*, **1961**, *156*, 67.
23). Laurent, T.C. *Biochem J.*, **1964**, *93*, 106.
24). Laurent, T.C. *Acta. Chem. Scand.*, **1964**, *18*, 1, 274.
25). Balazs, E.A. *U.S. Patent #* 4582 865, **1986**.
26). Balazs, E.A.; Leshchiner, A. *Chemical Abstracts*, **1986**, *104*, p.449.
27). Balazs, E.A. *ACS Polymeric Materials Science & Engineering Fall Meeting Proceedings*, **1990**, *63*, p. 689.
28). Siegel, R.A.; Falmarzian, M.; Firestone, B.A.; Moxley, B.C. *J. Cont. Release*, **1988**, *8*, 179.
29). Peppas, L.B.; Peppas, N.A., *J.Cont. Release*, **1987**, *8*, 267.

30). Vartak, H.G.; Rele, M.V.; Rao M.; Deshpande, V.V. *Anal Biochem*, **1983**, *133*, 260.
31). Shah, C.B.; Barnett, S.M. Paper in Press, *J.App.Poly.Sci.*, **1991**.
32). Tanaka, T.; Sun, S.T.; Hirokawa, Y.; Katayama, S.; Kucera, J.; Hirose, Y.; Amiya, T. *Nature*, **1987**, *325*, 26.
33). Peppas, L.B.; Peppas, N.A., *J. Cont. Release*, **1988**, *7*, 181.

RECEIVED September 6, 1991

Chapter 8

Hydrophobic Polyelectrolytes

Effect of Hydrophobicity on Buffering and Colloid Osmotic Pressure

Ronald A. Siegel and Jose M. Cornejo-Bravo

Departments of Pharmacy and Pharmaceutical Chemistry, School of Pharmacy, University of California, San Francisco, CA 94143–0446

Effects of hydrophobicity on the titration curves and colloid osmotic pressures of polyelectrolyte copolymers containing diethylaminoethyl methacrylate hydrochloride (DEA•HCl) were studied. Hydrophobic and hydrophilic comonomers respectively decrease and increase the "buffering pH" observed in the titration curves. The colloid osmotic pressure of the fully ionized polyelectrolytes follows the Donnan model if an osmotic coefficient is introduced. The osmotic coefficient increases with separation of the charges on the polyelectrolyte, as occurs by incorporating unionizable comonomers. Comparison of data with the "cell model" for polyelectrolytes, supplemented by the "additivity rule," indicates that these models provide a quantitative description in some regions, but only a qualitative description in others. Results obtained for these hydrophobic polyelectrolytes are discussed in terms of their suitability as osmotic agents in a self-regulating, mechanochemical insulin pump.

In recent years considerable attention has been paid to weak acidic or basic polyelectrolyte gels as environmentally sensitive components of drug delivery systems (*1-3*) and as mediators of mechanochemical energy conversion (*4,5*). In addition, these gels can function as ion-exchangers (*6,7*). The swelling properties of polyelectrolyte gels are determined by environmental pH and ionic strength (*2-4,8-13*), as well as specific ion concentrations.

In most environmentally-sensitive drug delivery applications, drug release is modulated by altering the permeability of the gel to the pharmacologic agent of interest (*1,2,14*). Recently, we proposed a "mechanochemical" pump (*15*), the concept of which is illustrated in Figure 1. A polybasic gel is confined between a rigid screen and an elastomeric diaphragm. On the other side of the diaphragm are two chambers, one containing water and the other containing drug. These chambers are connected to the outside of the pump via one-way check valves. Ionization of the gel in response to an increase in hydrogen ion concentration causes an increase in the osmotic pressure in the gel, and consequently in the pressure inside the pump. When this pressure exceeds the critical cracking pressure of the valve leading from the drug chamber, drug is expelled from the pump. Thereafter when the hydrogen ion concentration decreases, the gel contracts, lowering the pressure in the pump. When the pressure

0097–6156/92/0480–0131$06.00/0
© 1992 American Chemical Society

difference between the inside of the pump and the external medium is reduced below the (negative) cracking pressure of the valve leading to the water chamber, water will flow into the latter. Thus, the volume of drug solution ejected is exactly compensated by the volume of water imbibed.

The device as described above is sensitive to hydrogen ion concentration. While there are a few body spaces in which pH fluctuates, a more likely application may result from coupling the gel to the enzyme glucose oxidase, which converts glucose into gluconic acid. Thus the pump would respond to changes in blood glucose level, and could be used as an artificial pancreas if an insulin formulation is placed in the drug chamber.

Although the final application may involve an enzymatic conversion step, it will be important to characterize the response of a proton-sensitive polyelectrolyte system to changes in hydrogen ion concentration. The present contribution discusses preliminary work toward this end.

We have chosen to work with uncrosslinked, linear polyelectrolytes instead of crosslinked gels. This decision is due in part to difficulties we experienced in loading a gel into the small space between the screen and the diaphragm. If the gel was too large, excess pressure would develop in the pump. If the gel was too small, then it would have to expand to fill the space before it could exert a force. The latter problem is obviated by a liquid polyelectrolyte system, which always fills the space it occupies. In this case a membrane which is impermeable to the polyelectrolyte must be placed between the screen and the polyelectrolyte solution in order to prevent leakage of the linear polyelectrolyte. For reasons to be discussed below, we have chosen to work with hydrophobic polyelectrolytes which precipitate when in the unionized state.

Aside from the practical advantage discussed above, we believe that liquid polyelectrolyte systems may be advantageous for other reasons. First, the crosslinks in gels oppose osmotic expansion. Second, the osmotic expansion of crosslinked polymer systems should be slower than that of liquid systems. Therefore, liquid systems may be more efficient mechanochemical energy converters than gels.

The osmotic pressures that can be measured readily in these liquid polymer systems should be similar to the swelling pressures occurring in lightly crosslinked gels consisting of the same materials at similar volume fractions. Therefore, Donnan pressure data gathered for polyelectrolyte solutions will be useful in explaining the swelling behavior of polyelectrolyte gels in response to changes in the electrolyte composition of external solutions.

In this contribution we discuss the thermodynamic behavior of a series of hydrophobic copolymers of N,N-diethylaminoethyl methacrylate and other, nonionic comonomers. Following a brief discussion of synthesis procedures, we report results of potentiometric titrations. Then, colloid osmotic (Donnan) pressures are measured for fully charged polyelectrolytes as a function of polymer structure and concentration. A comparison of the latter results with a semiempirical model will be made.

Experimental

Materials. All monomers were distilled before use. The vinyl monomers methyl methacrylate (MMA), butyl methacrylate (BMA), 2-hydroxyethyl methacrylate (HEMA), N,N-diethylaminoethyl methacrylate (DEA) and the free radical initiators 2,2' azobisisobutyronitrile (AIBN) and ammonium persulfate were obtained from Polysciences, Inc. Water was double distilled and deionized using the Barnstead Nanopure System. Methanol (Fisher Scientific, A.C.S. grade), absolute ethanol (Gold Shield Chemical Company), ethylic ether (Fisher Scientific, A.C.S. grade) and sodium chloride (Fisher Scientific, A.C.S. grade), were all used as received.

Poly(DEA•HCl) synthesis. Poly(DEA•HCl) was synthesized according to the method of Shatkay and Michaeli (*16*). The monomeric salt DEA•HCl was formed by passing dry gaseous HCl through a solution of DEA in dry ethylic ether. After precipitation and drying steps, the DEA•HCl was polymerized in aqueous solution at 40° with ammonium persulfate as initiator. The resulting solution was dissolved in methanol and poly(DEA•HCl) was precipitated in acetone and dried.

p(DEA•HCl/BMA), p(DEA•HCl/MMA) and p(DEA•HCl/HEMA) copolymer synthesis. The copolymers were synthesized to a maximum conversion of 10%. The appropriate molar proportions of the monomers (100 g total weight) were dissolved in 250 ml of methanol. The polymerization was carried out at 70° using AIBN (0.5 w% with respect to the monomers) as initiator. After 20 minutes the polymerization was stopped by pouring the polymerization mixture into distilled water. The precipitated polymer was separated by filtration and then dissolved into a liter of dry ethylic ether. Dry gaseous HCl was passed through the solution with cooling and stirring. The copolymers precipitated and were subsequently dissolved in an excess of HCl. The resulting solution was precipitated in acetone. After drying in vacuum at room temperature to a constant weight, analysis yielded an N to Cl^{-} ratio of 1.00. Chemical analyses were also performed to determine the polymer composition following synthesis and precipitation.

Titration studies. Titration behavior of the copolymers was studied as follows. Precise aliquots of differing volumes 0.1M NaOH solution were added to 20 ml vials containing an amount of polymer equivalent to 0.165 mmol of DEA•HCl in 15 ml of a NaCl solution at a specified ionic strength. The pH in each vial was then recorded using a standard pH meter.

Colloid osmotic pressure studies. The colloid osmotic (or Donnan) pressure of the polyelectrolytes as a function of polyelectrolyte concentration was measured using a specially constructed Donnan cell with no volume flux. The cell, diagrammed in Figure 2, consists of two compartments. The top, reference solution compartment has a volume of 200 ml. The bottom, polyelectrolyte solution compartment, has a volume of 20 ml. The volume of the polyelectrolyte compartment was chosen to be sufficiently small to maintain ionic strength virtually constant in the reference compartment but sufficiently large to minimize error in the experiments due to any small changes in volume following pressure buildup. The two compartments are separated by a semipermeable cellulose acetate membrane (Wescan, Santa Clara CA), which is supported by metallic screens. The pressure is measured using a membrane pressure transducer (Ametek, Feasterville PA). The top compartment is stirred by a motorized impeller, and the bottom compartment is agitated with a magnetic stirrer.

Results

Titration Studies. Figure 3a displays potentiometric titration curves for p(DEA•HCl), p(DEA•HCl-co-HEMA) and the p(DEA•HCl-co-BMA) copolymers in 0.1 M NaCl solutions. The degree of neutralization, α, is defined as the amount of NaOH added divided by the initial equivalent concentration of DEA•HCl units in the solution. During the titration process a precipitated polymer phase appears at a particular degree of neutralization α, which is indicated in Figure 3 by arrows. When the precipitate is present, the polyelectrolyte solution acts as a very efficient buffer, with pH changing very little over a wide range of neutralization. For poly(DEA•HCl) the pH at which buffering occurs is 7.6. The buffering pH is increased when DEA is copolymerized with HEMA (a hydrophylic monomer). However, buffering pH is

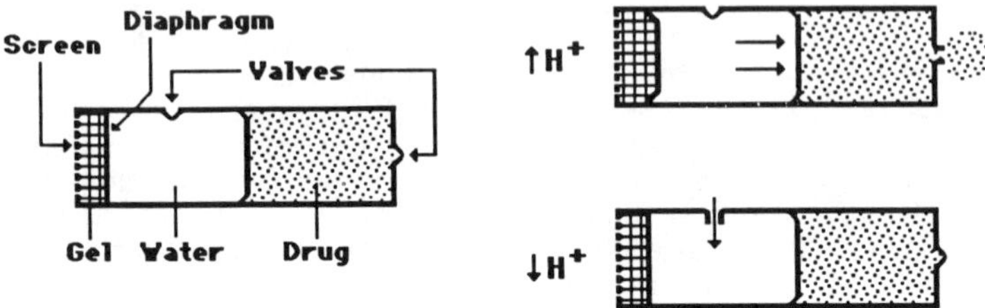

Figure 1. Schematic of a mechanochemical pump which responds to local hydrogen ion concentration changes, or indirectly to glucose level changes when glucose oxidase and catalase are coupled to the polyelectrolyte. Arrows point in direction of water flow.

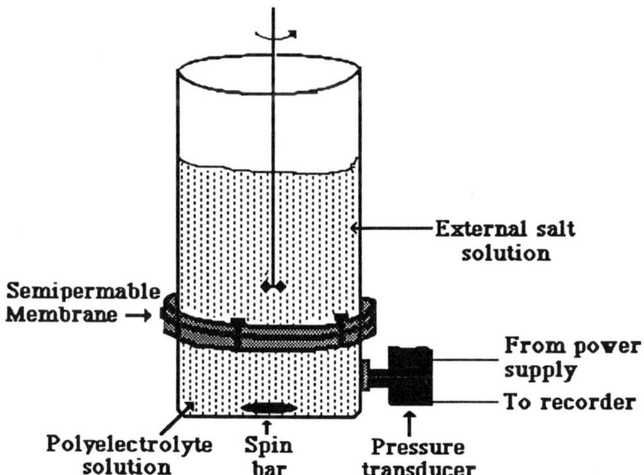

Figure 2. Schematic of Donnan osmotic pressure cell used in the present work.

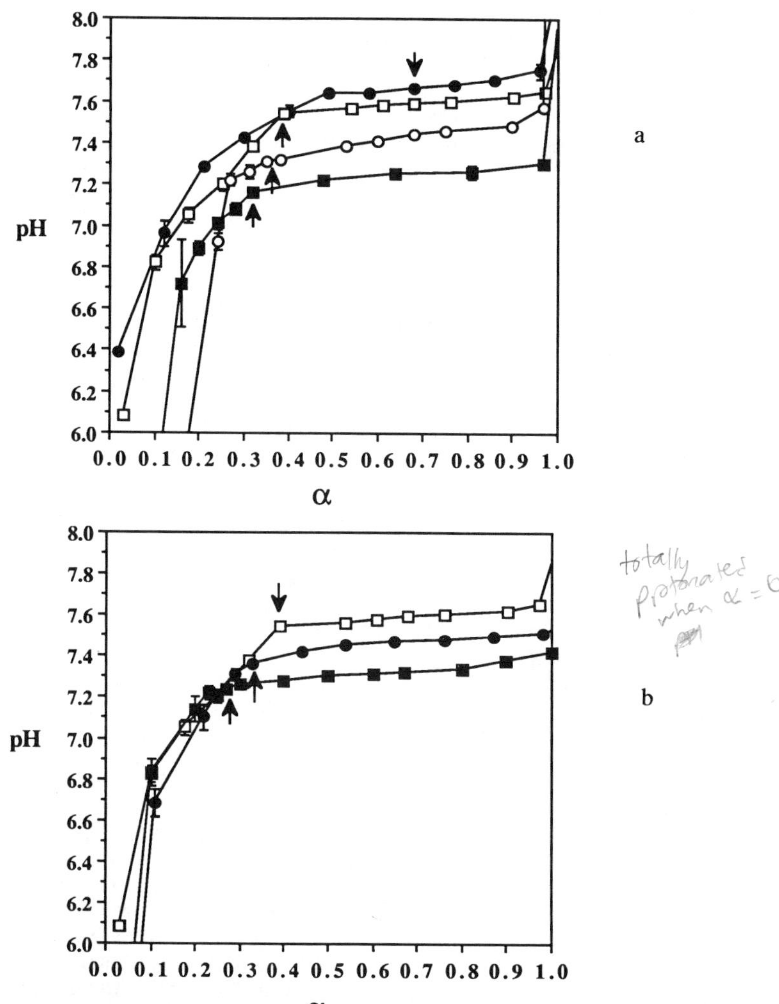

Figure 3. Potentiometric titration curves for copolymers in 0.1 M NaCl solutions.
a) (□) DEA•HCl homopolymer, (○) 98/2 DEA•HCl/BMA copolymer, (■) 80/20 DEA•HCl/BMA copolymer, (●) 73/27 DEA•HCl/HEMA copolymer.
b) (□) DEA•HCl homopolymer, (●) 81/19 DEA•HCl/MMA copolymer, (■) 56/44 DEA•HCl/MMA copolymer.

lowered when the proportion of butyl methacrylate (a hydrophobic monomer) in the copolymer is increased.

In Figure 3b similar data are displayed for p(DEA•HCl-co-MMA). As above, a decrease in the buffering pH with increasing hydrophobic (MMA) content is observed.

Figure 4 shows the effect of ionic strength on the titration curves of p(DEA•HCl-co-MMA) with 85% DEA content. A significant increase in the buffering pH is observed when the ionic strength is increased.

Colloid Osmotic Pressure Studies. Figure 5 shows the effect of the equivalent fixed charge concentration (C_M) on the colloid osmotic pressure, $\Delta\Pi$, for different copolymers of DEA•HCl and MMA, all measured at ionic strength 0.1 M. With increased content of MMA at fixed C_M, $\Delta\Pi$ increases. Also for a given copolymer composition, $\Delta\Pi$ increases parabolically with increasing C_M. The curves in Figure 5 are predictions of models which will be discussed below.

Discussion

Titration Data. The majority of studies of the titration of polyelectrolytes have involved polymers that are sufficiently hydrophilic to remain in solution at all degrees of neutralization. Generally, polyelectrolytes act as buffers in the sense that they stabilize the solution pH when amounts of acid or base are added that would, in the absence of the polyelectrolyte, lead to large swings in pH. However, when pH is plotted against the degree of neutralization, α, the curves are considerably steeper than is observed for monoprotic buffers of the same equivalent concentration (*17,18*). This phenomenon is explained by noting that the charge state of the polyelectrolyte changes with neutralization, leading to increased repulsion, or diminished attraction, of the neutralizing species by the titratable groups. For example, a hydrophilic, polyamine HCl salt will initially carry a high positive charge, which will strongly attract neutralizing hydroxyl ions. As the neutralization progresses, however, the polyelectrolyte charge diminishes and so will the attraction. Thus, the titration process exhibits negative cooperativity.

Certain water soluble but more hydrophobic polyelectrolytes exhibit more complex behavior. For example, poly(methacrylic acid) (*18-20*) and alternating copolymers of maleic anhydride and alkyl vinyl ethers (polysoaps) (*21,22*) show three phases in the titration. Starting from the totally protonated form (α=0), pH initially rises sharply with increasing neutralization by NaOH. In a critical range of α, however, pH flattens out, after which it rises steeply again. Within the critical range, the titration process exhibits positive cooperativity. The flattening has been attributed to structural transitions in the polyelectrolytes that occur within the critical range. Due to the increase in charge which overcomes the chain hydrophobicity, the polymer goes from a relatively compact configuration to a more swollen state, as demonstrated by dramatic changes in viscosity. During this transition process the polyelectrolyte coil shows virtually no "resistance" to neutralization: hence, added base neutralizes the coils instead of increasing the pH of the solution. When the equivalence point is achieved, a rapid rise in pH ensues with added base. Throughout the whole range of neutralization, the polyelectrolyte remains in solution.

Similar three-phase titration effects have been observed with poly[thio-1-(N,N-diethylaminoethyl ethylene], a hydrophobic polybase, and its partially quaternized derivatives (*22*). We consider here the case where such a polybase is initially in its HCl salt form. Neutralization by NaOH will proceed in a similar manner as discussed in the previous paragraph, except that the polymer will go from a charged to an uncharged state with increasing neutralization. Therefore, within the region in which the pH response flattens out, polymers will go from an extended state to a collapsed state (*22*).

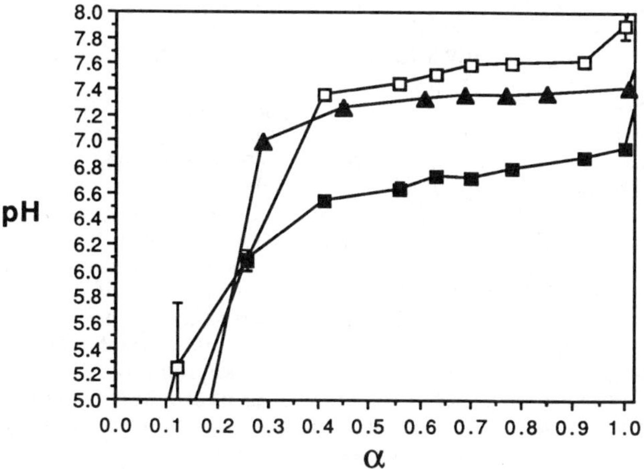

Figure 4. Potentiometric titration curves for 85/25 DEA•HCl/MMA copolymers in NaCl solutions with different ionic strengths (I). (■) no added salt, (▲) I=0.1 M, (□) I=0.2 M.

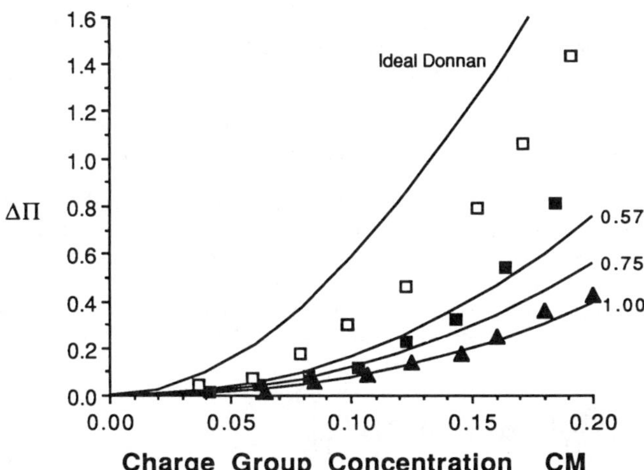

Figure 5. Donnan osmotic pressures measured for DEA•HCl/MMA copolymer solutions against 0.1 M NaCl external solutions. (▲) DEA•HCl homopolymer, (■) 75/25 DEA•HCl/MMA, (□) 57/43 DEA•HCl/MMA. All copolymer ratios were determined by elemental analysis. Curves are predictions based on Eqs. (9)-(11). Values of X_{DEA} used in calculations are indicated next to curves.

The copolymers of DEA with esters of methacrylic acid, studied in the present work, are hydrophobic and insoluble in the ionized state, but fully soluble in the ionized state. Starting in the fully ionized HCl form, and titrating with NaOH, we initially observe a steep pH dependence on α. However, in every case the titration curves flatten out. Unlike the cases discussed above, the point where the curves flatten coincides with the onset of polymer precipitation. Beyond that point, the curves remain flat until virtually all the polyelectrolyte is neutralized. This indicates superb buffering.

The titration behavior of these precipitating polyelectrolytes has a straightforward explanation. In the initial stages all chains are highly charged and prefer to remain in solution. However, when a critical neutralization α_{ppt} is reached, a phase separation, driven by the essential hydrophobicity of the polymer without charge, commences. When a polymer molecule enters the precipitate, it releases all its bound protons, thus neutralizing the added NaOH. This mechanism explains the unusually strong buffering behavior which persists until virtually all the polyelectrolyte is neutralized.

A consequence of the argument of the previous paragraph is that the pH at which precipitation and buffering occur should be lowered as the polyelectrolyte chains become more hydrophobic. The polymer/solvent/ion equilibria that are established are the result of competition between polymer/ion and polymer/polymer interactions. As the polymer interaction becomes stronger (greater hydrophobicity), a higher proton activity (lower pH) is needed to maintain the preference for ion binding (and hence keeping the polymer in solution) over polymer/polymer association (i.e. precipitation). This prediction is confirmed by the data in Figure 3. By including increasing amounts of hydrophobic BMA in the copolymer, the buffering pH is progressively reduced. On the other hand, including the relatively hydrophilic HEMA raises the buffering pH, presumably by destabilizing the precipitate.

Similar arguments account for the observation that an increase in ionic strength raises the buffering and precipitation pH. With increasing ionic strength, the affinity of the polymer for protons increases, since the electrostatic field due to the polyelectrolyte charge is shielded. Thus, lower proton chemical potentials are required to maintain the polyelectrolyte in solution and precipitation does not occur until a higher pH is achieved.

Based on these arguments, one might expect that the precipitation point should occur at lower degrees of neutralization as the polymer hydrophobicity increases. This expectation is confirmed in Figure 3.

With few exceptions, most researchers have avoided studying precipitating hydrophobic polyelectrolyte systems, presumably because the precipitation was considered to be a complicating factor. The most complete studies of which we are aware have been performed by Shatkay and Michaeli. These authors presented a simple thermodynamic analysis which is valid when the precipitated polymer phase is present. The main result is an equation which can be rearranged to the following form:

$$P(1-\overline{\alpha^*}) = \frac{d}{d\,pH}\left(-\log C_P + 0.4343\,\frac{\mu_0}{RT}\right) \quad (1)$$

where R is the gas constant and T the absolute temperature, P is the number of ionizable amine groups per polymer molecule, $1-\overline{\alpha^*}$ is the mean degree of ionization of the polymer molecules in the solution phase, C_P is the molar concentration of polymer in solution, and μ_0 is the chemical potential of the polymer in the precipitate. Shatkay and Michaeli investigated poly(DEA•HCl), measuring C_p and $1-\overline{\alpha^*}$ during the course of titration with NaOH. They were able to show that Equation (1) describes the data well, and they were further able to disregard the μ_0 term. The latter was possible because the chemical composition of the precipitate does not change during the titration process. Shatkay and Michaeli also showed that virtually no HCl was to be found in the precipitated phase, which is consistent with the notion of a competition between polymer/polymer and polymer/proton interactions.

A consequence of Equation (1) with constant μ_0 is that when the precipitate is present, large changes in the solution polymer concentration will occur with very little change in pH, since the product $P(1-\overline{\alpha}^*)$ is large. This explains the exquisite buffering capabilities of the precipitating polyelectrolyte systems. It should be noticed (Figures 3 and 4) that the slopes of the titration curves in the buffering region are somewhat larger for the copolymers than for the pure poly(DEA•HCl). We believe that this observation can be explained in the context of Equation (1). The copolymers are expected to exhibit some compositional heterogeneity, and hence variability in their hydrophobicity. During the process by which the polyelectrolyte molecules are neutralized by NaOH, the most hydrophobic polymer molecules will be the first to precipitate. Subsequently, less hydrophobic molecules will be added to the precipitate. Thus the chemical composition, and hence the chemical potential μ_0 will change with the course of titration. A sizeable value of $(1/RT)(d\mu_0/dpH)$ could conceivably offset the $P(1-\overline{\alpha}^*)$ term. Intuitively, as the solution phase becomes less hydrophobic as the titration proceeds, the attraction of the precipitate for the polymer remaining in solution becomes weaker, and a higher pH will be needed to precipitate the remaining polymer.

Colloid Osmotic Pressure Data. In the experimental configuration shown in Figure 2 , the polyelectrolyte is incapable of diffusing through the semipermeable membrane, but water and small ions are free to pass between the polyelectrolyte solution and the external salt solution. Due to the requirement of bulk electroneutrality in both phases, ion species will not be distributed identically between these phases at equilibrium. This situation represents a classical Donnan equilibrium.

Since the composition differs between the two phases, a colloidal osmotic, or Donnan pressure arises. This pressure may be explained in two complementary ways. In the first explanation the highly charged polyions repel each other, which encourages water flow into the polyelectrolyte compartment. This explanation has given rise to rigorous analyses based on the McMillan-Mayer solution theory (*23-25*), particularly for rod-shaped polyelectrolytes such as DNA. This theory requires calculation of a virial series, the coefficients of which depend on the mean forces of repulsion for pairs, triplets, etc., of polyelectrolyte molecules in the salt solution. Unfortunately, calculations beyond the second virial coefficients are prohibitive for most realistic models of polyelectrolytes. Accordingly, the McMillan-Mayer approach appears applicable only to solutions containing very low concentrations of polyelectrolyte.

In the second explanation for osmotic pressure it is noticed that the electroneutrality requirement implies that the total concentration of microions in the polyelectrolyte solution phase must exceed the total microion concentration in the outer solution. This concentration difference leads to an osmotic pressure (*26*). This explanation appears to be less rigorous than the McMillan-Mayer approach, but it can be applied over much larger polyelectrolyte concentration ranges, and is simpler to implement. The ensuing discussion will be based on the second explanation.

In the following primed quantities refer to the outer solution phase, while unprimed quantities refer to the polyelectrolyte solution phase. Thus c_i and c_i' are the molar concentrations of the i'th ionic species in those phases, respectively, and a_i and a_i' are the corresponding activities. The subscript i can take on the following "values": Na^+, Cl^-, "s" for the neutral salt NaCl, and "M" for the charged amine groups attached to the polyelectrolyte molecules. The quantity we seek is the Donnan osmotic pressure $\Delta\Pi$, which may be generally defined by

$$\Delta\Pi = RT(\phi\Sigma c_i - \phi'\Sigma c_i') \tag{2}$$

where and ϕ and ϕ' are osmotic coefficients. The summation in Equation (2) is taken over all microion species and ignores the molar contribution of the polyelectrolyte molecules, which is substantially lower than the molarities of the microions.

In a Donnan equilibrium, the salt activities in the two phases are equal. This implies the general relation

$$(a_{Na+})(a_{Cl-}) = (a'_{Na+})(a'_{Cl-}) \tag{3}$$

We first consider the ideal case where concentrations and activities are equal, and the osmotic coefficients ϕ and ϕ' are unity. Electroneutrality in both phases leads to the following relations:

$$a'_{Na+} = a'_{Cl-} = c_s' \ ; \ a_{Na+} = c_s \ ; \ a_{Cl-} = c_s + c_M \tag{4}$$

Combining Equations (3) and (4) yields, for the ideal case,

$$c_s(c_s+c_M) = (c_s')^2 \tag{5}$$

Since c_s' and c_M are specified in the experiment, c_s can be determined. Then in view of Equations (2) and (4),

$$\Delta\Pi = RT(2c_s-2c_s'+c_M) \tag{6}$$

From Figure 5 it is clear that the ideal model greatly overestimates the observed Donnan pressures. This discrepancy is readily explained by noticing that strong electric fields surround the polyions; these fields attract the counterions (in this case, Cl^-) to the polyions. Thus the counterions are not completely osmotically active, and their activity is reduced. A more detailed model is required to explain the data.

The approach we will discuss next is based on the semiempirical "additivity rule" for polyelectrolytes in salt solutions (*27-31*). Let π_p be the osmotic pressure of the counterions (here, Cl^-) to the polyelectrolyte in a solution with no added salt. Let π_s be the osmotic pressure of a salt solution in the absence of polyelectrolyte. Then the additivity rule states that the osmotic pressure of the combined polyelectrolyte/salt solution will be

$$\pi = \pi_p + \pi_s$$

and the Donnan pressure will be

$$\Delta\pi = \pi_p + \pi_s - \pi_s' \tag{7}$$

Moreover, the activity of the counterion species (Cl^-) is the sum of the activity of the counterions in the salt-free solution, $(a_{Cl-})_p$ and the counterion activity in the polymer-free salt solution, $(a_{Cl-})_s$, i.e.

$$a_{Cl-} = (a_{Cl-})_p + (a_{Cl-})_s \tag{8}$$

It remains to calculate the terms in Equations (7) and (8). For simplicity we ignore nonideality effects for the salt, so that π_s=2RTc_s and π_s'=2RTc_s'. To make further progress we utilize the "cell model" of polyelectrolyte solutions (*28-33*). In this model polyelectrolyte molecules are represented as charged cylindrical rods surrounded by their counterion atmosphere. Rods in the same locale are assumed to be aligned in parallel, forming a hexagonal lattice. The justification for this picture is that when highly charged, polyelectrolytes are sufficiently stretched that they may be considered to be locally rod-like. Also, the local lattice-like arrangement represents a minimum energy configuration. Clearly, this picture will apply only locally--whole polyelectrolyte molecules are not rods, and the orientation of the rodlike portions must vary in different regions of the solution.

With this geometric specification microion distributions are calculated using the Poisson-Boltzmann equation (*24-26, 28-34*), and the osmotic pressure is calculated applying the ideal van't Hoff expression for regions in which the computed electric field vanishes.

The details of the cell model will not be discussed here, as they are reviewed extensively elsewhere (*29-33*). We will simply summarize the pertinent results. First, in the absence of added salt, the osmotic coefficient ϕ_p and activity coefficient f_p $[(a_{Cl^-})=f_p(c_{Cl^-})]$ of the counterions are identical. Again assuming that the salt ions in both the polyelectrolyte and outer solutions behave ideally, Equations (3) and (8) may be therefore combined to obtain

$$c_s[c_s+(1-\alpha)\phi_p c_M] = (c_s')^2 \tag{9}$$

where $1-\alpha$ is the fraction of the tertiary amine groups that are ionized. The Donnan osmotic pressure of Equation (7) is then given by

$$\Delta\pi = RT[2c_s-2c_s'+(1-\alpha)\phi_p c_M] \tag{10}$$

The second result of the cell model is the specification of ϕ_p. The parameters needed to calculate ϕ_p are V_p, the polyelectrolyte volume fraction, and λ, the linear charge density of the polyelectrolyte chains. The latter is defined by

$$\lambda = (1-\alpha)e^2/\varepsilon bkT$$

where e is the protonic charge, ε is the solution dielectric constant (taken to be that of bulk water), b is the average distance between charge groups along the chain backbone, and k is the Boltzmann constant. For the vinylic copolymers under study and at room temperature, $\lambda=2.83(1-\alpha)X_{DEA}$, where X_{DEA} is the mole fraction of DEA units in the copolymer. To calculate ϕ_p the following equations are used:

$$\phi_p = (1-\beta^2)/2\lambda, \text{ where } \lambda = (1-\beta^2)/[1+\beta\coth(\beta\gamma)] \qquad \lambda < \gamma/(\gamma+1)$$
$$\phi_p = (1+\beta^2)/2\lambda, \text{ where } \lambda = (1+\beta^2)/[1+\beta\cot(\beta\gamma)] \qquad \lambda > \gamma/(\gamma+1) \tag{11}$$

where $\gamma=-(\ln V_p)/2$. Thus, we first solve for β in terms of λ and γ and then for ϕ_p in terms of β.

Figure 6 displays calculated osmotic coefficients ϕ_p for the counterions in the absence of salt. The polymer volume fraction term V_p was taken to be $0.24C_M$, based on density measurements of the copolymers. The MMA component was assumed to contribute little to the polymer volume, since the molecular weight of an MMA unit is about half that of a DEA unit, and since in all cases DEA is the majority comonomer. A weak but definite trend of increasing ϕ_p with C_M is observed. This trend is due, of course, to the increase in V_p. With increasing V_p polymer "rods" become more closely spaced, and the attraction of a counterion to a particular rod will be reduced due to attractions to neighboring rods, leading to greater freedom of movement and

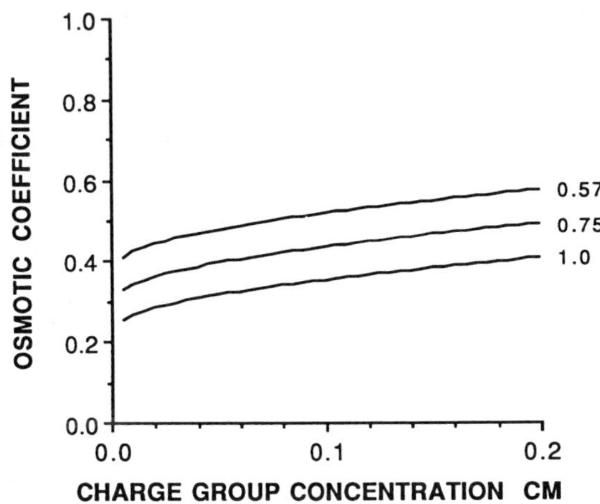

Figure 6. Osmotic coefficients ϕ_p calculated for DEA•HCl/MMA copolymers in salt-free solutions. Values of X_{DEA} used in calculations are indicated next to curves.

hence a higher osmotic coefficient. More dramatic differences are seen between the curves for the different values of X_{DEA}. This will be reflected, of course, in the value of b and hence λ. As charge groups become more separated by MMA groups, the electric field around the chain weakens and the osmotic coefficient will increase.

Figure 5 presents a comparison of predictions based on Equations (9)-(11) against experimentally determined values, as measured above. For $X_{DEA}=1.0$ the theory predicts well the data up to $C_M=0.20$ M. For $X_{DEA}=0.75$ theory diverges from experiment for $C_M>0.10$ M. For $X_{DEA}=0.57$ the theory consistently underestimates the experimental values of $\Delta\pi$, although the values for $C_M\leq0.06$M are in reasonable agreement.

At present we are unable to explain the success of the model in some regions and its failure in others. However, some discussion of the present data within the context of previous studies is useful. In a review article Katchalsky et al. (*29*) display osmotic coefficients for various polyacid solutions at varying degrees of ionization α, with no added salt. In order to account for the data, the charge density parameter λ for a given polyelectrolyte had to be adjusted to a value that is higher than would be inferred from the structure of the chain backbone. For polyacrylic and polymethacrylic acid, it was necessary to increase λ by a factor of two. For stiffer chain polyelectrolytes such as carboxymethylcellulose and alginates, λ was increased by a factor of 1.5. While no explanation was provided, it was conjectured that dielectric saturation in the vicinity of the charge groups could account for this discrepancy.

In contrast, we find that no modification of λ is needed to fit the osmotic pressure data of the DEA/MMA copolymers for $X_{DEA}=1.0$ and 0.75, and at sufficiently low concentrations C_M. In all our studies $\alpha=0$. For $X_{DEA}= 0.57$ one would have to decrease λ to force the theory to fit the data, but this modification seems implausible. It is noteworthy that Shatkay and Michaeli (*35*) inferred values of osmotic coefficients for poly(DEA) ($X_{DEA}=1.0$) which are approximately twice the values estimated for polyacrylic and polymethacrylic acids. These estimates are in a direction consistent with our findings. We speculate that the DEA monomer may be less prone to the dielectric saturation effects mentioned above since the DEA group is bulkier than a carboxylic acid, so that the electric field near the interface between the DEA group and water will be weaker than the corresponding field at the carboxylate-water interface. Of course, this explanation cannot account for the lack of agreement between theory and experiment at higher polyamine concentrations, and for $X_{DEA}=0.57$.

Implications for the Mechanochemical Pump.

The experiments reported in this paper were designed to investigate the suitability of DEA copolymers for the mechanochemical pump described earlier. The results indicate that somewhat hydrophobic copolymers of DEA will be superior to the poly(DEA) homopolymer. Most obviously, adding spacer comonomers between the amines will increase the osmotic efficiency, due to the higher resulting osmotic coefficients. Also, it is necessary to incorporate the hydrophobic component in order to lower the buffering region below physiologic pH (7.4). Otherwise the polymer will be almost completely ionized at physiologic pH, and it will not show much sensitivity to protons supplied to it by the glucose oxidase reaction.

We believe that it will be most useful to set the buffering pH just below the range of pH fluctuations that one might expect in a diabetic. Doing so will prevent the device from being triggered by acidotic episodes. Thus, during normoglycemic periods the internal pH will be essentially 7.4 and the polymer will be precipitated. After a rise in blood sugar, one may expect a slight delay due to the need to accumulate sufficient protons to lower the local pH to the point where the polymer will start to go into solution and become osmotically active. We are in the initial stages of investigating the kinetics of conversion of glucose to protons, the acidification and dissolution of the precipitated polymer (as well as the reverse process) and the development of osmotic pressure. The rates of these steps will be influenced by device geometry, and probably by the presence of physiologic buffers such as phosphate and bicarbonate.

From previous discussions it should be evident that the precipitating polybases are probably superior to more hydrophilic polybases that remain in solution throughout a titration. The latter are negatively cooperative in their ionization behavior. Therefore, as more protons are provided by the enzyme reaction, they will become progressively less efficient in generating osmotic pressure. Said another way, one can expect a large pH drop within the device due to the progressive unwillingness of the polyelectrolytes to adsorb protons, and this will discourage diffusion of new protons into the device. This problem is reduced substantially in the precipitating system, since the pH inside the device changes little with progressive ionization. Osmotic pressure can be developed with scarce change in pH.

Conclusions

The results demonstrate that the buffering pH of DEA•HCl copolymer can be "tuned" by changing the content of hydrophobic or hydrophilic nonionic monomers. This observation together with the data from osmotic experiments indicate that hydrophobic copolymers of DEA•HCl can be used as osmotic agents for a self regulating mechanochemical insulin pump.

The present work is incomplete in a number of ways. First, it should be noted that the titration experiments were conducted at high dilution, whereas the osmotic pressure data was gathered at much higher concentrations in order to achieve pressures that will be useful for the proposed pump. A more consistent study would also check the titration behavior at higher concentrations. Moreover, such titrations should be carried out in a Donnan cell. Second, the studies were conducted at room temperature with ionic strength set at 0.1M by NaCl. More realistic studies should be conducted at body temperature, and in phosphate and bicarbonate buffered media which reflect the contents of plasma. Finally, the present experiments concerned equilibria. The present do not address kinetics, which will be extremely important in designing an implantable insulin pump which must respond rapidly to changes in blood glucose levels.

Acknowledgments

This work was supported in part by grants from the National Institutes of Health (DK 38035), The Whitaker Foundation, and the California Biotechnology Research and Education Program.

Literature Cited

1. Siegel, R.A.; Falamarzian, M.; Firestone, B.; and Moxley, B.C. *J. Controlled Release* **1988,** *8*, 179-182.
2. Pradny, M.; and Kopecek, J. *Makromol. Chem.* **1990,** *191*, 1887-1897.
3. Brannon-Peppas, L.; and Peppas, N.A. *Biomaterials* **1990,** *11*, 635-644.
4. Kuhn, W.; Hargutay, B.; Katchalsky, A.; and Eisenberg. H. *Nature*, **1950,** *165*, 514-516.
5. Kuhn, W.; Ramel, A.; Walters, D.H.; Ebner G.; and Kuhn, H.J. *Fortschr. Hochpolym.-Forsch.*, **1960,** *1*, 540-592.
6. Helfferich, F. *Ion Exchange*; McGraw-Hill, New York, NY, 1962.
7. *Ion Exchange*; Marinsky, J., Ed.; Marcel Dekker: New York, NY, 1966, Vol. 1.

8. Michaeli, I.; and Katchalsky, A. *J. Polym. Sci.*, **1957**, *23*, 683-696.
9. Hasa, J.; Ilavsky, M.; and Dusek, K. *J. Polym. Sci.*, **1975**, *13*, 253-274.
10. Ricka, J.; and Tanaka, T. *Macromolecules* , **1984**, *17*, 2916-2921.
11. Siegel, R. A.; and Firestone, B. A. *Macromolecules*, **1988**,*21*,3254.
12. Firestone, B. A.; and Siegel, R. A. *Polym. Commun.* **1988**, *29*, 204.
13. Grimshaw, P.E.; Nussbaum, J.H.; Yarmush, M.L.; and Grodzinsky A.J. *J. Chem. Phys.*, **1990**, *93*, 4462-4472.
14. Kost, J.; Horbett, T.A.; Ratner, B.D.; and Singh, M. *J. Biomed. Mater. Res.*, **1984**, *19*, 1117-1133.
15. Siegel, R. A; and Firestone, B. A. *Journal of Controlled Release* **1990**, *11*,181.
16. Shatkay, A; and Michaeli, I. *J. Phys. Chem.* **1966**, *70*, 3777.
17. Katchalsky, A., Mazur, J. and Spitnik, P. *J. Polym. Sci*, **1957**, *13*, 513-532.
18. Katchalsky, A.; Shavit, N.; and Eisenberg, H. *J. Polym. Sci.*, **1954**, *13*, 69-84.
19. Leyte, J.C.; and Mandel, M. *J. Polym. Sci.*, **1964**, *2*, 1879-1991.
20. Noda, I.; Tsuge T.; and Nagasawa, M. *J. Phys. Chem.*, **1970**, *74*, 710-719.
21. Dubin, P.; and Strauss, U. *J. Phys. Chem.*, **1970**, *74*, 2842-2847.
22. Vallin, D.; Huguet, J.; and Vert, M. *Polym. J.*, **1980**, *12*, 113-124.
23. Hill, T.L. *An Introduction to Statistical Thermodynamics*; Dover, New York, NY, 1986.
24. Hill, T.L. *Disc. Faraday Soc.*, **1956**, *21*, 31-45
25. Stigter, D.; *Biopolymers*, **1977**, *16*, 1435-1448.
26. Overbeek, J. Th.G. *Prog. Biophys. Biophys. Chem.*, **1956**, *6*, 57-84..
27. Alexandrowicz, Z. *J. Polym Sci.* **1960**, *43*, 337-349.
28. Alexandrowicz, Z. *J. Polym Sci.* **1962**, *56*, 97-114.
29. Katchalsky, A.; Alexandrowicz, Z.; and Kedem, O. In *Chemical Physics of Ionic Solutions*; Barradas, C., Conway, R.G., Eds.; Wiley: New York, NY, 1966; Chapter 17.
30. Katchalsky, A. *Pure Appl. Chem.* **1971**, *26*, 327-373.
31. Marinsky, J. in *Ion Exchange*; Marinsky, J., Ed.; Marcel Dekker: New York, NY, 1966, Vol. 1.
32. Lifson, S.; Katchalsky, A. *J. Polym. Sci.*,**1954**, *13*, 43-55.
33. Rinaudo, M in *Polyelectrolytes*, Selegny, E., Ed.; D. Reidel: Dordrecht, Holland, 1974, 157-193.
34. Sharp, K. and Honig, B. *J. Phys. Chem.*, **1990**, *94*, 7684-7692.
35. Shatkay, A. and Michaeli, I. *J. Polym. Sci.: Part A-2*, **1967**, *5*, 1055-1059.

RECEIVED August 23, 1991

Chapter 9

Gel Structural Heterogeneity, Gel Permeability, and Mechanical Response

A. Silberberg

Weizmann Institute of Science, Rehovot 76100, Israel

It is shown that in consequence of the additional correlation between segments implied by a crosslink whether of finite or infinite lifetime heterogeneities must of necessity occur in dilute gel systems in steady state. In systems where the blob chain length g, calculated from the overall volume fraction ϕ of the network substance, is larger than the mean chain length between crosslinks g_o, calculated from the overall composition, the chemically imposed heterogeneities will dominate. At higher overall concentrations, $g < g_o$, however, the built in heterogeneities tend to be drowned out by the random short-lived contacts. How these heterogeneties affect permeability and plateau modulus are discussed and illustrated in the case of aqueous polyacrylamide gels made by copolymerisation.

Our objective is two-fold. Firstly we wish to provide arguments which will demonstrate the existence, necessarily, of heterogeneities in a chemically cross-linked system, whether or not the crosslinks are permanent. Secondly we wish to show that these heterogeneities will tend to be overwhelmed by physical contacts as the overall segments concentration increases and the volume effectively excluded by a segment declines with solvent power. This is today very clearly appreciated in the case of the single polymer chain, in consequence perhaps of the tremendous emphasis placed on this second point in the early work of Flory and his discovery of the Theta point, the temperature at which associative attractions between polymer segments, in a given solvent medium, just cancel the self volume repulsive contact potential between those segments (*1*). The mean square radius of gyration then shows ideal Gaussian behavior and for infinitely long chains the Theta point is the point of total self collapse. Without an inherent length scale the statistical segment of the ideal Kuhn random coil which is

0097–6156/92/0480–0146$06.00/0
© 1992 American Chemical Society

derived from the mean square end-to-end distance by dividing it by the real contour length of the chain would thus be zero.

Also a three-dimensional infinite network tends to give the total collapse of the network at the Theta point. Only because with real chains there are finite dimensions associated with the network component does one, therefore, away from the Theta point, obtain a macroscopically sized structure with finite open spaces which can and will be filled with solvent when an open contact between network and a solvent source is created.

Binary contacts between segments a certain distance apart along the contour of an isolated homopolymer chain in solution will occur uniformly with the same probability at all segment locations (near-to-end segments exempted). In the swollen network, however, such contacts become dependent upon the location of the segment with respect to the crosslinks. The network is constrained by possessing built-in chemical crosslinks, and this constraint enhances the probability that segments neighboring the crosslink on different strands and a certain contour length apart will contact each other in space. Chemical crosslinks, even in the most ideal network (a fully satisfied crosslink functionality and equal strand length between crosslinks), thus impose a heterogeneous structure.

It is particularly wrong to conceive of the strands between crosslinks as the equivalent of the "blobs" which de Gennes uses to define that section of a chain in a semidilute solution which "sees" only its own segments to the exclusion of all other segments in the system (*2*). Probably not even in swelling equilibrium is this precisely true. It certainly does not apply in the general case. Blobs are not meant to have fixed ends. They are only statistically located. Any segment of the chain has an equal statistical chance to be an "end segment" of a blob. Only when the blob length in a semidilute solution of similar concentration to the gel is much smaller than the mean strand length between crosslinks will the gel and the semidilute solution tend toward similar behavior in certain physical situations.

The Structure of Gels

In a semidilute solution, if we ignore the small number of segments at or near the ends of the entangled linear chain macromolecules, every segment has the same chance to stand in contact with any of the other segments which surround it. Predominantly, however, the interactions are confined to a regime which de Gennes calls the "blob", composed of the nearby segments of the segment's own chain. The segments so involved with each other in the blob form a section g segments long in the chain and shield each other from interaction with the cloud of other segments further out. In particular it can be stated that if two segments from different molecules are in contact the next such contact along the chain will, because of shielding, be some g segments away at that instant. The semi-dilute system is not free of structure, but it is the same structure statistically for each segment and is one constantly changing with time. Averaged over time the surroundings of any particular segment are the same.

Not so in a chemically linked network ! Even if by some clever chemistry each f-functional monomer making up the crosslink is surrounded by f-strands of exactly equal length, there will now be a clearly marked difference between segments. In a network of topologically fixed crosslinks, it becomes impossible to ignore the special status of the segments near the ends of the strands, i.e. the segments which are close to the crosslink itself. One can again define a characteristic structural element, but one quite different from the single blob, a star composed of the crosslink with its f surrounding strands each one conceptually cut in half. There is overlap between these units and they are entangled with each other. It is reasonable to assume that the status of each such entity is statistically the same. No longer, however, in a network, is each individual chain segment equivalent. The only equivalent structures are the crosslinks and the stars they define. In the ideal case these stars will fill space uniformly and will generate locally equivalent conditions.

In the hands of a less skilled chemist, one, for example, who simply co-reacts monofunctional and difunctional, or f-functional, monomers in the solvent medium using radical initiation polymerization, a much more heterogeneous structure will result. The multi-functional monomer, as a rule, reacts at a higher rate and the incorporation of each higher than mono-functional monomer creates a potential crosslink point in the polymer growing on the initiator molecule. The growing chain for some time, or even forever, may have unused functionalities along its contour. On the whole, however, these sites are much more accessible to the growing reactive end than a chance contact with a monomer unit coming in from solution. The chain will, hence, predominantly tend to branch within itself and create an array of local, highly concentrated gel regions in the early stages of the reaction. A much shorter chain length between crosslinks than the overall ratio of the two monomer units would suggest is thus created in this early phase. Only when the accessible chances of self crosslinking within the growing gel particles are all used up and only a few monomer units remain in solution will the chains, which grow, reach out beyond the core across the intervening solvent space and interact with another core particle, thus finally forming a macroscopic network of relatively concentrated core elements and a rather dilute connecting gel. To the eye the gel may still seem perfectedly homogeneous since the core particles are generally much smaller than the wave length of visible light.

Note that in this case as well, it is possible to define a structural repeat unit. No star this time, but a unit centered on the core gel particle. The number of replicas, however, is now not determined by the number of f-functional monomers, but, approximately, by the number of growth centers in the polymerization reaction, i.e. by the number of initiator molecules.

Other methods of crosslinking create other forms of local heterogeneities even though, on the macroscopic scale uniform systems are generated, as a rule. For example crosslinking a semi-dilute polymer solution by radiation tends to produce locally crosslinked systems of chains which may contract faster around their new center of mass than they will crosslink to other polymer molecules. Some crosslinks between developing cores will arise, but the local concentration differences introduced in the early stages of the reaction will tend to become consolidated and may even grow in emphasis until all the polymer molecules are linked to each other. It is again reasonable to believe that one can define a core of high local concentration which one can use as the center of a constructional element.

The semi-dilute solution is of course also a gel of sorts. The system is characterized by the fact that at any instant there are a given number of chance contacts between the different chains, generally one or more per macromolecule. The lifetime of these contacts will, however, be extremely short. Not only are the places where the contacts were established changing almost instantly, but also their total number is fluctuating about some mean. It is this rapid change in the structural arrangement which allows us to think of the solution as intermolecularly smooth and randomly disposed. Only the correlations along the chains, the blobs, reflect the macromolecular nature of the solute and even here the position of the blobs along the macromolecule fluctuates with time, Were we to freeze in, instantly, a particular state of a semi-dilute solution and then crosslink the resulting immobilized structure by forming a permanent bond at each point of contact in the system, we would destroy the larger than blob isotropy and each chemical crosslink would be surrounded by spherical shells of declining segment concentration.

Even reversible so-called physical gels will not be absolutely uniform. Reversible gels are semi-dilute polymer solutions where a minority of contacts between points on the polymer chains possess appreciable longer lifetimes than others. Such long lived contacts can only be achieved if the polymer involved is chemically or structurally nonuniform, i.e. is a copolymer containing a majority of units which are well soluble and exclude each other and a small minority of units which tend to associate strongly. The solvent is a better than Theta system for the first kind of unit and a worse than Theta system for the other. Since the long-lived contacts can arise only between certain chemically determined sites, not every purely conformationally determined contact pair becomes a crosslink. The tendency to produce long- and short-lived contacts distorts conformations, but since the rate of turnover of both kinds of crosslink is finite, time does tend to establish considerable uniformity in a reversible gel of a random copolymer. While at any instant local high points of concentration will arise, there is turnover and all similarly located reactive sites on the polymer chains can participate. Rheologically we have here an interesting interplay of relaxation times, the lifetimes of the contacts and the conformational relaxation times of the connected segments stand in competition and tend to undercut each other (*3*).

It should also be recalled that in certain important cases of reversible gel formation, the crosslinks are the result of strand pairing or even multistrand local crystal formation. The crosslink in such cases is a cooperative structure which makes use of selected structural regions in the copolymer where locally two or more sterically complementary sections form good association complexes and are prevented from growing too long by breaks in structural complementarity in the polymer chain. Such "crystallite" crosslinks can obviously be of very long lifetime and can impose a much more marked local structural inhomogeneity on the system.

Only occasionally will visible light show these inhomogeneities since they tend be on a smaller scale than the average polymer coil size and thus much below the wave length.

Permeability

The freedom with which solvent moves through a matrix of macromolecular chains is exquisitely sensitive to the number and distribution of the segments blocking its path. Let us assume that the number of matrix segments, and, therefore, the volume they occupy on the average in a unit volume of the system, is maintained constant. Only the arrangement of these segments is altered. One extreme case we can imagine has the segments uniformly distributed in space, another has them bunched in one region leaving the remainder of the unit volume very sparsely occupied. In view of hydrodynamic shielding, the second arrangement turns out to be more permeable by far.

How does hydrodynamic shielding work? The answer is its long-range nature. Even if we have one region essentially free of segments, the other where the segments are more crowded will still find them separated by large distances if the overall segment concentration is low. Thus, even in the "crowded" region random segment-segment contacts occur only rarely. Beyond a certain quite low concentration, however, the segments begin, hydrodynamically, to shield each other and flow in the intervening fluid is essentially stopped. The flow will preferably go through regions where very few segments stand in its way. It is this region which determines the permeability. Since resistance to flow is more a matter of the number of polymer segments which are exposed to the flow than the aspect area between them, the removal of a segment from the low concentration region and its "burial" in the hydrodynamic "shadow" of another in the high concentration part raises permeability although, overall, the segment concentration is the same. In other words the more segments one can pack into an already essentially impermeable region increases the permeability of the system. This packing of segments, as we have pointed out, occurs spontaneously, if the number of crosslinks which hold the segments together in the dense region is high while the total number of segments in the unit volume is kept the same. The permeability, at a fixed overall segment concentration, should thus go up as the number of local-segment - crowding-co-monomers is being increased. This increase of permeability as the number of crosslinks goes up at constant overall segment concentration is precisely what was found to happen in the case of aqueous polyacrylamide gels made by the copolymerization procedure. The effect is illustrated in Figure 1. Note that the system of lowest permeability at each overall volume fraction is the semi-dilute solution. Though the overall polymer volume fraction is kept constant, the permeability increases by much more than an order of magnitude as the fraction of the number of difunctional monomers is increased.

Note, however, that the effect decreases as the overall concentration increases. Let us understand the reason. If, as in the present case, the crosslinker monomers are difunctional the mean number g_o of segments in a strand between crosslinks is given by:

$$g_o = \phi_M/2\phi_D \ , \tag{1}$$

where ϕ_M is the volume fraction of the monofunctional monomer and ϕ_D is the volume fraction of the difunctional monomer. The overall volume fraction is given by:

$$\phi = \phi_M + \phi_D \ . \tag{2}$$

In a semi-dilute solution of volume fraction ϕ there are no long-lived chemical crosslinks. The blob, however, defines a self-contained length of chain of g segments, each of volume a^3, within whose domain there will on the average be one contact between two segments from different chains. The volume fraction of segments within the blob equals the segment volume fraction present overall in the system, so that, since water in our case is a close to Theta solvent for polyacrylamide, we can estimate the blob size g from:

$$\phi = ga^3/(gaA)^{3/2} = 1/g^{1/2}s^{3/2} \tag{3}$$

where $(gaA)^{1/2}$ is the root mean square end-to-end distance of the blob in a Theta-solvent given that A is the Kuhn statistical element and that:

$$s = A/a \tag{4}$$

is the number of segments in the statistical element.

Let us now compare such a system with a uniform gel of volume fraction ϕ where there are g_o segments precisely in the strands between crosslinks. We have two possibilities. In the semi-dilute system at concentration ϕ the "strand size" g, the mean length of chain between segments in contact, is either smaller or larger than g_o. Either the number of chemical crosslinks is negligible as compared with the number of inter-chain, larger than "blob" contacts produced by the semi-dilute system conditions in the gel, or g_o is smaller than g in which case the opposite is true. Combining equations (1)

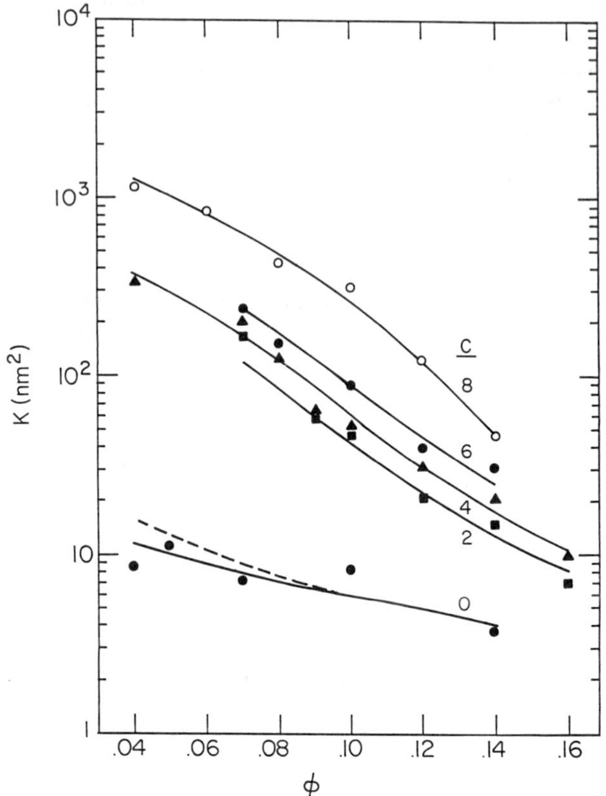

Figure 1: Permeability coefficient K versus overall volume fraction ϕ of polymer segments in the gel at various values of C, the percentage of difunctional monomer in the gel system.

$C = 100\phi_D/\phi$ where ϕ_D is the volume fraction of difunctional comonomer in the system. C=0 corresponds to the non-crosslinked case, the semi-dilute solution where K was computed from sedimentation data (*6*). The data on K for C > 0 are taken from publications (*4-7*). The dotted line is computed by means of the Kuhn pearl necklace model for sedimentation of linear chains assuming a chain diameter of d=1.4 nm.

and (3) yields a relation, valid under close to Theta conditions:

$$g_o/g = (s^3\phi^2/2)[(\phi/\phi_D)-1] \quad (5)$$

which in the case of aqueous polyacrylamide gels can decide this question.

Fortunately we have a value for s. The permeability of the polyacrylamide system, in terms of the two phase model we described earlier (*4*), is constrained by the condition of internal equilibrium with respect to the solvent in the concentrated and dilute gel phase. Matching this condition to give agreement with the measured permeability fixes the parameter s (*4*). It was found that s is 1.5 and with this value we can conclude from equation (5) that g begins to be smaller than g_o in regions where ϕ is in excess of about 0·15.

In the region of ϕ larger than 0.15 and not too large concentrations of the difunctional co-monomer the blob size g is smaller than g_o and thus dominates the number of effective crosslinks. In that region the distribution of the chemical crosslinks ceases to determine gel structure. The effective number of crosslinks is now dominated by the blob size which sets the size of chain which can function as a strand in terms of ideal network concepts. As Figure 1 shows the permeabilities of all the gels tend toward the value set by the "permeability" of the semi-dilute solution as the overall concentration is increased.

The Plateau Storage Modulus

Further evidence for the heterogeneity of gel structure is provided by our data on the storage rigidity modulus (*5*). The storage rigidity modulus of aqueous polyacrylamide gels in the ϕ, ϕ_D range of Figure 1 was measured for radial frequencies from about 1 to 10^3 radians/sec. The modulus tends to be independent of frequency over some decades in the lower frequency range and then begins to rise. We will discuss the plateau value as given in Table I. The plateau value of the rigidity modulus characterizes the long range correlations in the gel and is a measure of the number of effective independent crosslinks ν in the network in terms of the classical formula for the storage modulus G'_p in the plateau region:

$$G'_p = f\nu kT \quad , \quad (6)$$

where ν is the number of crosslinks of effective functionality f per unit volume, k is the Boltzmann constant and T is the absolute temperature (*1*). In equation (6) we include in ν all the chemically recognizable entities that function as crosslinks and treat f as a measure of the number of strands which start on the crosslink and actually establish a physical link with one of the neighboring crosslink structures.

Table I

G'_p
(in kdynes/cm^2 as taken from (*5*))

$\phi =$		0.04	0.07	0.10	0.12	0. 16
$100\ \phi_D/\phi$ =	1	0.31	2.40	5.50	8.60	12.3
	2	0.65	4.60	10.3	16.0	25.0
	4	1.50	9.10	20.9	30.0	50.0
	6	1.66	8.80	22.0	31.5	57.0
	8	1.08	5.10	16.0	24.0	48.0

We have pointed out that it is appropriate, in the case of the polyacrylamide gels made by *in situ* copolymerization, to treat the crosslink as the concentrated gel particle core which arises around each initiator molecule in the early stages of the copolymerization process. The effective functionality f is a measure of the actually completed chain links between core particles in the last stages of the reaction. Many such cores do not link into the network and constitute dangling ends. Hence only a small minority of cores contribute to G'_p .

In Table I we give our results for G'_p (*5*) when in all cases the initiator concentration [I] to which ν is proportional was 5.5 mM. Assuming that T=300°K and putting ν equal to [I] (in terms of molecular crosslinks per unit volume), we can calculate f as shown in Table II. Since ν is overestimated by equating it to [I], f, which ideally should be 1 or larger, is expected to be underestimated and is indeed less than 1 in all cases. Clearly however, it tends towards larger values as the gel concentration is increased. On the other hand it is established that G'_p is scaled by [I]. Only the value of ν so estimated gives the right order of magnitude for G'_p.

Table II

f
(calculated from G'_p/[I])

$\phi =$		0.04	0.07	0.10	0.12	0. 16
$100\ \phi_D/\phi$ =	1	0.004	0.034	0.079	0.120	0.180
	2	0.009	0.066	0.150	0.230	0.360
	4	0.021	0.130	0.300	0.430	0.710
	6	0.024	0.130	0.310	0.450	0.810
	8	0.015	0.073	0.230	0.340	0.690

Let us, for example, consider another estimate for ν. Let us calculate G'_p on the basis that the number of chemical crosslinks of functionality 2 is a measure for ν. The difunctional monomer concentration in our case (Table I) ranges from $\phi_D = 4x10^{-4}$ to $1.3x10^{-2}$. When f=1 and the molar volume of acrylamide is 80 ml, we can calculate from equation (6) that G'_p should range from 250 to 8200 kdynes/cm^2. These estimates (see Table I) are 2 to 3 orders of magnitude too large.

The data in Table I, however, show an interesting correlation with gel composition. This correlation is illustrated in Table III where we have calculated $G'_p/\phi^2\phi_D$. Considering the relatively big changes in G'_p, the $G'_p/\phi^2\phi_D$ values are remarkably constant. Only for combinations of ϕ and ϕ_D where we can expect semi-dilute solution behavior to dominate the gel (ϕ larger than 0.15) does the value of $G'_p/\phi^2\phi_D$ begin to decline.

Table III
$G'_p/\phi^2\phi_D$
(in (dynes/cm^2)x10^{-8})

ϕ =		0.04	0.07	0.10	0.12	0.16
100 ϕ_D/ϕ =	1	4.8	7.0	5.5	5.0	3.0
	2	5.1	6.7	5.2	4.7	3.0
	4	5.9	6.6	5.2	4.4	3.1
	6	4.3	4.3	3.7	3.0	2.3
	8	2.1	1.9	2.0	1.7	1.5

For the semidilute case under close to Theta conditions G'_p should scale with ϕ^3. We have, therefore, also calculated G'_p/ϕ^3 in Table IV. A proportionality of gel modulus to ϕ^3 independent of ϕ_D (in the case of a Theta environment) is predicted when we assume that in the semi-dilute case:

$$\nu \sim \phi/g \ , \qquad (7)$$

substitute the expression (3) for g into (7) and use the result in (6). Note from Table IV that the relationship G'/ϕ^3 is indeed approximately constant in the range of ϕ, ϕ_D pairs which correspond to the dominance of semi-dilute character, i.e. the case where g is smaller than g_o.

What, then, is the significance of the proportionality to $\phi^2\phi_D$ demonstrated in Table III? The combination $\phi^2\phi_D$ is a measure of the number of cases where two mono-

Table IV

G'_p/ϕ^3
(in (dynes/cm^2)x10^{-6})

$\phi =$		0.04	0.07	0.10	0.12	0.16
$100\,\phi_D/\phi$ =	1	4.8	7.0	5.5	5.0	3.0
	2	10.2	13.4	10.3	9.3	6.1
	4	23.4	26.5	20.9	17.4	12.2
	6	25.9	25.7	22.0	18.2	13.9
	8	16.9	14.9	16.0	13.9	11.7

functional monomers, not in the same growing core, have reacted with the same difunctional monomer and have thus established a crosslink between cores. Hence $\phi^2\phi_D$ is a measure of fv and

$$G'_p \sim fv \sim \phi^2\phi_D \quad (8)$$

We can thus expect $G'_p/\phi^2\phi_D$ to be constant in the region of the ϕ, ϕ_D plane where g_o is smaller than g.

Note that these gels were not studied under conditions of swelling equilibrium with water. Even in the permeability experiments the gel volume was constrained (*4-7*).

Polyelectrolyte Gels

A network studded with groups which are capable of ionization is even more markedly a heterogeneous system, although to my knowledge no systematic study has yet been performed.

Where, in the uncharged case, the internal equilibrium of the system depended upon the chemical potential of the solvent only, there is now also an electrical potential field set up which depends upon the segment distribution of the network and the distribution of charge along it. In addition to the chemical potential of the solvent the electrochemical potential of the counter ions must be constant throughout the gel. The effects of segment concentration distributions on the solvent chemical potential are compensated by a distribution of hydrostatic pressure in the solvent (*8*). The gradients in hydrostatic pressure are generated (and balanced) mechanically by the stress distribution in the network. Stresses in a polymeric network arise from coil expansion and, in the case of a polyelectrolyte network, also from the partially shielded coulomb interactions between the charged sites.

The expansion of the network and the local stresses which deformation induces, the distribution of charge and the electrical potential which sign and density of charge determines must be such that the resulting solvent chemical potential and counterion electrochemical potential distributions are uniform. Segment concentration distribution and counterion concentration distribution will thus be complex.

As is well known, in the presence of high added neutral salt concentrations, polyelectrolyte systems tend to conform to the usual case of a polymer/solvent system. Hence also the behavior of charged polymeric networks will reduce to the previous case of a non-uniform network and a single solvent and show the characteristics of heterogeneous structure mainly then when the added salt concentration is high. This is one limiting case.

In the other limiting case, where there is no added salt, much depends upon the local concentration of fixed charges since the fixed charge distribution determines the distribution of the counterions within the gel. Any tendency to smooth out the counterion concentration and thus reduce its effect on solvent chemical potential gradients would require the creation of a very strong gradient in the electrical potential close to the polymer chain, i.e. would be a consequence of counterion condensation. Polyelectrolyte gels, much as polyelectrolyte systems generally, will very strongly reflect any built-in organizational heterogeneity of the network and adjust the counterion concentration accordingly rather than the other way around. To the extent that the "fixed" charges are derived from weakly ionizing groups some smoothing out can take place by charge redistribution. Highly charged networks on the other hand are subject to counterion condensation effects if the groups involved are strongly ionizing.

In polyelectrolyte gels perhaps the most interesting cases arise when the fixed charge concentration, provided locally by a homogeneous network, is of the same order of magnitude as the neutral salt concentration outside the gel or, in the case of a very heterogeneous gel, if the fixed charge concentration in the concentrated gel matches the salt concentration in the spaces of low network concentration. In these and other such cases, Donnan equilibrium strongly distorts the counter- and co-ion distribution and very large hydrostatic pressure differences can be induced. The regions of high fixed charge concentration will be stretched to nearly maximum extent and the load bearing capacity of the gel will be very high (*9*).

Phase Transitions

As is well known polymer solution can undergo phase separations forming two systems in equilibrium - one almost free of polymer, the other highly concentrated. The effect in a large measure is a consequence of the energy of interaction between polymer segments and solvent molecules and will thus occur in the gel as well. In a heterogeneous gel system, both the dilute and the concentrated phase will reversibly collapse to an even more concentrated gel phase. Heterogeneity of the chemical crosslink distribution imposes constraints which cause the gel sample to buckle and to form very strange shapes. The collapse patterns which are produced thus provide information about the in-built structural limitations (*10*).

Conclusion

We stress that heterogeneity in gels is to be expected in all cases. Its effects become more and more pronounced the lower the overall network segment concentration and the larger the relative number of chemical crosslinks chemically or statistically introduced. The spectrum of lifetimes of these contacts produces interesting rheological effects. Only when network segment concentration is high (larger than, say, 0.15 in volume fraction) will chains overlap and will interpenetration begin to smooth out the in-built differences in organization.

Literature Cited

(1) Flory, P.J. *Principles of Polymer Chemistry*, Cornell Univ. Press, Ithaca, N.Y., 1953.

(2) de Gennes, P.-G. *Scaling Concepts in Polymer Physics*, Cornell Univ. Press, Ithaca, N.Y., 1979.

(3) Silberberg, A., Hennenberg, M. *Nature* **1984**, *312*, 746-748.

(4) Weiss, N., Van Vliet, T., Silberberg, A. *J. Polymer Sci., Polymer Physics*, **1979**, *17*, 2229-2240.

(5) Weiss, N., Silberberg, A. *British Polymer Journal*, June **1977**, 144-150.

(6) Weiss, N., Silberberg, A. In: *Hydrogels for Medical and Related Applications;* Andrade, J.D., Ed., ACS Symposium Series, Number 31; American Chemical Society: Washington, D.C., 1976, pp. 69-79.

(7) Weiss, N., Van Vliet, T., Silberberg, A. *J. Polymer Sci., Polymer Physics*, **1981**, *19*, 1505-1512.

(8) Silberberg, A. *Macromolecules*, **1980**, *13*, 742-748.

(9) Silberberg, A. *Biorheology*, **1988**, *25*, 303-318.

(10) Tanaka, T., Sun, S-T., Hirokawa, Y., Katayama, S., Kucera, J., Hirose, Y., Amiya, T. *Nature* **1987**, *325*, 796-798.

RECEIVED September 9, 1991

Chapter 10

Swelling Behavior and Elastic Properties of Ionic Hydrogels

W. Oppermann

Institute of Physical Chemistry, Technical University Clausthal, Arnold-Sommerfeld-Strasse 4, W–3392 Clausthal-Zellerfeld, Germany

The swelling of ionic hydrogels in aqueous salt solutions as well as the elastic properties of such gels are discussed on the basis of molecular thermodynamics. The major factors governing the swelling equilibrium are identified and judged with respect to their significance. It is shown that the use of non-Gaussian chain statistics is indispensable in order that the model calculations shall agree with basic experimental observations.

Networks made of strongly hydrophilic polymers are of great technical importance because such systems can absorb large amounts of aqueous fluids. Basic requirements to be met by a superabsorbent material are large fluid uptake and high mechanical strength of the gel. These fundamental physical properties depend on the precise molecular structure of the network and on the characteristics of the substances (polymer and fluid) involved.

A theoretical estimate of the gel properties can, in principle, be obtained from a thermodynamic and statistical mechanical treatment of the system. The fundamental problem is to devise a model which is simple enough to deal with, yet realistic enough to bring out all the essential features of the material.

Molecular theories of rubber elasticity have been developed for fifty years. They describe with reasonable accuracy the dependences of the amount of swelling and the elastic modulus on molecular structure, as long as non-ionic, hydrophobic networks are considered which swell in organic solvents (*1*,*2*). If one has to deal with ionic polymer networks and aqueous salt solutions as solvents, the situation is far more complicated for at least two reasons: the degrees of swelling achieved are more than one order of magnitude larger, and the intermolecular interactions now comprise major coulombic, hydrogen-bonding, and polar contributions difficult to quantify.

In this chapter, the theoretical framework used to predict the properties of extremely swollen, ionic hydrogels is discussed. Special emphasis is placed on the necessity to use non-Gaussian chain statistics in order that the model predictions shall agree with basic experimental observations. Therefore, we will start with recalling briefly the typical properties of superabsorbent materials.

0097–6156/92/0480–0159$06.00/0
© 1992 American Chemical Society

Typical Behavior of Superabsorbent Polymer Networks

Figure 1 shows the dependence of the equilibrium degree of swelling, q, on the concentration of NaCl in the swelling medium, c_s. The network considered here consists of poly(trimethylammoniumethyl methacrylate chloride) (PTMAC), being prepared via a crosslinking copolymerization with 1 mol-% N,N'-methylene-bis-acrylamide in 10% solution (•) and 20% solution (▲) (*3*). Similar behavior is observed with various other kinds of ionic hydrogels and is well documented in the literature (*4-6*). A theoretical treatment must be able to predict the up to tenfold decrease of the degree of swelling with increasing salt concentration as well as the plateaus below $c_s \approx 10^{-3}$ mol/l and above $c_s \approx 1$ mol/l.

The second important experimental observation, which seems to be less widely known, is the dependence of the modulus of the gel on the degree of swelling. Figure 2 gives two examples of this dependence (*7*). The shear modulus, G, of a gel made of partially (32%) hydrolyzed polyacrylamide and crosslinked by 2.5 mol-% (•) and 1.25 mol-% (▲) N,N'-methylene-bis-acrylamide in 15% solution is plotted versus the degree of swelling on double logarithmic scales. At low swelling ratios, $q \lesssim 20$, the modulus decreases with increasing dilution, as expected on grounds of the Gaussian elasticity theory. At still higher degrees of swelling, however, the modulus rises markedly. This behavior, which has been observed with quite a number of different types of ionic gels, is an indication of the limited extensibility of the chains (*6*). The network stiffens when an appreciable portion of the chains become highly extended. This enhancement of the modulus due to non-Gaussian behaviour of the chains is a highly desirable effect when such gels are considered a potential material for superabsorbents.

It must be expected from a theoretical approach that the swelling ratios obtained in different salt solutions and the course of the modulus on swelling are consistently predicted.

Thermodynamics of Swelling

Thermodynamic equilibrium between a gel phase (g) and the surrounding liquid phase (l) is established when the chemical potentials $\mu_i(g)$ and $\mu_i(l)$ of all substances i occuring in both phases are identical (*1,8*):

$$\Delta\mu_i = \mu_i(g) - \mu_i(l) = 0 \ . \tag{1}$$

This relationship has to be valid for all components which exist in both phases. In particular, for a hydrogel swollen in an aqueous salt solution, water (component 1) and salt (component 3) are the diffusible species. Hence,

$$\Delta\mu_1 = 0 \ , \tag{1a}$$

$$\Delta\mu_3 = 0 \ . \tag{1b}$$

The equilibrium condition with respect to water can be met by reducing or increasing the concentration of water within the gel. Equation 1a, therefore, directly determines the equilibrium degree of swelling.

Equation 1b requires the concentrations of salt inside and outside the gel to be adjusted such that $\Delta\mu_3 = 0$. This condition is generally achieved by the diffusion of salt into the gel. As the chemical potential of water depends

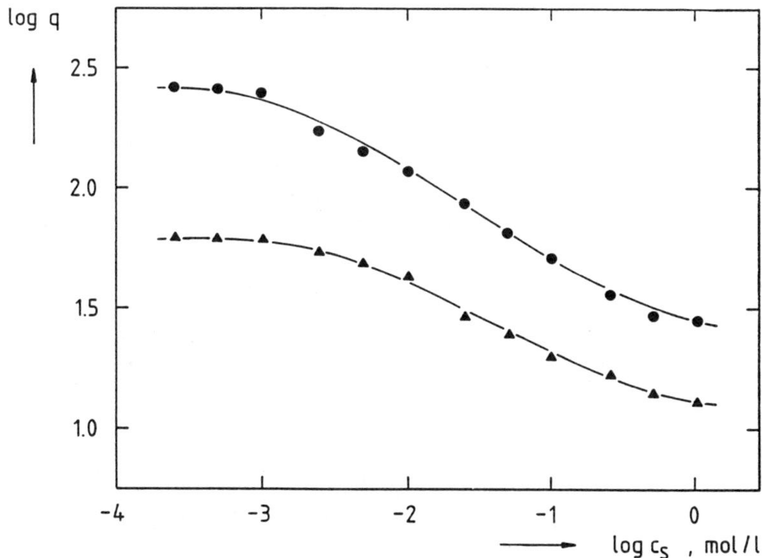

Figure 1. Dependence of the equilibrium degree of swelling, q, on salt concentration, c_S, for PTMAC-networks prepared in a 10% solution (●) or a 20% solution (▲).

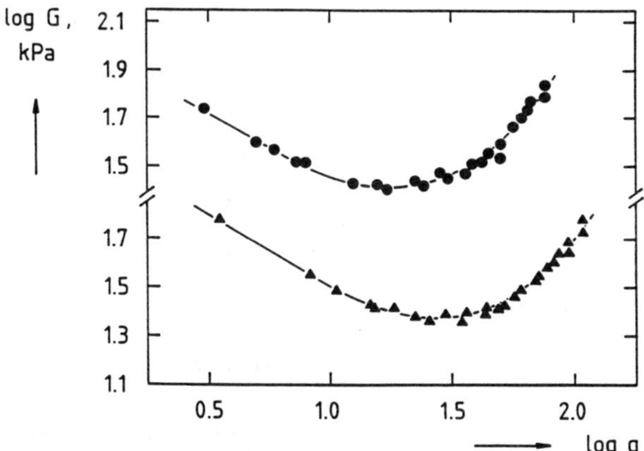

Figure 2. Shear modulus, G, of networks made of partially hydrolyzed polyacrylamide as a function of the degree of swelling,
● : 2.5 mol-% crosslinker, ▲ : 1.25 mol-% crosslinker.

markedly on salt concentration, equation 1b is needed and it is assumed that the respective equilibrium is established in order to calculate $\Delta\mu_1$.

Usually, $\Delta\mu_1$ is assumed to be composed of three additive terms:

$$\Delta\mu_1 = \Delta\mu_1^{elast} + \Delta\mu_1^{pol} + \Delta\mu_1^{ion} . \quad (2)$$

The two terms $\Delta\mu_1^{pol}$ and $\Delta\mu_1^{ion}$ are mixing terms, the former referring to the mixing of water and polymer and the latter to the mixing of water and the diffusible ionic species due to the presence of salt in the swelling medium. We are treating the mixing terms according to standard approaches which have been extensively used in the literature (*1,8-11*).

Ion/Water-Mixing Contribution to the Chemical Potential. $\Delta\mu_1^{ion}$ is due to the fact that the ion activity inside the gel is larger than outside. To simplify the treatment, concentrations are used instead of activities. This approximation may be justified even at rather high ionic strengths because the mean ion activity coefficients (for free ions) inside and outside the gel are probably in the same range. The expression obtained for $\Delta\mu_1^{ion}$ is:

$$\Delta\mu_1^{ion} = - R\,T\,V_1\,\Delta c_{tot} \;;\quad \Delta c_{tot} = \Sigma c_i(g) - \Sigma c_i(l) , \quad (3)$$

where the summations include all mobile ions only. V_1 is the molar volume of the solvent, R is the gas constant, and T is temperature.

The difference in concentration is caused by the fact that the charges fixed on the polymer, which are confined to the gel phase, require an equivalent number of counterions to stay within the gel because of neutrality. These counterions will always be there in excess of additional ions, which may diffuse into the gel, but can do so only in equivalent portions of positively and negatively charged species.

To calculate Δc_{tot}, one proceeds from equation 1b which takes the simple form:

$$c_+(g)\; c_-(g) = c_+(l)\; c_-(l) . \quad (4)$$

(The treatment is limited to 1,1-electrolytes for simplicity.)

The concentration of ions in the liquid phase which is usually fixed by experimental conditions is termed c_s. Equating the number of free counterions due to the gel forming polyelectrolyte to $\rho/(q\,M_2)$, where ρ is the density of the gel, M_2 the molar mass of the polyelectrolyte per free counterion, and q the degree of swelling, one obtains:

$$\Delta\mu_1^{ion} = 2\,R\,T\,V_1 \left(c_s - \sqrt{c_s^2 + (\rho/2qM_2)^2}\right). \quad (5)$$

It should be noted that for a 100% ionized polyelectrolyte, M_2 is generally not the molar mass of the repeat unit. An appreciable fraction of the counterions cannot be treated as being free because there are strong electrostatic binding effects between macroions and counterions. According to Manning's counterion condensation theory, only the uncondensed counterions would be free. Manning's theory provides an equation to calculate the fraction of free counterions, but the theory is devised for an infinitely diluted system (*12,13*). Hence, it seems to be inappropriate to apply it quantitatively to calculate M_2. We therefore prefer to treat M_2 as an adjustable parameter and use Manning's theory to obtain some estimate about what magnitude of M_2 might be reasonable.

Polymer/Water-Mixing Contribution to the Chemical Potential. For the solvent/polymer mixing term, we use the Flory-Huggins relationship with χ as the interaction parameter.

$$\Delta\mu_1^{pol} = R\,T\,(\ln(1-q^{-1}) + q^{-1} + \chi\,q^{-2}) \quad (6)$$

Because of the complex polymer/water mixing effects, the applicability of equation 6 in aqueous systems is highly questionable. More realistic models which distinguish between hydrogen-bond donating sites, hydrogen-bond accepting sites, and dispersion-force interaction sites and therefore lead to orientation dependent interaction potentials, have recently been developed by Prausnitz and collaborators (*5,14,15*). These approaches have successfully been used to predict phase diagrams with lower critical solution temperatures. However, these models are much more complicated mathematically and contain three adjustable exchange-energy parameters.

For polyelectrolyte gels which swell to very large swelling ratios, $\Delta\mu_1^{pol}$ is usually small compared with $\Delta\mu_1^{ion}$. Under these conditions, the exact dependence of $\Delta\mu_1^{pol}$ on the degree of swelling is rather insignificant. Therefore, we have decided still to use the Flory-Huggins equation as a first-order approach.

To visualize the significance of the mixing terms of the chemical potential and their dependence on the degree of swelling, the results of some model calculations are shown in Figure 3. The calculations are based on the following set of parameters: $\chi = 0.47$, $V_1 = 18$ ml/mol, $\rho = 1$ g/cc, $M_2 = 600$ g/mol. Curve 1 represents the polymer/solvent mixing term. The others show the ionic mixing term in the case of different salt concentrations in the swelling medium. It is obvious that the ionic term is by far the dominating factor. The polymer/solvent mixing contribution is significant only at the very low degrees of swelling and in rather concentrated salt solutions. Hence, the use of a simple approach for $\Delta\mu_1^{pol}$ is justified.

Elastic Contribution to the Chemical Potential. The elastic term $\Delta\mu_1^{elast}$ refers to the change of the chemical potential of the water due to the dilation of the network. To express $\Delta\mu_1^{elast}$ as a function of the degree of swelling and of the network parameters, one has to proceed from a particular network theory. The Gaussian approach is certainly the most common and widespread starting point. In terms of this theory, one obtains (*2,16*):

$$\Delta\mu_1^{elast}(\text{Gaussian}) = \nu\,R\,T\,V_1\,q^{-1/3}\,q_c^{-2/3}. \quad (7)$$

Here ν is the network density (molar number of network chains per unit volume of the unswollen sample), q the actual degree of swelling and q_c the degree of swelling at which the formation of the network has taken place (reference state).

Equation 7 differs from the original Flory-Rehner equation by having the term $(q^{-1/3}-q/2)$ replaced by $q^{-1/3}$ only, as obtained from the phantom network approach (*17*). The difference is rather small, especially with high degrees of swelling.

As was shown in the brief review of experimental results, the treatment based on Gaussian chain statistics is obviously no longer valid in the case of highly swollen gels. Calculations using non-Gaussian chain statistics are much more difficult and some assuptions have to be made to arrive at a tractable model. We base our calculations on the inverse Langevin approximation and assume affine displacement of the crosslinks. As we are only interested in swelling deformation, which is isotropic, this assumption seems to be justified even at high deformations, contrary to the case when uniaxial deformations are considered.

The entropy of a single chain having an end-to-end distance r and a maximum end-to-end distance r_{max} is given by equation 8 (*2*):

$$S = \text{const} - k\, n \left(\frac{r}{r_{max}}\, \beta + \ln\frac{\beta}{\sinh \beta}\right), \tag{8}$$

where $\beta = \mathcal{L}^{-1}(r/r_{max})$, $\mathcal{L}^{-1}$ being the inverse Langevin function. n is the number of statistical segments per chain with each segment having the length l. r_{max} is then equal to n·l.

When purely entropic elasticity is assumed, summation over all chains in the network and multiplication by −T gives the free energy of the network, whose partial derivative with respect to the number of moles of water is the chemical potential of the water. Expressing the latter via the degree of swelling leads to the relationship:

$$\Delta\mu_1^{elast} = -\nu\, T\, V_1\, N_L \left(\frac{dS}{dq}\right)_{T,P}, \tag{9}$$

which, when $(dS/dq)_{T,P}$ is analytically calculated, finally results in (*18*):

$$\Delta\mu_1^{elast} = {}^1\!/_3\, \nu\, R\, T\, V_1\, q^{-2/3}\, q_c^{-2/3}\, q_{max}^{1/3}\, \mathcal{F} \tag{10}$$

with $\mathcal{F} = \beta + (r/r_{max} + 1/\beta - \coth\beta)\, \beta^2 \sinh^2\beta/(\sinh^2\beta - \beta^2)$.

The maximum degree of swelling, q_{max}, is related to the maximum end-to-end distance of a single chain via:

$$r/r_{max} = (q/q_{max})^{1/3}. \tag{11}$$

The number of statistical segments n per chain was substituted as:

$$n = r_{max}^2/\langle r^2\rangle_o = (q_{max}/q_c)^{2/3}. \tag{12}$$

In doing so it was assumed that the end-to-end distance of the network chains equals the mean-square end-to-end distance of free chains under the conditions of network formation, i.e. at q_c.

To illustrate the significance of the application of non-Gaussian chain statistics, Figure 4 shows the term $\Delta\mu_1^{elast}/RT$ as a function of the degree of swelling. The dashed curve represents the result of the calculation according to the Gaussian approach, equation 7, and the other three curves show the prediction of equation 10 which is based on non-Gaussian statistics. The parameters chosen were $q_c = 3.33$ (crosslinking was achieved in a 30% solution), and $q_{max} = 250$ (n = 18), 500 (n = 28), and 1000 (n = 45), respectively. The other parameters are the same as used for Figure 3.

At low degrees of swelling, all curves coincide because the state of deformation of the chains is far from the fully extended one. At moderate and

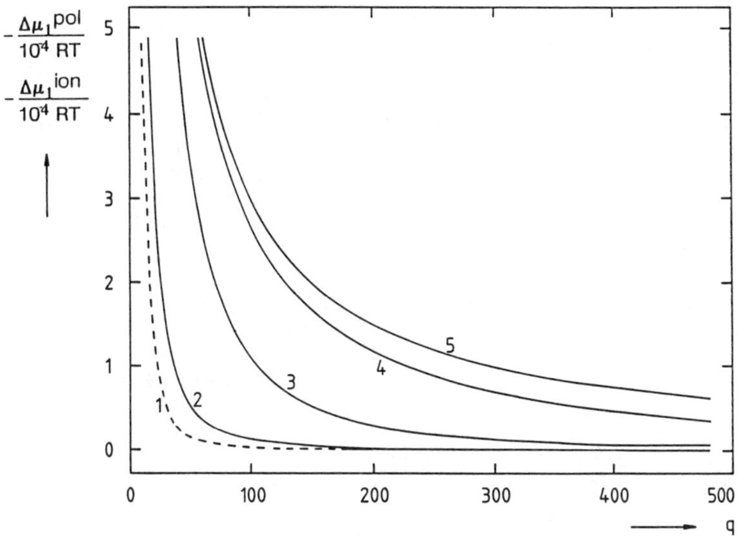

Figure 3. Mixing terms of the chemical potential of water as a function of the degree of swelling. 1: Flory-Huggins term, $\Delta\mu_1^{pol}$; 2-5: $\Delta\mu_1^{ion}$ at salt concentration, c_s, 0.1 M (2), 0.01 M (3), 0.001 M (4), 0 (5).

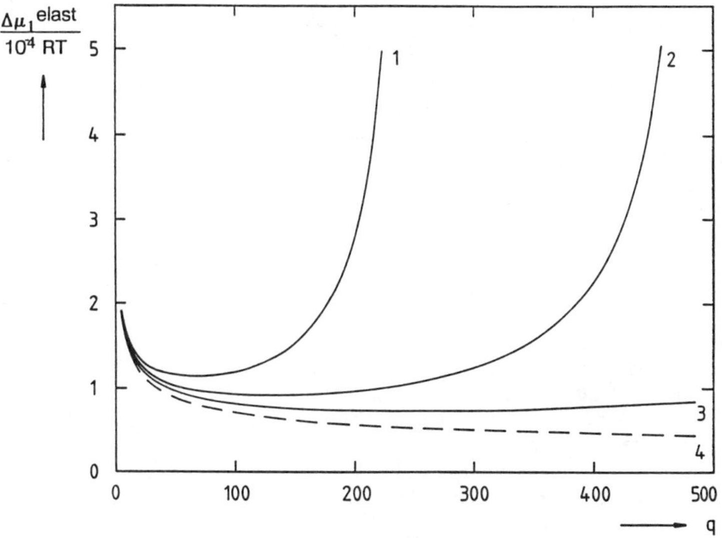

Figure 4. Elastic term of the chemical potential of water as a function of the degree of swelling. 1: q_{max} = 250, 2: q_{max} = 500, 3: q_{max} = 1000, 4: Gaussian theory.

particularly at high degrees of swelling, however, there is a marked difference between the Gaussian and the non-Gaussian curves. This fact has extensive influence on the swelling properties.

The use of non-Gaussian chain statistics for considering the swelling of polymer networks is not new. The various approaches described in the literature are similar in that they are based on a suitably truncated series expansion of the inverse Langevin function (*19-21*). Such a truncation can give rise to appreciable errors when the chain extension comes close to the limit. One formula even contains a correction factor which is largest at low swelling ratios and approaches unity at high swelling ratios (*22-23*). This representation seems to be physically unsound.

Swelling Equilibrium in Salt Solutions. To avoid the hardly assessable consequenses of truncations and approximations, we prefer to use the inverse Langevin function in full. This procedure has the disadvantage that one cannot solve equation 1a with respect to the degree of swelling. There are, however, analytical expressions for all three terms of the chemical potential and q can be found by iteration. This procedure corresponds to finding the intersection of the $\Delta\mu_1^{elast}$-curve with the $-\Delta\mu_1^{mix}$-curve ($\Delta\mu_1^{mix} = \Delta\mu_1^{pol} + \Delta\mu_1^{ion}$), as shown in Figure 5. For a family of curves resulting from different salt concentrations in the swelling medium, the dependence of the degree of swelling on salt concentration is obtained.

In Figures 6 - 8, the results of such calculations are presented. These graphs are used to elucidate the influence of the various parameters upon which the calculations are based. In all cases the network density ν is set to 0.04 mol/l and q_c is 3.33. Furthermore, the standard set of parameters comprises a χ-value of 0.47, a maximum degree of swelling of 500 (n=28), and a molar mass per free counterion, M_2, of 600 g/mol. The consequence of the variation of one of these three parameters is shown in each graph.

In Figure 6, χ is varied between 0.3 and 0.5. It is seen that this change influences the degree of swelling only at rather high salt concentrations. However, this is the regime of greatest technical importance. With a concentration of c_s = 0.15 mol/l (physiological), q changes from 28 to 53 when χ is lowered from 0.5 to 0.3. This dependence demonstrates that it is not only the ionic charges which are responsible for the degree of swelling, but also the hydrophilicity of the polymer.

Figure 7 shows the influence of the non-Gaussian properties of the chains on swelling. q_{max} is set to 250, 500, and 1000. This parameter is important when very high degrees of swelling are reached, namely in solutions of low salinity.

Finally, Figure 8 shows the effect of the degree of ionization on q. The number picked for M_2 is relevant over the whole range of salt concentrations with the major influence being observed at medium concentrations. At high salt concentrations all curves approach the same degree of swelling, which is determined by the χ-parameter.

Elastic Modulus of Highly Swollen Gels

It is obvious that the shape of the theoretical swelling curves can be manipulated widely by adjustment of the corresponding parameters to match any experimental data. Also, the same curves could possibly be obtained by different combinations of parameters. A crucial test whether or not a particular set of parameters is meaningful can be obtained by comparing the theoretical

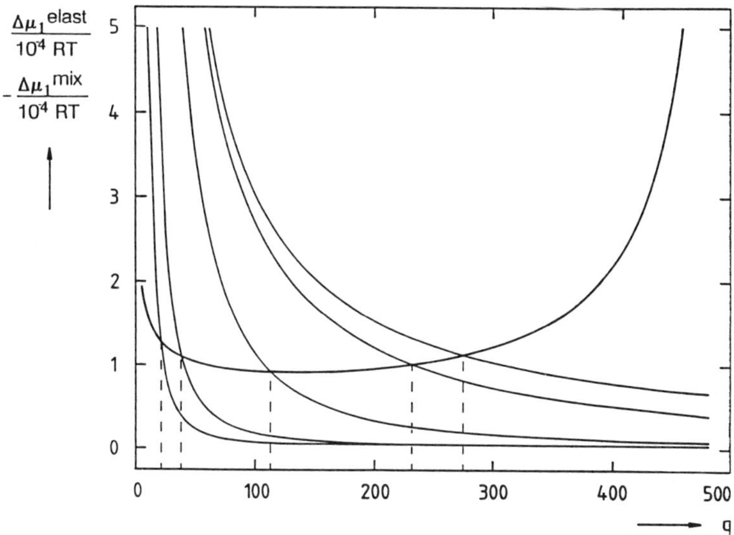

Figure 5. Determination of the equilibrium degree of swelling from the relationship $\Delta\mu_1^{elast} + \Delta\mu_1^{mix} = 0$.

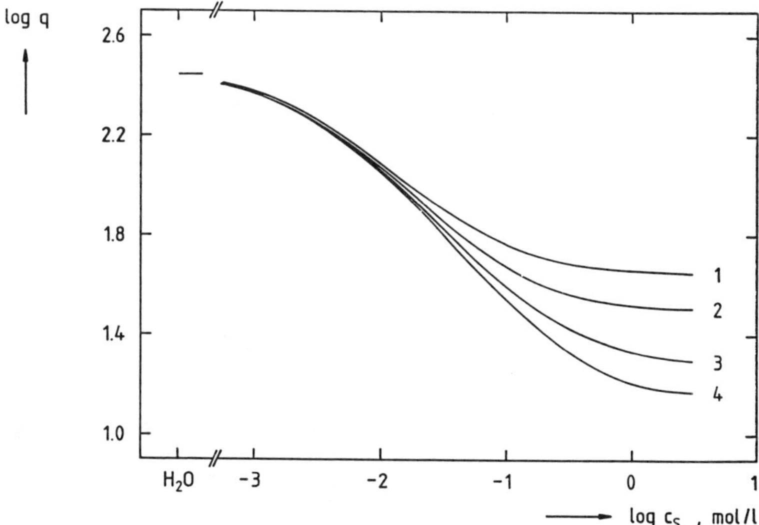

Figure 6. Theoretically calculated dependence of the degree of swelling on salt concentration. ν = 0.04 mol/l, M_2 = 600 g/mol, q_{max} = 500, χ = 0.3 (curve 1), 0.4 (curve 2), 0.47 (curve 3), 0.5 (curve 4).

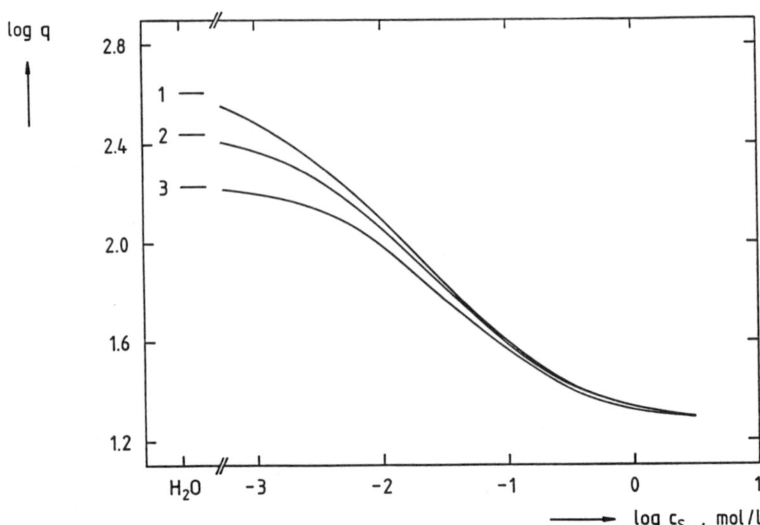

Figure 7. Theoretically calculated dependence of the degree of swelling on salt concentration. ν = 0.04 mol/l, M_2 = 600 g/mol, χ = 0.47, q_{max} = 1000 (curve 1), 500 (curve 2), 250 (curve 3).

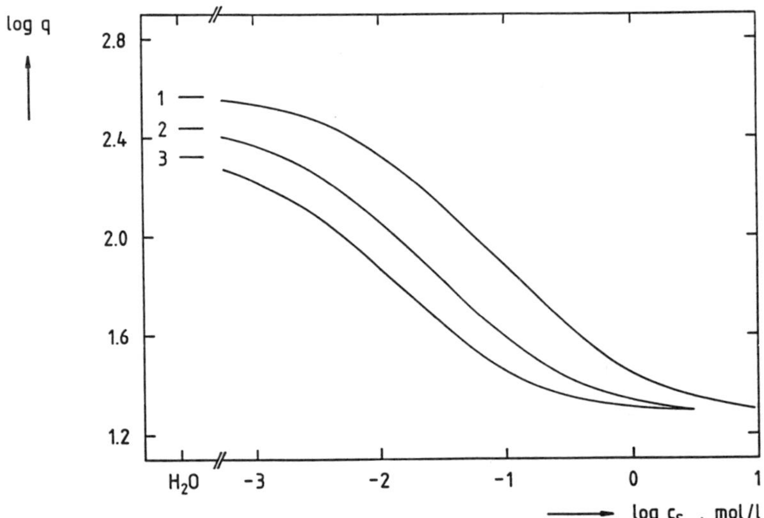

Figure 8. Theoretically calculated dependence of the degree of swelling on salt concentration. ν = 0.04 mol/l, χ = 0.47, q_{max} = 500, M_2 = 300 g/mol (curve 1), 600 g/mol (curve 2), 900 g/mol (curve 3).

prediction for another physical property with experimental observations. As stated in the beginning, the elastic modulus can serve this purpose. The shear modulus, G, is derived within the same theoretical framework as the elastic part of the chemical potential, $\Delta\mu_1^{elast}$. An equation for G has been obtained on the basis of non-Gaussian chain statistics by Smith (*24*). Phrased in our notation and expressed as a function of the degree of swelling, it reads:

$$G = \nu R T q^{-1/3} q_c^{-2/3} [\beta^2 \sinh^2\beta/(\sinh^2\beta - \beta^2) + \beta r_{max}/r]/6, \qquad (13)$$

where the bracketed term times the factor 1/6 is a correction factor due to non-Gaussian properties. The two parameters which enter into equation 9 for $\Delta\mu_1^{elast}$, namely ν and q_{max}, also determine the elastic modulus. When the same numbers as used to calculate the curves in Figure 4 are employed, the relationship depicted in Figure 9 is obtained for the dependence of G on the degree of swelling. This theoretical result can be compared with the experimental observations as pictured in Figure 2. Obviously, the Gaussian approximation (dashed curve) is inadequate. On the other hand, the upturn of the modulus at high degrees of swelling is well accounted for when the non-Gaussian approach is used. Thus, the experimentally established course of the modulus with increasing swelling gives the strongest and definite hint that it is indispensable to apply a non-Gaussian network theory. When one proceeds accordingly, the theoretical treatment can give a correct description of the dependences of the degree of swelling on salt concentration and of the elastic modulus on the degree of swelling.

Some limitations of the model also need to be mentioned. When the non-Gaussian chain statistics are applied to a network, it was implicitely assumed that all network chains are of the same length. It is not quite clear how a broad chain-lenth distribution would influence the results. Secondly, specific

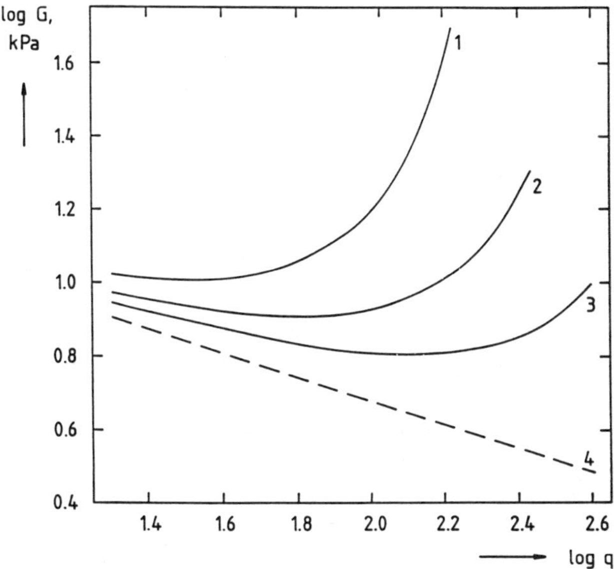

Figure 9. Theoretically calculated dependence of the shear modulus on the degree of swelling. ν = 0.04 mol/l, curve 4: Gaussian theory, curves 1-3: non-Gaussian theory with q_{max} = 250 (1), 500 (2), 1000 (3).

interactions between counterions and macroions are not considered. Such interactions often occur with divalent counterions, for instance between Ca^{2+} and poly(acrylic acid). Therefore, an application of the model to the swelling of ionic hydrogels in electrolyte solutions containing divalent ions may fail in some cases.

Acknowledgments

Financial support for this project by Chemische Fabrik Stockhausen GmbH, Krefeld, is gratefully acknowledged.

Literature Cited

1. Flory, P. J. *Principles of Polymer Chemistry*; Cornell University Press: Ithaca, NY, 1953.
2. Treloar, L. R. G. *The Physics of Rubber Elasticity*; Clarendon Press: Oxford, 1975.
3. Oppermann, W. *Angew. Makromol. Chem.* **1984**, *123/124*, 229.
4. Hooper, H. H.; Baker, J. P.; Blanch, H. W.; Prausnitz, J. M. *Macromolecules* **1990**, *23*, 1096.
5. Hooper, H. H.; Blanch, H. W.; Prausnitz, J. M. In *Absorbent Polymer Technology*; Brannon-Peppas, L.; Harland, R. S., Eds.; Elsevier Science Publ.: Amsterdam, 1990, 203.
6. Anbergen, U.; Oppermann, W. *Polymer* **1990**, *31*, 1854.
7. Oppermann, W.; Rose, S.; Rehage, G. *Brit. Polym. J.* **1985**, *17*, 175.
8. Tanford, C. *Physical Chemistry of Macromolecules*; John Wiley & Sons, Inc.: New York, 1967.
9. Harsh, D. C.; Gehrke, S. H. In *Absorbent Polymer Technology*; Brannon-Peppas, L.; Harland, R. S., Eds.; Elsevier Science Publ.: Amsterdam, 1990, 103.
10. Ricka, J.; Tanaka, T. *Macromolecules* **1984**, *17*, 2916.
11. Brannon-Peppas, L.; Peppas, N. A. In *Absorbent Polymer Technology*; Brannon-Peppas, L.; Harland, R. S., Eds.; Elsevier Science Publ.: Amsterdam, 1990, 67.
12. Manning, G. S. *J. Chem. Phys.* **1969**, *51*, 924 and 934.
13. Manning, G. S. *Biophys. Chem.* **1978**, *9*, 65.
14. Prange, M. M.; Hooper, H. H.; Prausnitz, J. M. *AIChE J.* **1989**, *35*, 803.
15. Beltran, S.; Hooper, H. H., Blanch, H. W.; Prausnitz, J. M. *J. Chem. Phys.* **1990**, *92*, 2061.
16. Dusek, K.; Prins, W. *Adv. Polym. Sci.* **1969**, *6*, 1.
17. Flory, P. J.; Erman, B. *Macromolecules* **1982**, *15*, 800.
18. Anbergen, U.; PhD Thesis, Clausthal 1991.
19. Ilavski, M. *Polymer* **1981**, *22*, 1687.
20. Kovac, J. *Macromolecules* **1978**, *11*, 362.
21. Konak, C.; Bansil, R. *Polymer* **1989**, *30*, 677.
22. Peppas, N. A.; Lucht, L. M. *Chem. Eng. Commun.* **1984**, *30*, 291.
23. Peppas, N. A.; Barr-Howell, B. D. In *Hydrogels in Medicine and Pharmacy*; Peppas, N. A., Ed.; CRC Press, Boca Raton, FL, 1986, Vol. 1, 27.
24. Smith, T. L. *J. Polym. Sci., Polym. Symp.* **1974**, *46*, 97.

RECEIVED July 29, 1991

Chapter 11

Complexation of Polymeric Acids with Polymeric Bases

A. B. Scranton[1], J. Klier[2], and C. L. Aronson[1]

[1]Department of Chemical Engineering, Michigan State University, East Lansing, MI 48824
[2]Dow Chemical Company, Central Research, Midland, MI 48674

The complexation of poly(ethylene glycol) with poly(methacrylic acid) was studied in dilute aqueous solutions and in covalently crosslinked hydrogels. Complexation in dilute solution was detected by NMR relaxation time and Nuclear Overhauser Effect (NOE) experiments at lower poly(ethylene glycol) molecular weights than previously reported. NOE experiments revealed intimate contact between the ethylene groups of the poly(ethylene glycol) and the α-methyl groups of poly(methacrylic acid). The swelling of crosslinked gels containing covalently attached poly(ethylene glycol) and poly(methacrylic acid) was highly sensitive to pH, temperature, solvent type and network composition. Large transitions in swelling in response to changes in pH and solvent composition were observed, and these transitions were attributed to complexation and decomplexation induced by the changes in surrounding conditions.

Polymer complexes are formed by the association of two or more complementary polymers, and may arise from electrostatic forces, hydrophobic interactions, hydrogen bonding, van der Waals forces or combinations of these interactions (1-4). Due to the long-chain structure of the polymers, once one pair of complementary repeating units associate to form a segmental complex, many other units may readily associate without a significant loss of translational degrees of freedom. Therefore the complexation process is cooperative, and stable polymer complexes may form even if the segmental interaction energy is relatively small (1,4). The formation of complexes may strongly effect the polymer solubility, rheology, conductivity and turbidity of polymer solutions. Similarly, the mechanical properties, permeability and electrical conductivity of the polymeric systems may be greatly affected by complexation.

0097–6156/92/0480–0171$06.00/0
© 1992 American Chemical Society

In this contribution we present results from experimental investigations of complexing polymer systems containing poly(ethylene glycol) (PEG) and poly(methacrylic acid) (PMAA). These species form hydrogen-bonded complexes which are stabilized by hydrophobic interactions. Dilute solutions of PEG and PMAA in water and crosslinked polymer networks containing both species are considered. Nuclear magnetic resonance spin-spin relaxational experiments and nuclear Overhauser enhancement studies were performed to provide information about the underlying complexation processes. Crosslinked poly(methacrylic acid-g-ethylene glycol) networks exhibited swelling transitions in response to changes in pH and solvent composition. These networks contain both complex-forming constituents covalently linked to one another and swelling is largely regulated by control of the complexation. The experimental results were explained in terms of a statistical thermodynamic description of the complexation equilibrium. For completeness, a review of previous experimental and theoretical investigations of electrostatic polyelectrolyte complexes and hydrogen-bonded complexes is given below.

Polyelectrolyte Complexes. Many polymer complexes form as a result of electrostatic forces. Polymeric acids may form complexes by proton transfer to complementary polymeric bases, resulting in poly(cation):poly(anion) pairs. For example, synthetic polycarboxylates may form such ionic complexes with poly(ethylene imine) (5). Similarly, the salts of polymeric acids (e.g. anionic polyelectrolytes) may associate via electrostatic interactions with polyelectrolytes of opposite charge (6,7). A simple example of this type of complex is the association of poly(sodium styrene sulfonate) with poly(vinyl trimethyl ammonium chloride). In addition to the inter-polymer complexes, intra-polymer complexes of polyampholytes have been studied (8,9). In these systems complexes may be formed between oppositely charged moieties on the same polymer chain, or even on the same monomer unit. Finally, simple polyelectrolytes in solution may be linked together by multivalent ions to form gels or coacervates (10,11).

Electrostatic interactions are considerably stronger than most secondary binding interactions. Therefore, electrostatic polyelectrolyte complexes exhibit unique physical and chemical properties. These complexes have been divided into four classes based upon the possible combinations of strong and weak polyelectrolytes (1). The properties and potential applications of polyelectrolyte complexes have recently been reviewed (1-3). Although the complex stability is dependent upon such variables as charge density, solvent, ionic strength, pH and temperature; polyelectrolytes of high charge density may form relatively insoluble complexes.

Hydrogen-Bonded Complexes. Compared to complexes between complementary polyelectrolytes, complexes formed by hydrogen bonding between complementary polymeric Lewis acids and bases (1-4) typically dissociate under a wider variety of conditions. Hydrogen bonding occurs between a Lewis acid containing an electron deficient proton and a Lewis base containing a lone pair of electrons. Hydrogen bonds are distinctly directional and specific, and are more localized than any other type of weak intermolecular interaction (12). In associating polymer systems the equilibrium between complexed (hydrogen bonded) and uncomplexed polymers in solution may be highly sensitive to surrounding conditions, such as pH, temperature, solvent composition and polymer concentration. Molecular structural parameters such as molecular weight and polymer architecture may also influence the complexation equilibrium.

The properties of hydrogen-bonded complexes are reviewed below, focusing on the relationships among environmental conditions or molecular parameters and the physical properties of polymer solutions and gels. The properties of complexes of poly(ethylene glycol) (PEG) with poly(acrylic acid) (PAA) or poly(methacrylic acid) (PMAA) will be emphasized since this system has been extensively studied in the literature, and was used as a model system for the work reported in this paper. PEG is a Lewis base by virtue of the electron lone pairs in the ether linkage, while PMAA and PAA are Lewis acids. Since the PEG:PMAA complexes increase in strength as the temperature is increased, they are believed to be stabilized by hydrophobic interactions.

Experimental Analysis of Hydrogen-Bonded Complexes. The properties of dilute aqueous solutions of PEG and PMAA or PAA have been studied by a number of investigators using viscometry, turbidimetry and potentiometric titration (1-4,13-17). These studies have shown that the complexed polymer fraction depends strongly upon the molecular weight of the complementary polymers. In experiments with PAA or PMAA of high molecular weight, a critical PEG molecular weight was identified below which no complexes were observed. A critical molecular weight of about 2000 was reported for the PEG:PMAA system, while a value of around 6000 was reported for the PEG:PAA system (14-16). Several investigators (1-4,14-17) found that both atactic PAA and PMAA complexed with PEG in a 1:1 repeating unit molar ratio while isotactic PAA formed complexes with a 2:3 carboxylate:ether molar ratio. Finally, all investigators found that the complex stability increased with decreasing pH, although there are discrepancies in the value of the critical pH for complexation. For example, based upon turbidimetry Ikawa (17) reported a critical pH for PMAA:PEG complexation of 3.0 in contrast to a value of 5.6 obtained previously using viscometry and potentiometry.

Several authors have investigated the hydrophobic stabilization of PEG:PMAA complexes. The earliest evidence for hydrophobic stabilization in aqueous media arose from the temperature dependence of the complex stability. PMAA:PEG complexes increase in strength with increasing temperature (14,15), whereas those of PAA:PEG do not change appreciably. These trends suggest that the PEG:PMAA complexes are hydrophobically stabilized, perhaps due to the α-methyl group which is present on PMAA. Further evidence of hydrophobic stabilization of PMAA:PEG complexes was reported by Ikawa et al. (17). For PMAA systems the critical chain length of PEG increased significantly when the solvent was changed from water to a water/methanol mixture; while it remained relatively constant for the PAA:PEG system. Similarly, Osada and collaborators (16,18) found the temperature dependence of complex stability to be opposite in ethanol/water mixtures from that in pure water, due to disruption of hydrophobic interactions in the former case. These results were supported by measurements by Papisov et al. (19) who found that complexation of PMAA and PEG in water is endothermic with a positive entropy change.

Recently the complexation of PAA and PMAA with PEG were studied using fluorescent spectroscopic techniques (20-26). Morawetz and coworkers (24,25) studied PAA association with PEG using PAA labeled with dansyl chromophores. These chromophores exhibit a fluorescent intensity change and spectral shift when moved from a hydrophilic to a hydrophobic environment. Large peak intensity changes upon addition of PEG to dansylated PAA revealed that the chromophore ends up in an environment largely devoid of water. Frank and collaborators (20-23) used fluorescence spectroscopy to investigate the complexation of pyrene end-labeled PEG with PMAA and PAA. Examination of the pyrene excimer to monomer ratio of sparsely tagged PEG allowed intramolecular end-to-end contact to be characterized, while experiments with fully tagged PEG provided information about both intra- and intermolecular contacts. These studies confirmed previous observations that complexation is highly dependent upon chain length, facilitated by low pH, reduced by neutralization of the acid and reduced by the addition of methanol. However, in contrast to earlier results, complexes of the PEG with PMAA and PAA were detected for PEG molecular weights as low as 1850 and 4200, respectively (20-23).

Investigations of the complex stoichiometry by fluorescence spectroscopy yielded mixed results. Based upon studies using the previously described chromophores, Morawetz and collaborators (24,25) and Frank and collaborators (20-23) found the extent of complexation in the PEG:PAA system increased as the number of acid moieties was increased past a 1:1 molar ratio. In fact, for low molecular weight PAA, the formation of intermolecular excimers plateaued at the PAA:PEG repeating unit ratio of around 3:1 (23). In contrast, Heyward and Ghiggino (26) used fluorescence polarization studies of acenaphthylene labeled PAA to show that complexation is maximized for the 1:1 molar ratio of repeating units.

The affinity of PMAA and PAA for polymeric Lewis bases can depend quite strongly upon the structure of the base. For example, PMAA will bind more strongly with poly(vinyl pyrollidone) (PVP) than with PEG (1-4,27). In fact, PVP will displace PEG of similar molecular weight from a complex with PMAA. Similarly the Lewis acid poly(itaconic acid monomethyl ester) will form a hydrogen-bonded complex with PVP, but not with PEG. Other polymeric Lewis bases with which poly(carboxylic acids) will form complexes include poly(acrylamide), poly(-dimethoxyethylene), poly(vinyl methyl ether), poly(vinylbenzo-18-crown-6) and poly(vinyl alcohol).

Applications of Hydrogen Bonded Complexes. Complexes of PEG and PMAA have been used to design novel polymeric systems. For example, PEG complexes with PMAA have been used to form membranes with controlled permeability (28). Addition of PEG to water-swollen PMAA membranes at low pH gave rise to large reversible shrinkage of the membranes. The shrinkage increased with PEG concentration and with temperature. PEG with a molecular weight as low as 2000 gave rise to shrinkage (29), while addition of alcohols or base to shrunken membranes broke the complexes and resulted in swelling. Osada and coworkers (30,31) used these membrane systems as chemical valves with controllable permeability. The porous PMAA membranes were fixed in frames so that when shrinkage and swelling were induced, the pores would open and close, thereby regulating solute flux. A similar concept was studied by Nishi and Kotaka (32). End-linked PEG gels were swollen with acrylic acid, which was subsequently polymerized. The resulting interpenetrating networks showed pH responsive swelling and permeability similar to that described above. The flux or solute macromolecules was varied by changing the pH of the surrounding medium.

Theoretical Models of Polymer Complexation. The complexation of complementary polymers has been modeled theoretically by Kabanov and collaborators (4,15,33). In this analysis the total free energy of complexation was divided into two contributions: one arising from the specific interactions between complexing functional groups, and a second arising from configurational changes of the system upon complexation. The authors typically assumed that in the complexed state, the polymer of the shortest chain length (the PEG) was completely bound with all repeating units participating in hydrogen bonds. This assumption greatly simplified the calculation of the number of system configurations in the complexed state. According to these models the extent of complexation can change very abruptly near critical values of polymer molecular weight or free energy of complexation due to cooperative effects in the complexation process. Calculations based on the model also agreed at least qualitatively with other experimentally observed trends. For example, the equilibrium bound fraction depended strongly upon chain length and stable complexes formed even with weak segmental interactions.

Several authors have modeled the association of biological polymers (34-37). The various models differ in the manner in which the polymer conformation in the complexed state is considered. One approach incorporates the assumption that a complexed polymer has only one possible conformation; that in which all repeating units are bound to the complementary polymer. Examples include models for the complexation of poly- and oligo-nucleotide (34,35). In contrast, models of the double-stranded helix complexes of long-chain nucleic acids have included the possibility of loops in the conformation of a complexed chain (36,37). In these models, which are reviewed by Poland and Scheraga (37), it is typically assumed that the loops are fairly long (> 20 units) when calculating the loop entropy.

Mathematical techniques based on sequence generating functions have been proposed for efficient formulation of the statistical mechanical partition functions of polymeric systems. For example, Lifson (38) reported a procedure for evaluating the canonical partition function of long-chain polymers with repeating units existing in two or more distinct states, while Eichinger et al. (39) reported a generating function technique for the evaluation of the partition functions of a long polymer chain absorbed onto a planar surface. Both of these methods were applicable in the limit of infinite chain length. Scranton et al. (40) used a similar generating function technique to describe the complexation thermodynamics of free and graft oligomers with complementary polymers. Since the long chain assumption was not required, this analysis was applicable to short and intermediate chain lengths.

In summary, polymeric acids may form complexes with a number of polymeric Lewis bases (including PEG) in water. The extent of complexation increases with polymer concentration, polymer molecular weight and with reductions in pH. The extent of complexation may increase, decrease or remain essentially unchanged with changing temperature depending upon the contribution of hydrophobic interactions to the complex stability. Recent spectroscopic measurements suggest that complexes may take place at lower molecular weights than previously determined using viscometry, turbidimetry or titration. In addition, some complexes previously thought to form with a 1:1 repeating unit stoichiometry may not do so in every case. Certain detection techniques may be more sensitive to complexation than others, perhaps accounting for the discrepancies in critical chain length and complex stoichiometries.

Experimental

Safety Considerations. After washing, the polymers used in these experiments present essentially no safety hazards. In fact, PEG is commonly used in cosmetic and pharmaceutical applications. However the reactants used to synthesize the polymers present potential safety hazards. For example, the monomers methacrylic acid, ethylene glycol dimethacrylate, tetraethylene glycol dimethacrylate and methoxy poly(ethylene glycol) methacrylate as well as the initiators ammonium persulfate and sodium bisulfite are irritants of the eyes, skin and respiratory system, and may act as skin sensitizers. These reactants were always weighed and handled with gloves, lab coat and safety glasses in a fume hood. Furthermore, the monomer species are sensitive to ultraviolet light and may rapidly polymerize at elevated temperatures. Therefore, the monomers must be stored at shaded areas at prescribed temperatures. The solvents acetone, ethanol and methanol can be toxic. Eye and skin contact as well as inhalation should be avoided (especially methanol). Furthermore, these solvents are highly flammable and should be kept away from spark sources.

Synthesis of Polymer Networks. The copolymer networks were synthesized by the free-radical copolymerization of methacrylic acid (MAA) (Polysciences, Warrington, PA) with methoxy poly(ethylene glycol) methacrylate (PEGMA) (Polysciences) in the presence of small amounts of a crosslinking agent, ethylene glycol dimethacrylate (EGDMA) (Aldrich, Milwaukee, WI) or tetra(ethylene glycol) dimethacrylate (TEGDMA) (Polysciences). Methoxy poly(ethylene glycol) methacrylate with pendent PEG chains of molecular weight 200 and 400 were used in this study (henceforth denoted PEGMA-200 and PEGMA-400, respectively). The monomer mixture was diluted with a 50:50 (w/w) mixture of ethanol/water to a final concentration of 40 or 50 wt%. Polymerization took place in sealed 5 ml polyethylene vials of diameter 1 cm at 40 °C for at least 24 hours and was initiated by 0.025 wt% ammonium persulfate (Aldrich) and sodium bisulfite (Aldrich) based on monomer weight. The resulting crosslinked polymer network contained a backbone of poly(methacrylic acid) and grafts of poly(ethylene glycol).

The polymer cylinders were sliced into disks with a thickness of approximately three millimeters. The crosslinked polymer disks were then purified by washing with distilled water for at least one week, followed by dialysis against distilled water with a pH of approximately 4 for at least 48 hours. The freshly cut disks turned cloudy upon placing them in the distilled water, but cleared up after washing for several days. During this washing period the swelling solution was replaced with fresh distilled water daily. The dialysis against pH 4 water ensures that any cations contained within the network will be exchanged for protons. Before this step, sodium and ammonium ions could be present in the networks due to the initiators.

Equilibrium Swelling. Swelling studies were performed to determine the effects of pH, network composition and solvent composition on the equilibrium degree of swelling. For the pH studies the washed polymer disks were equilibrated in acidic solutions of the appropriate pH by changing the solution twice daily for two weeks. Equilibrium swelling in basic solutions was accomplished by preequilibrating the gels at a pH of 11 for two days, followed by equilibration in the basic aqueous solution of the desired pH. The gels were allowed to swell in sealed containers for two weeks. The temperature was controlled to within one Celsius degree. Swelling in salt and alcohol solutions was performed under air, while swelling in HCL and NaOH solutions between pH 5.5 and 8.5 was performed under nitrogen. After swelling the samples were weighed and dried to constant weight in vacuo. Finally, the dried samples were weighed so the equilibrium solvent weight fraction could be determined.

A similar procedure was used for the studies as a function of the solvent composition. Swelling solutions were prepared by mixing an appropriate solvent with distilled water to desired proportions. Two distinct cosolvents were investigated - methanol and acetone. For each cosolvent swelling studies were performed in solutions with concentrations ranging form 0 wt% to 100 wt% in increments of 10 wt%. The washed polymer disks were placed in sealed jars containing solutions of desired composition, and were allowed to swell until measurements on consecutive days indicated no change in the swollen mass. After swelling, the gel weights were determined in air, the gels were dried to constant weight in vacuo at 60°C, and the dried gels were weighed to determine the equilibrium solvent weight fraction.

NMR Sample Preparation. Monodisperse samples of PEG (Polysciences) with molecular weights ranging from 200 to 20,000 as well as polydisperse samples of larger molecular weight were used as received. The molecular weights and polydispersities of the PEG samples used in this study are listed in Table I in the Results and Discussion section. NMR samples were prepared by dissolving appropriate amounts of the PEG in D_2O (99.9%, Cambridge Isotope Laboratories).

Uncrosslinked PMAA for NMR samples was synthesized at 40°C by reacting 20 vol% MAA monomer in water with 0.50 wt% of both sodium bisulfite and ammonium persulfate based upon the total mass. The polymer product was dialyzed with D_2O to exchange $H^+$with D^+, and the resulting polymer was dried in vacuo. The resulting uncrosslinked PMAA was used for the NOE and relaxational studies of PEG complexed with PMAA. Samples containing both species were prepared by dissolving appropriate amounts of each constituent in D_2O. These samples were typically prepared by first forming dilute solutions of each polymeric species separately, then mixing the solutions in appropriate proportions.

Spin Echo NMR Experiments. ^1H NMR relaxational studies were conducted using a VXR-300 spectrometer (Varian, Palo Alto, CA) located in the MAX T. Rogers NMR facility at Michigan State University or a Gemini-300 spectrometer (Varian) at The Dow Chemical Company (Midland, Michigan). These experiments were performed with a controlled temperature of 25°C, using a transmitter frequency of 299.949 MHz. The 180 degree pulse was measured for each sample and varied between 30.0 and 31.8 microseconds. The delay between successive pulses was at least fifteen seconds (T_1 inversion recovery was less than two seconds). At least 16 transients were added before Fourier transformation. Spin-spin relaxation times were measured by spin-echo using a Carr Purcell Meiboom Gill (CPMG) pulse sequence (42) with a Levitt Freeman 180 degree composite refocussing pulse (43). The sequence is shown below.

$$90^\circ\ (\tau\ 180^\circ\ 2\tau\ 180^\circ\ \tau)_n\ \text{acquire}$$

The CPMG pulse sequence effectively removed relaxational effects due to magnetic field heterogeneity and greatly limited the effects of diffusion, while the composite pulse was used to remove pulse imperfections (43) and off-resonance effects. The NMR tube was not allowed to spin during the spin echo experiments. The proton NMR spectrum of PEG is very simple with all of the ethylene protons occurring in one peak. For each polymer sample the intensity of the ethylene peak was measured as a function of total time for relaxation (equal to $4n\tau$).

Nuclear Overhauser Effect Measurements. The complexation of PMAA with excess PEG of high molecular weight was studied using pre-steady state NOE. The PEG concentration was an order of magnitude higher than that of PMAA to ensure that nearly all of the PMAA would exist in the complexed state. PEG of high molecular weight was used to ensure that the complex would be soluble in water. A D_2O solution of 0.01 wt% PMAA and 0.09 wt% PEG of molecular weight 5 million (Aldrich Chemical Company) was prepared by first forming dilute solutions of each polymeric species separately, then mixing the solutions in appropriate proportions.

The pre-steady state proton NOE measurements were performed using an XL200 NMR spectrometer (Varian). The pulse sequence consisted of presaturation of the PEG ethylene peak, a 2 msec equilibration delay, a 90° pulse for acquisition and a three second delay for relaxation. The presaturation time was varied between 100 msec and 1 second. T_1 values were measured using an inversion recovery pulse, and the longest T_1 was 477 msec. Therefore, the 3 second delay was more than adequate for complete spin-lattice relaxation. Two spectra were acquired simultaneously using a pulse interleaving procedure. A sequence of 8 pulses in which the PEG ethylene was saturated and interleaved with 8 pulses in which the decoupler irradiated "off resonance" on the opposite side of the PMAA peaks. For each spectrum 240 transients were added before Fourier transformation. In the spectra the PMAA α-methyl and methylene peaks were well resolved. The two spectra were subtracted, and the integrated intensity in the difference spectrum was divided by the intensity in the off resonance spectrum to obtain the value of the NOE enhancement factor.

Results and Discussion

Swelling Studies. Figure 1 contains a plot of the equilibrium solvent weight fraction as a function of pH for PMAA-g-PEG networks. Here PEGMA-200 and MAA were mixed to give a 60:40 EG:MAA molar ratio of repeating units, and TEGDMA was present at 2 wt% of the total monomers. The monomers were then diluted to 40 wt% with the ethanol/water mixture before polymerization (see Synthesis of Polymer Networks). When swelling was conducted in salt-free aqueous solutions (curve 1), a low degree of swelling was observed under acidic conditions, and a high degree of swelling (over 99 wt.% water) was observed under basic conditions. This behavior can be explained in terms of the effect of pH on the underlying complexation equilibrium of PEG with PMAA. Under acidic conditions PMAA exists in a neutral (protonated) state, and can form hydrogen-bonded complexes with PEG. When complexes are formed the network exhibits a low degree of swelling because the hydrophilic moieties (ether and carboxylic acid groups) are bound with each other rather than with water. If the pH is basic, the carboxylic acid is not protonated, the complexes cannot form, and the network exhibits a much larger degree of swelling.

Curve 2 in Figure 1 illustrates the effect of ionic strength on the swelling behavior. It is important to note that the pH being monitored in the experiment is that in the swelling solution external to the polymeric gels. For swelling in water which contains no dissolved salts, the swelling transition occurs at an external pH of 7.0. If the experiment is performed in water containing 0.1 molar NaCl (curve 2), the transition pH is shifted to a value of approximately 4.4. The effect of the dissolved salt can be explained by the Donnan equilibrium of the system. In the absence of sodium ions virtually all of the protons remain inside the polymer network giving rise to a low internal pH and a relatively high external pH. At high ionic strength sodium ions in the bath will exchange for protons inside the polymer gel. The final equilibration of protons inside and outside the network will result in a decrease in the external pH relative to the case of low ionic strength. It is interesting to note that while increasing the ionic strength of the surrounding solution results in a shift in the pH of the swelling transition, it has a small effect on the magnitude of the swelling transition as characterized by the equilibrium solvent weight fraction.

Figure 2 contains a plot of the equilibrium solvent weight fraction as a function of pH for pure PMAA networks. These polymer networks were formed by polymerizing MAA in the presence of TEGDMA crosslinking agent (2 wt% of the total monomers). The monomers were diluted to 40 wt% with the ethanol/water mixture before polymerization. As shown in the figure, the PMAA networks exhibit a swelling transition from an equilibrium solvent weight fraction of approximately 90 wt% under acidic conditions, to over 99.5 wt% under basic conditions. This swelling transition may be attributed to the ionization of the PMAA. It is interesting to note that the magnitude of the swelling transition for the pure PMAA networks is much smaller than that of PMAA-g-PEG networks formed under identical conditions (Figure 1). This result supports the idea that in the PMAA-g-PEG system network swelling is regulated by polymer complexation and cannot be explained solely by the ionization of PMAA.

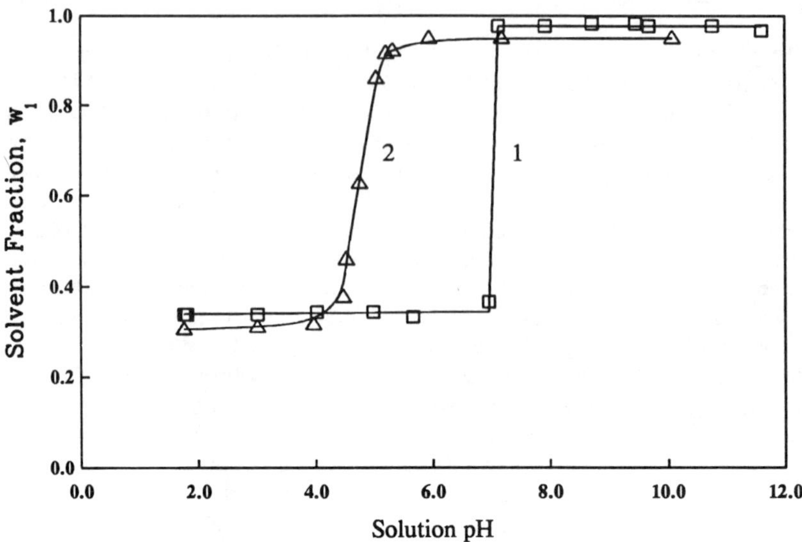

Figure 1. Equilibrium solvent weight fraction, w_1, in P(MAA-g-PEG) networks as a function of solution pH. Curve 1: swelling in salt-free aqueous solutions; curve 2: swelling in aqueous 0.1 molar NaCl solutions. (Adapted from ref. 41)

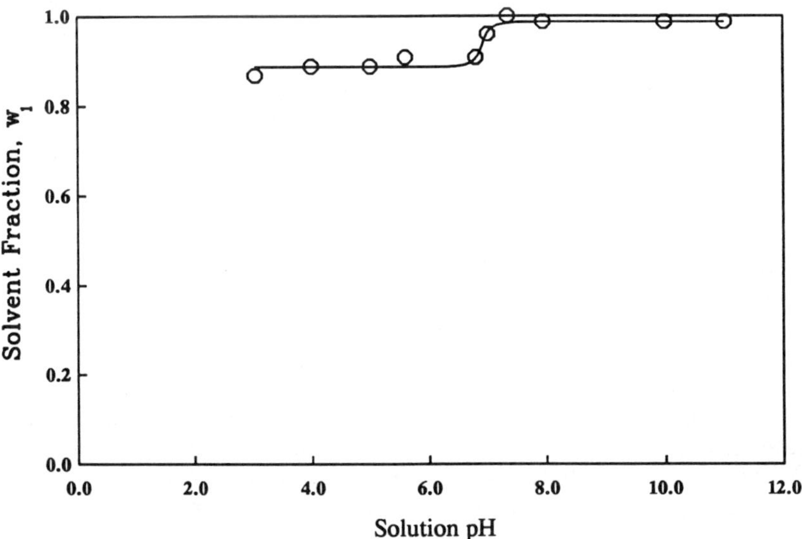

Figure 2. Equilibrium solvent weight fraction, w_1, in PMAA networks as a function of solution pH.

Figure 3 illustrates the effect of the composition of the polymeric network on the equilibrium solvent weight fraction. A series of copolymer networks ranging in composition from pure PEGMA to pure PMAA were synthesized and were swollen in methanol as well as acidic and basic aqueous solutions. Again, PEGMA-200 was used to create the PEG grafts, 2 wt% of TEGDMA was used as a crosslinking agent, and the monomers were diluted to 40 wt% with the ethanol/water mixture before polymerization. Figure 3 shows plots of the equilibrium degree of swelling in each of these solvents as a function of the mole fraction of methacrylic acid in the polymer network. Curve 1, which corresponds to swelling in pH 4 water, illustrates that although the homopolymer networks exhibited a high degree of swelling, typically near 90 wt% water, the copolymer swelling was sharply lower, exhibiting a minimum near a 1:1 molar ratio of EG/MAA. Swelling studies in methanol (curve 2) indicate that the polymer networks exhibit a relatively high degree of swelling between 85 and 90 wt% solvent regardless of MAA mole fraction. In a basic solvent of pH 9 (curve 3), the swelling is very high and increases as the MAA content of the networks is increased.

The results shown in Figure 3 are consistent with the idea that the network swelling properties are regulated by polymer complexes. The equilibrium swelling minimum in the vicinity of a 1:1 repeating unit molar ratio for swelling in pH 4 water may be attributed to a maximum in the extent of complexation at that point. As discussed earlier, acidic conditions promote the formation of complexes because the carboxylic acid moieties are present in the protonated state. Methanol and base are both known to disrupt PEG:PMAA complexes. Curves 2 and 3 illustrate this point since the swelling is uniformly high for these conditions. Furthermore, the small increase in the equilibrium solvent weight fraction with MAA content under basic conditions (curve 3) may be attributed to the increasingly ionic nature of the polymer chains.

The swelling minimum in the vincinity of a 1:1 repeating unit molar ratio is consistent with complex stoichiometries reported by previous investigators (1-4,14-17) based upon viscometry, tubidimetry, and potentiometric titration. Intrinsic viscosity minima and turbidity maxima have been observed for aqueous PMAA/PEG solutions at a 1:1 repeating unit molar ratio. These results have been attributed to the formation of compact polymer complexes with the acidic moieties of the PMAA hydrogen bonded to the ether groups of PEG in a 1:1 ratio. According to this picture, deviations from a 1:1 stoichiometry lead to free polymer chains or increased hydration of the complex (due to the presence of uncomplexed hydrophilic units within a complexed chain). Either of these occurrences would lead to increases in net solubility, hydrodynamic radius, and intrinsic viscosity. Similarly, when aqueous solutions of PMAA were titrated with PEG, pH increases were observed until a 1:1 base molar stoichiometry was attained. The titration with PEG causes the acidic groups to be bound in complexes, making them unavailable for dissociation.

Like the intrinsic viscosity minima and turbidity maxima reported previously, the swelling minima in the present systems suggest that the maximum manifestation of the polymer complexation occurs for a 1:1 EG:MAA repeating molar ratio. However, these observations leave significant questions about the fundamental nature of the complexes, and do not conclusively establish that the complexes form in a strict 1:1 stoichiometry with each acidic moiety bound with an ether linkage. It is possible that the swelling minima arise from a nonspecific combination of hydrophobic interactions and hydrogen bonds, and are not necessarily accompanied by a stoichiometric EO:MAA ratio. Similarly, the observed pH effects in the potentiometric titrations may be due to the low dielectric constant of the hydrophobic complexes and not to specific EO:MAA interactions. Further experiments will be required to unambiguously resolve these questions.

Figure 4 illustrates the effect of solvent on the swelling characteristics of the PMAA:PEG networks. In this figure, data are shown for swelling in aqueous mixtures of acetone (curve 1) and methanol (curve 2). Here PEGMA-400 and MAA were mixed to give a 50:50 EO:MAA molar ratio of repeating units, and EGDMA was present at 1 mol% of the total monomers. The monomers were then diluted to 50 wt% with the ethanol/water mixture before polymerization. As shown in Figure 4, both organic solvents exhibit an equilibrium weight fraction of just over 0.3 in the limit of infinite dilution (0 weight fraction) and a value of around 0.8 for the pure solvent. This trend is consistent with the idea that solvents with Lewis base character may compete with PEG for the hydrogen bonding sites on PMAA. The effectiveness of a solvent in breaking the complexes is probably related to its ability to disrupt the hydrophobic interactions as well as compete for the hydrogen bonding sites. The swelling results at intermediate compositions indicate that acetone effectively breaks the complexes at lower concentrations than methanol. The sigmoidal nature of both plots indicates that the incremental effectiveness of the solvent in breaking the complexes (as indicated by the slope of the curve) passes through a maximum at an intermediate concentration, between 0.2 and 0.5 weight fraction.

NMR Relaxational Studies. Nuclear magnetic relaxation may be resolved into two components - relaxation along an axis parallel to the external magnetic field and that in the plane perpendicular to the field. The former is called spin-lattice or longitudinal relaxation, while the latter is called spin-spin or transverse relaxation. These nuclear magnetic relaxations are characterized by the exponential time constants T_1 and T_2 respectively. Relaxation ultimately arises from fluctuating magnetic fields experienced by the nuclei as they interact with other molecules while undergoing random thermal motion. Fluctuating magnetic field components of the proper frequency lead to nuclear magnetic relaxation. Because the frequencies of the fluctuating fields are dependent upon the mobility of the nuclei, so are the relaxation times, T_1 and T_2. In general T_2 decreases monotonically as the mobility decreases, while T_1 passes through a minimum. This behavior is due to the fact that transverse relaxation relies upon low-frequency contributions to the spectral density function for efficient relaxation, while longitudinal relaxation relies upon higher frequency contributions (44,45).

The spin-spin relaxation times of dilute PEG solutions were determined using spin echo pulse sequences. These experiments were performed on solutions of 0.1 wt% PEG in D_2O. The data for the intensity of the ethylene peak as a function of the total time for relaxation were fit by nonlinear regression to a single decaying exponential, thereby providing a value for T_2. In all cases the error in the value of T_2 was

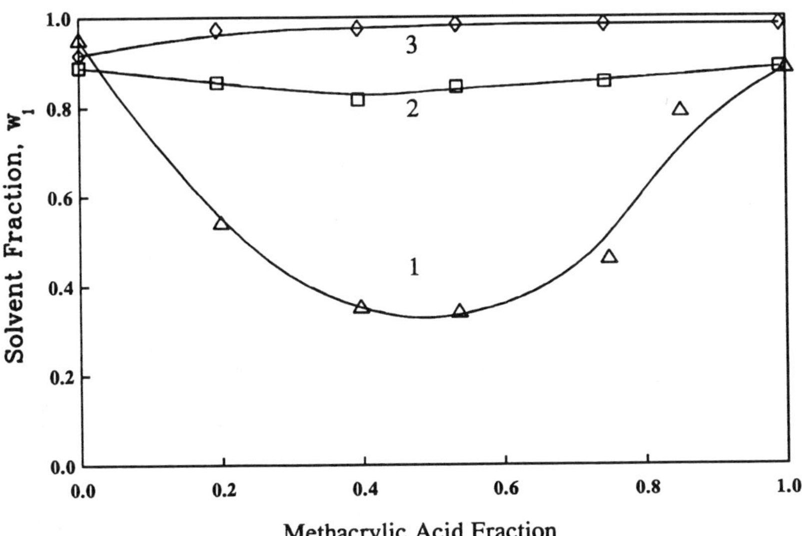

Figure 3. Equilibrium solvent weight fraction, w_1, in P(MAA-g-PEG) networks as a function of MAA mole fraction. Curve 1: swelling in pH 4 water; curve 2: swelling in methanol; curve 3: swelling in pH 9 water. (Adapted from ref. 41)

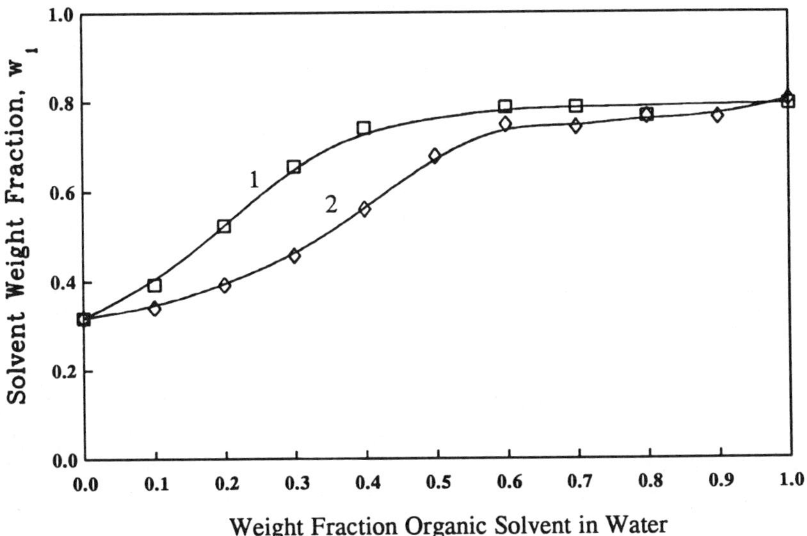

Figure 4. Equilibrium solvent weight fraction, w_1, in P(MAA-g-PEG) networks as a function of the weight fraction of organic solvent in water. Curve 1: swelling in acetone/water mixtures; curve 2: swelling in methanol/water mixtures.

less than 5%. Results for a series of experiments with different PEG molecular weights are shown in Table I. These results serve as a basis of comparison for the subsequent experiments using complexed PEG. As expected, the value of the spin-spin relaxation time decreased as the molecular weight of the PEG chain was increased. This trend is explained by an increase in the low frequency contribution to the spectral density function as the PEG mobility decreases. As illustrated in table I, the incremental decrease in the relaxation time with increasing molecular weight is most pronounced for the smallest molecular weights. This trend probably arises from a diminishing incremental decrease in mobility with increasing chain length for long polymer chains.

Table I. Spin-Spin Relaxation Times of PEG Ethylene Protons

Molecular Weight	Polydispersity	Relaxation Time (sec)
62	1.00	3.70
202	1.05	0.99
600		0.64
1500	1.05	0.61
5000	1.05	0.55
1,000	1.04	0.48
20,000	1.13	0.47
5,000,000		0.45

Relaxational studies were also performed on PMAA and on PEG:PMAA complexes. Solutions of 0.1 wt% PMAA and 0.01 wt% PEG were formed by dissolving appropriate amounts of these polymers in D_2O, while solutions of PEG:PMAA complexes were formed by dissolving 0.1 wt% PMAA and 0.01 wt% PEG in the same tube. Again, values of the spin-spin relaxation time were obtained by fitting the data for the peak integrals as functions of the time for relaxation to a single decaying exponential. The relaxation times of both the α-methyl and methylene protons on PMAA were less than 10 msec, while values for PEG are shown in Table II.

The relaxation time for complexed PEG chains is expected to be shorter than that for uncomplexed chains because complexation effectively decreases the mobility of the chains, thereby enhancing the low frequency contributions to the spectral density function. In fact, if irreversible complexes were formed, the T_2 value for the complexed chains would likely approach that of the PMAA protons (approximately 10 msec), an order of magnitude lower than the values for uncomplexed PEG (over 500 msec). However the complexes are reversible and the PEG chains undergo rapid exchange relative to the NMR relaxation timescale. Therefore the observed relaxation time is a weighted average of the values for the complexed and uncomplexed chains (46). As shown in Table II, although the presence of the PMAA has a relatively small effect on the relaxation time of PEG of molecular weight 1500, it has a pronounced effect on the T_2 value of PEG of molecular weight 5000. These results suggest that the spin-spin relaxation time is indeed sensitive to the formation of complexes, and that there is a measurable chain length effect on the complex stability between PEG molecular weights of 1500 and 5000. For the 1500 molecular weight PEG, the ten percent decrease in T_2 from 570 msec to 511 msec provides evidence for complexation even at this relatively short chain length.

Table II. T_2 Values for Complexed and Uncomplexed PEG

Molecular Weight	Uncomplexed (msec)	Complexed (msec)
1500	570	511
5000	530	310

The experimental results may be explained in terms of the statistical thermodynamic description of the complexation equilibrium reported by Scranton et al. (40). These authors considered the complexation thermodynamics of dilute solutions of free and graft oligomers with complementary polymers. The total free energy change upon complexation was divided into two contributions which were considered separately. An internal contribution to the free energy accounted for conformational degrees of freedom as well as the segmental binding interactions, while an external contribution accounted for the configurational (translational) degrees of freedom. The external contribution could be evaluated from combinatorial considerations, while the internal canonical and grand canonical partition functions were formulated in terms of sequence generating functions for trains, tails and loops. A simple random walk model was used to evaluate the statistical weights of these generating functions. Details of these calculations may be found in reference 40.

In agreement with the experimental results reported here and in the literature the theoretical simulations revealed a marked dependence of the equilibrium bound fraction on the segmental binding free energy and the PEG chain length. Moreover, simulation results indicated that covalently attaching the complementary polymers to one another promotes the formation of complexes primarily due to a decrease in the configurational entropy change upon complexation. If the complementary polymers are covalently attached to one another, they are effectively immobilized and are in close proximity prior to complex formation. This result corroborates the NOE experiments which indicate the lack of a critical chain length for the graft copolymers. Other simulation results indicate the loops may be important in the conformation of a complexed oligomer, and the conformational averages for the ungrafted case asymptotically approach those for the grafted case as the segmental binding free energy, polymer concentration and PEG chain length increase.

Nuclear Overhauser Effect Studies. Proton nuclear Overhauser effect (NOE) studies were conducted to provide insight into the molecular characteristics of PEG:P-MAA complexes. NOE is useful for structural determination in solution because its presence indicates that two nuclei are undergoing dipole-dipole cross relaxation, and are therefore in close proximity to one another. In proton NMR the presence of a nuclear Overhauser effect indicates the two protons are within 5 angstroms of one another (47). Information about the spatial distance between the interacting protons can be obtained from the initial growth rate in a pre-steady-state NOE experiment (48). Here the NOE growth rate is higher for protons which are closer together. Klier et al. (41) demonstrated the sensitivity of proton NOE experiments for detecting PEG:PMAA complexes in systems containing free and grafted PEG. These investigators measured the NOE growth rate in the PEG ethylene peak when the PMAA α-methyl was irradiated. NOE was detected under complex promoting conditions, but not under conditions which break the complex. Due to the existence of a NOE, complexes were detected for PEG molecular weights as low as 1500.

In this study NOE was used to determine the relative importance of the PMAA α-methyl and methylene groups in the hydrophobic stabilization of the complexes. The experiments were performed by irradiating the ethylene peak on the PEG, and then measuring the NOE growth rates in both the α-methyl and the methylene on the PMAA. The NMR sample contained 0.09 wt% PEG of molecular weight 5,000,000, and 0.01 wt% PMAA in D_2O. Excess PEG was used to ensure that the PMAA would exist in the complexed state, while PEG of high molecular weight was used to keep the complexes soluble in water.

The NOE results are shown in Figure 5. The figure includes plots of the NOE evolution for both the PMAA α-methyl and methylene peaks upon irradiation of the PEG ethylene peak. The initial NOE growth rate for the α-methyl is higher than that of the methylene. This result indicates that in the complexed state the ethylene protons of the PEG are spatially closer to the α-methyl protons than to the methylene protons, and suggests that the α-methyl group plays a larger role than the methylene in the hydrophobic stabilization of the PMAA:PEG complexes. Therefore, these studies provide spectroscopic evidence for the importance of the α-methyl group, and supports previously reported conclusions based upon pH and viscosity measurements.

Conclusions

The complexation of poly(ethylene glycol) with poly(methacrylic acid) was studied in dilute aqueous solutions and in covalently crosslinked hydrogels. Crosslinked polymer gels of P(MAA-g-EG) exhibited responsive swelling behavior over a wider range of PEG molecular weights than previously reported. In contrast to ungrafted polymer systems, no lower limit on PEG molecular weights was found for the graft copolymers. This behavior was attributed to a smaller configurational entropy change upon complexation of covalently attached grafts as compared to independent polymer chains. The sensitivity of complexation to pH and solvent composition resulted in gels which exhibited transitions in swelling over narrow pH and solvent composition ranges. The ability of an organic solvent to disrupt complexes in these systems is probably due to a combination of its ability to disrupt hydrophobic interactions and to compete with the PEG ether oxygens for hydrogen-bonding sites. Investigation of solvent effects in swelling is being extended to strong Lewis base solvents and poly(acrylic acid-g-ethylene glycol) systems.

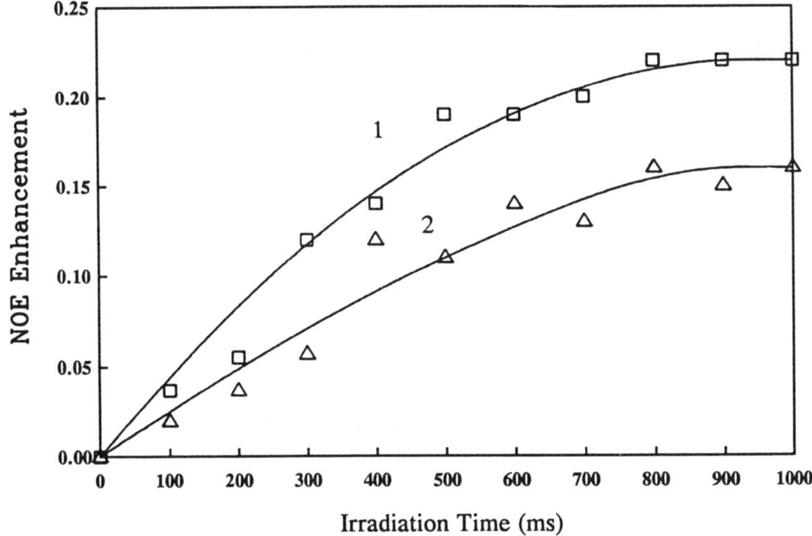

Figure 5. Negative NOE enhancement of PMAA protons upon irradiation of PEG ethylene protons as a function of irradiation time. Curve 1: α-methyl; curve 2: methylene.

The nature of the complexation equilibria were studied using NMR spectroscopy. Proton nuclear Overhauser effect experiments and relaxational studies indicated that complexation between PEG and PMAA may occur at lower PEG molecular weights than previously reported, and at quite low concentrations. However, the effects of complexation on the spin-spin relaxation time are much more pronounced as the PEG molecular weight is increased, suggesting a strong molecular weight dependence upon complex stability. Consequently, discrepancies among critical molecular weights, concentration effects and complex stoichiometries reported in the literature may be due to differences in the sensitivities of the experimental techniques. Complexation probably takes place outside of the molecular weight and stoichiometry limits measured by viscometry and sedimentation as indicated by more sensitive spectroscopic techniques. Other NOE results shown here indicate that in the complex an intimate contact exists between the α-methyl groups of PMAA and the ethylene groups of PEG. These results affirm the importance of hydrophobic stabilization of the complexes.

Acknowledgments

This work was supported in part by a Du Pont Young Faculty Grant to ABS and the State of Michigan Research Excellence Funds. In addition, the authors wish to thank Kimberly Hammonds and Stephanie Smith for their help with the swelling studies in organic solvents.

Literature Cited

1. Tsuchida, E.; Abe, K. *Adv. Polym. Sci.* **1982,** 45, 1.
2. Bekturov, E.; Bimendina, L.A. *Adv. Polym. Sci.* **1980,** 41, 100.
3. Osada, Y. *Adv. Polym. Sci.* **1987,** 82, 1.
4. Kabanov, V.A.;. Papisov, I.M. *Vysokomol Soyed.* **1979,** A21, 243.
5. Abe K.; Tsuchida, E. *Makromol. Chem.* **1975,** 176, 803.
6. Kabanov, V.A.; Zezin, A.B., *Pure Appl. Chem.* **1984,** 56, 343.
7. Michaels, A.S.; Mir, L.; Schneider, N.S.; *J. Phys. Chem.* **1965,** 69, 1447.
8. McCormick, C.L.; Johnson, C.B. *Macromolecules* **1988,** 21, 694.
9. Monroy Soto, V.M.; Galin, *J.C. Polymer* **1984,** 25, 254.
10. Kolawule, E.G.; Mathieson, S.M. *J. Polym. Sci., Polym. Chem.* **1977,** 15, 2291.
11. Chatterjee, S.K.; Gupta, S.; Sethi, K.R. Ange. *Makromol. Chem.* **1987,** 147, 133.
12. Vinogradov, S.N.; Linnell, R.H. *Hydrogen Bonding* Van Nostrand Reinhold, New York, NY, 1971.
13. Bailey, F.E.; Lundberg, R.D.; and Callard, R.W. *J. Polym. Sci.* **1964,** A2, 845.
14. Antipina, A.D.; Baranovsky, V.Y.; Papisov, I.M.; Kabanov, V.A. *Vysokomol. Soyed.* **1972,** A14, 941.
15. Baranovsky, V.Y.; Litmanovich, A.A.; Papisov, I.M.; Kabanov, V.A. *Eur. Polym. J.* **1981,** 17, 696.
16. Osada, Y.; Sata, M. J. Polym. Sci., *Polym. Lett.* **1976,** 14, 129.
17. Ikawa, T.; Abe, K.; Honda, K.; Tsuchida, E. *J. Polym. Sci., Polym. Chem.* **1975,** 13, 505.

18. Osada, Y.; *J. Polym. Sci., Polym. Chem.* **1979,** 17, 3485.
19. Papisov, I.M.; Baranovsky, V.Y.; Serfgieva, Y.I.; Antipina, A.D.; Kabanov, V.A. *Vysokomol. Soyed.* **1974,** A16, 1133.
20. Oyama, H.T.; Tang, W.T.; Frank, C.W. *Polym. Mater. Sci. Prepr.* **1987,** 57, 13.
21. Oyama, H.T.; Tang, W.T.; and Frank, C.W. *Macromolecules* **1987**, 20, 474.
22. Char, K.; Frank, C.W.; Gast, A.P.; Tang, W.T. *Macromolecules* **1987,** 20, 1833.
23. Oyama, H.T.; Tang, W.T.; Frank, C.W. *Macromolecules* **1987,** 20, 1839.
24. Chen, H.L.; Morawetz, H. *Eur. Polym. J.* **1983,** 19, 923.
25. Bednar, B.; Morawetz, H.; Schaeffer, J.A. *Macromolecules* **1984,** 17, 1636.
26. Heyward, J.J.; Ghiggino, K.P. *Macromolecules* **1989,** 22, 1159.
27. Abe, K.; Koide, M.; Tsuchida, E. *Macromolecules* **1977,** 10, 1259.
28. Osada, Y. *J. Polym. Sci., Polym. Chem.* **1977,** 15, 255.
29. Osada, Y. *J. Polym. Sci., Polym. Lett.* **1980,** 18, 281.
30. Osada, Y.; Takeuche, Y. *J. Polym. Sci., Polym. Lett.* **1981,** 19, 303.
31. Osada, Y.; Honda, K.; Ohta, M. *J. Membr. Sci.* **1986,** 27, 327.
32. Nishi, S.; Kotaka, T. *Macromolecules* **1986,** 19, 978.
33. Papisov, I.M.; Litmanovich, A.A. *Vysokomol. Soyed.* **1977,** A19, 716.
34. Magee, W.S.; Gibbs, J.H.; Zimm, B.H. *Biopolymers* **1963,** 1, 133
35. Damle, V.N. *Biopolymers* **1970,** 9, 353.
36. Poland, D. *Cooperative Equilibria in Physical Biochemistry* Oxford University Press, Oxford, 1978.
37. Poland, D.; Sheraga, H.A. *Theory of Helix-Coil Transitions in Biopolymers* Academic Press, New York, NY, 1970.
38. Lifson, S. *J. Chem. Phys.* **1964,** 40, 3705.
39. Eichinger, B.E.; Jackson, D.M.; McKay, B.D. *J. Chem. Phys.* **1986,** 85, 5299.
40. Scranton, A.B.; Klier, J.; Peppas, N.A. *J. Polym. Sci., Polym. Phys.,* **1991,** 29, 221.
41. Klier, J.; Scranton, A.B.; Peppas, N.A. *Macromolecules* **1990,** 23, 4944.
42. Carr, H.Y.; Purcell, E.M. *Phys. Rev.* **1954,** 94, 630.
43. Levitt, M.H.; Freeman, R. *J. Magn. Res.* **1981,** 43, 65.
44. Abragam, A. *Principles of Nuclear Magnetism* Oxford University Press: New York, NY, 1961.
45. Slichter, C.P. *Principles of Magnetic Resonance* Springer-Verlag, New York, NY, 1990.
46. Sanders, J.K.M. and Hunter, B.K. *Modern NMR Spectroscopy* Oxford University Press: New York, NY, 1987.
47. Wurthrich, K. *Science* **1989,** 243, 45.
48. Neuhaus, D; Williamson, M.P. *The Nuclear Overhauser Effect in Structural and Conformational Analysis* VCH Publishers: New York, NY, 1989.

RECEIVED August 19, 1991

Chapter 12

Monitoring Polyelectrolyte Mobility by Gel Electrophoresis

Evangelia Arvanitidou, David Hoagland, and David Smisek

Department of Polymer Science and Engineering, University of Massachusetts, Amherst, MA 01003

Electrophoretic methods are applied to studies of probe chain motion in gels, exploiting free solution measurements of the same motion to isolate gel matrix effects. Trends observed for gels, particularly in relationships between mobility and chain length and between mobility and chain topology, support a three state entanglement depiction, each state characterized by its own chain transport mechanism. These include *excluded volume* (unentangled regime), *entropic barriers* (weakly entangled regime), and *reptation* (strongly entangled regime). Intermediate entanglement provides the best fractionation, and evidence is presented that spatial variations in chain entropy control motion whenever the probe size approximately matches the gel mesh spacing. The backbone charge density, potentially an uncontrolled variable for synthetic samples, is shown to exert negligible influence on the mobility of highly charged probe molecules.

Gel electrophoresis has long been recognized as the method of choice for fractionating and analyzing mixtures of high molecular weight biological molecules (*1*). Despite the method's widespread acceptance, until recently only minor efforts had been made to develop a comprehensive understanding of electrophoretic transport mechanisms at the molecular level. Fundamental research was finally spurred in the early 80's with the discovery that pulsed field gel electrophoresis can fractionate enormous, chromosome-sized DNA chains (*2*). Fully exploiting pulsed field methods requires a sophisticated understanding of molecular transport in gel media, and the conspicuous lack of such understanding initiated a wave of research activity that continues unabated. Recognizing the richness and universality of electrophoretic transport mechanisms, and realizing that applications in the synthetic polymer field were notably lacking, we began a systematic study of the electrophoresis of model polymers about 5 years ago; these polymers serve as model materials in the sense that we select chemical structures that are well-suited for the testing of proposed transport mechanisms. In the absence of a basic molecular description for even the simplest electrophoretic techniques, we have focused our attentions on the least complicated cases: electrophoresis in homogeneous gels and solvents under steady, low applied fields. The present contribution will summarize our most recent findings in this area.

0097–6156/92/0480–0190$06.25/0
© 1992 American Chemical Society

The term electrophoresis is used to describe the migration of a charged solute under the action of an electric field. At steady state, the driving force due to the electric field balances the drag exerted on the solute by the surrounding medium. The resulting steady motion is characterized by the electrophoretic mobility μ, defined as the ratio of the velocity u to the field strength E:

$$\mu = u/E \tag{1}$$

In the low field limit, μ is independent of E. At higher fields, μ may vary with E, reflecting the potential of strong applied fields to alter solute friction. As a random coil polymer stretches and orients in the field direction with increasing E, for example, μ will rise since these processes reduce drag. Pulsed field electrophoretic methods rely entirely on the nonlinear nature of $\mu(E)$, with the nonlinearity greatly accentuated by the presence of a strongly confining gel. Operation in the nonlinear $\mu(E)$ range with steady fields, however, significantly reduces the sensitivity of mobility to solute structure. Our research program has thus been directed almost entirely toward experiments performed at very low applied fields ($\lesssim 2$ V/cm), incapable of perturbing equilibrium polymer structure and verified in each case to produce mobilities independent of E. The penalty for following this strategy is a lengthy run time (~ 2 days) for high molecular weight samples (i.e., those with molecular weights above 10^6).

If a system is diluted in charged solute and E is small, μ can be related to the solute diffusion coefficient in the same environment,

$$\mu \sim DQ \sim DN \tag{2}$$

where D is the tracer diffusion coefficient, Q is the total solute charge, and N is the degree of polymerization; a prefactor involving the ionic strength and the density of ionic groups along the backbone has been neglected. Essentially the classical Einstein relationship for ion mobility, equation 2 is derived under the implicit assumption that long range intramolecular hydrodynamic interactions are the same for both diffusion and electrophoresis; in fact, long range hydrodynamic interactions are absent during electrophoresis of flexible macromolecules (reasons to be discussed later), so the proper diffusion coefficient for equation 2 is the one evaluated in the absence of hydrodynamic interactions (i.e., using the Rouse model). The information contained in μ is analogous, within a factor involving the charge density and ionic strength, to the information contained in D. Most importantly, this equation provides a means of relating the molecular weight dependences of μ and D. Therefore, electrophoresis provides a convenient means to study diffusion phenomena in a variety of environments ranging from gels to polymer solutions to porous media.

Support media (gels) were originally introduced in electrophoretic methods to suppress convection currents arising from Joule heating. Investigators later recognized that the same media might also be responsible for electrophoretic separations based on molecular size. Agarose and poly(acrylamide) gels are the most common separation media, mainly because of their simple preparations and high mechanical strengths. Properly prepared gels contain few, if any, bound ionic groups; if such groups are present, solvent flow created by the electrical forces on mobile counterions can produce solute convection (i.e., electro-osmosis) greatly in excess of solute motion by electrophoresis. Agarose possesses large pores (~10-100 nm) (*3,4*) and is consequently employed to study larger macromolecules such as DNA restriction fragments. Poly(acrylamide) gels, with smaller pores (*5*), are used for separating proteins and sequencing short DNA fragments. Agarose is a natural polysaccharide, isolated from red algae, with a low percentage of bound, negatively charged functional groups. Gelation occurs as aqueous solutions are cooled to room temperature; the mean pore size, as well as the highly distributed pore size distribution, depends strongly on agarose concentration. Poly(acrylamide) gels are the product of polymerization of acrylamide in the presence of a multifunctional cross-linker such as N,N'-methylene-bis-acrylamide. The gel network is formed at room temperature in the presence of the buffer solution selected for the electrophoresis experiment. Poly(acrylamide) gels are normally prepared at higher gel concentrations (3.0-20.0 wt. %) than agarose gels (0.1-4.0 wt. %).

Many polymers routinely analyzed by gel electrophoresis can also be studied by aqueous size exclusion chromatography (SEC). Electrostatic exclusion from the pores of the support, coil expansion due to intramolecular electrostatic repulsion, and adsorption on support walls are a few of the difficulties that can make aqueous SEC less attractive than gel electrophoresis (*6*). Moreover, velocity gradients in the chromatography experiment create band-broadening effects not present in electrophoresis. The resolution of an electrophoretic separation can thus be much higher than one produced chromatographically; for example, 500 basepair DNA chains are routinely baseline resolved from 499 basepair DNA chains by electrophoresis in sequencing gels. Even under the most ideal circumstances, chromatographic methods for high molecular weight polymers are restricted to much lower resolutions (*6,7*). In fact, with synthetic polymers of molecular weight less than 10^5, gel electrophoresis can theoretically provide discrete molecular weight distributions, i.e., with each degree of polymerization resolved into a separate band. Efforts to attain this unprecedented separation quality are currently being pursued for anionically polymerized poly(styrene) samples. The final advantage of electrophoresis is its molecular weight range; separations are obtained for molecules ranging in size from dimers to whole chromosomes with 10^7 repeat units. The negative features of the electrophoretic approach, as opposed to SEC, are the need for solute charge and solubility in a high dielectric constant solvent such as water. In some cases an inappropriate solute, such as the poly(styrene) just mentioned, can be readily modified to provide a charged analog with the original backbone structure and length.

Numerous biopolymer applications of gel electrophoresis have been developed, with new methods reported regularly in journals such as *Electrophoresis*. The most common and perhaps most important applications are the molecular weight fractionation of proteins in poly(acrylamide) gels (*8-10*) and DNA fragments in agarose gels (*11-13*). Determination of the molecular weight distributions of *synthetic* polyelectrolytes by gel electrophoresis was first attempted by Chen and Morawetz (*14*). They described poly(acrylamide) gel electrophoresis of low molecular weight poly(styrene sulfonate) [PSS], adapting conventional protein sizing techniques. Recently, Smisek and Hoagland (*15*) discussed methods to measure PSS molecular weight in agarose gels, a procedure more akin to fractionation of large DNA fragments. In their initial studies with synthetic polymers, both groups demonstrated that by using monodisperse polyelectrolyte standards, a calibration curve could be developed between molecular weight and mobility. This curve, in conjunction with the measurement of the mobility of an unknown of the same chemical type, permitted determination of molecular weight and its distribution by comparison. This approach sidesteps the need for a molecular level understanding of the mechanism of molecular weight fractionation. Without a fuller understanding, however, further quantitative applications in the synthetic polymer field will remain problematical since only a handful of charged polymers with monodisperse standards are available.

Although electrophoresis is simple to perform, many parameters are known to affect the mobility (gel concentration and type, ionic strength, electric field strength, chain stiffness). Our goal is to understand, at a molecular and pore size level, the influence of each parameter. The most novel feature of this work is the use of model synthetic polymers as the probe species. By using model solutes and gels, we attempt to focus directly on the mechanisms of electrophoresis. Efforts to develop satisfactory model gels, however, have not been entirely successful; we believe that access to model gels (or alternative media) will eventually become a major research issue in the electrophoresis field.

Experimental

Electrophoresis in poly(acrylamide) gels is performed in a vertical apparatus using an Ephortec 500-V power supply operating at constant voltage. After each run, the gel is stained with a pH 4.0, 0.01 wt. % methylene blue solution for 15 minutes and destained with distilled water for several hours; stained bands are then scanned using an ISCO Model 1312 densitometer operated at 580 nm. Poly(acrylamide) gels are prepared from a 40% stock solution of acrylamide/bis-acrylamide (29:1 ratio) mixed with 25 ml TEMED, 600 μl of 10 wt. % ammonium persulfate, 10 ml 0.01 M sodium phosphate dibasic buffer (I=0.03 M), and water. The volumes of water and of the acrylamide/bis-acrylamide stock solution are varied to control gel concentration. Since polymerization of acrylamide is inhibited by oxygen, gels are poured between two 19.5 cm x 21.5 cm glass plates. The gel thickness is dictated by rigid spacers inserted between the plates and is typically about 1.0 cm. Polymerized gels are allowed to solidify for 30 minutes at room temperature before use. Gel concentrations range from 3.0 to 7.0 wt. %.

Agarose gels are prepared by dissolving FMC SeaKem agarose in a 0.01 M sodium phosphate dibasic buffer solution at a high temperature (95 C). After cooling, the solution is cast in a 15 cm x 15 cm tray, and electrophoresis is performed in the horizontal mode (submarine cell from BioRad). Agarose concentrations vary from 0.3 to 0.9 wt. %. The staining procedure is the same as for poly(acrylamide) gels; in a few cases, with DNA as the probe species, ethidium bromide is exchanged for methylene blue in the staining step, and DNA bands are recorded photographically over a UV transilluminator. Further experimental details on run procedures were presented in an earlier publication (*15*).

Electrophoresis of PSS in the absence of gel, termed free solution electrophoresis, is performed with a Coulter Electronics DELSA electrophoretic light scattering apparatus. Samples are prepared at a series of concentrations spanning the dilute and semidilute regimes, and the true dilute solution mobility is obtained by extrapolating mobility data to zero polymer concentration. The DELSA instrument permits redundant scattering measurement at four angles using two crossed incident beams to create an optical grating at the scattering volume. The scattering volume is located at the stationary plane of the electrophoresis chamber to eliminate extraneous solute convection arising from electro-osmosis. The rate of motion of the charged solute is determined from the frequency spectrum of the fluctuating scattered light intensity; the fluctuation occurs as charged solute moves through the grating. Unless the same velocity is measured at all angles, the data are not reported here. At low molecular weight and/or at low ionic strength, the scattering contrast decreases and the mobility becomes unobtainable by this method. In a few cases, the free solution mobility has been verified by direct measurement on a dilute solution in a capillary electrophoresis apparatus (ISCO).

A range of narrow molecular weight distribution PSS samples are available ($4{,}000<M_w<2{,}000{,}000$), either purchased from Pressure Chemical or prepared from monodisperse PS standards (Polysciences, Polymer Laboratories and Toyo Soda) according to Vink's (*16*) procedure. When Vink's procedure is applied, the starting poly(styrene) concentration must be well below the polymer's overlap concentration C^* to reduce or eliminate the fraction of cross-linked chains in the sulfonated product. For electrophoresis, PSS is dissolved into the appropriate buffer at concentrations about 50 ppm, well below C^* at all conditions and molecular weights; experience shows that overly concentrated samples (although still dilute when compared to C^*) become obvious during staining, with the stained region for these samples extending outside the borders of the sample lane and distorting the appearance of the band. DNA samples are purchased from BRL Life Technologies and studied under the same conditions as with PSS.

Results and Discussion

Free Solution Mobility. To discern the influence of the gel matrix on the mobility, an important control step is to explore the mobility in absence of a gel medium. In this case, assuming the field is low, the two most relevant variables are the probe molecular weight and the solvent ionic strength. The dependence of PSS mobility on these variables is plotted in Figures 1 and 2. A key result is immediately apparent, that with all other conditions fixed, the mobility in absence of gel is independent of molecular weight (or more accurately, of degree of polymerization N). This observation is in accord with many previous investigations (*17-21*), and a qualitative physical explanation has emerged in the literature (*22*). Superficially the molecular weight independence appears strange, since from equation 2 this result is consistent with D inversely proportional to chain length N, a dependence at odds with well-documented diffusion data for PSS and similar flexible coil polymers in aqueous solvents. The classical Rouse model predicts the necessary N dependence for D, of course, but only by neglect of intramolecular hydrodynamic interactions between distant chain segments.

To rationalize this behavior, one must account properly for the dynamics of the counterions clustered around the charged chain backbone. In an electrophoresis experiment, the electrical forces exerted on solvent by these ions are equal and opposite to the forces exerted on solvent by the backbone-bound ionic groups; this balance is a consequence of the zero net charge of the polyelectrolyte and its surrounding counterion cloud. At short length scales, the electrical forces are not in balance, with the chain and its counterions dragged by the electric field in opposite directions. The Debye length κ^{-1} ($\kappa = [2e^2I/\varepsilon kT]^{1/2}$, where e is the electron charge, I is the solvent ionic strength, and ε is the solvent dielectric constant) provides the length scale over which the local excess of counterions decays to zero; this length also characterizes the distance over which the hydrodynamic forces induced by the electric field's action on the polymer backbone can be transmitted. Inter- and intramolecular hydrodynamic forces imparted by a backbone segment are "screened" from segments located at distances much greater than the Debye length, and consequently free solution electrophoresis becomes a problem in local polymer dynamics. In other words, at length scales greater than κ^{-1}, the polymer coil is "free-draining," and the Rouse approach is indeed correct. Because a large flexible polyelectrolyte expands to keep distant segments statistically separated by distances greater than order κ^{-1} (thereby reducing electrostatic energy), screening by counterions in dilute solution is extremely efficient. Hydrodynamic friction at short length scales (less than the Debye length) does reflect the local segment geometry, which presumably is cylindrical. The interplay of forces described here is not specific to polyelectrolyte electrophoresis and has been well-documented in textbooks and papers that discuss electrophoresis of colloidal particles (*23*).

Figure 2 demonstrates that the mobility of a flexible polyelectrolyte decreases sharply with increasing ionic strength. This behavior is again associated with the counterions, which are only weakly associated with the chain at low ionic strengths: the counterion cloud size scales linearly with κ^{-1}. With an increase in ionic strength, the ion cloud compresses toward the backbone, and the electric forces acting on this cloud are more effectively transmitted by solvent to the chain. Since these forces are in the direction opposite to the forces acting directly on the chain, the chain mobility decreases as ions cluster more closely. Again, this effect is well documented in the electrophoresis literature for colloidal particles.

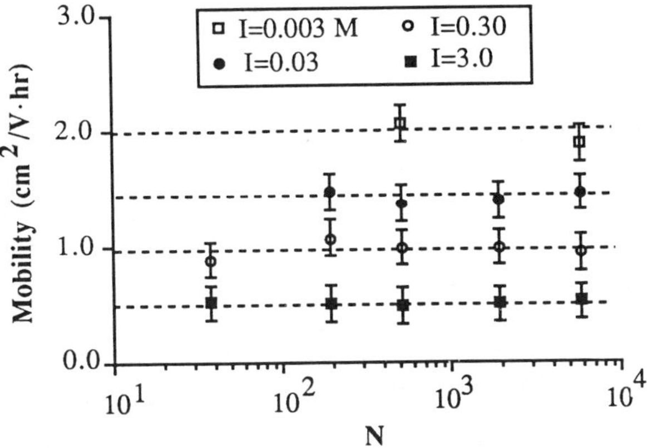

Figure 1. Chain length dependence of the free solution mobility for PSS; I is the ionic strength.

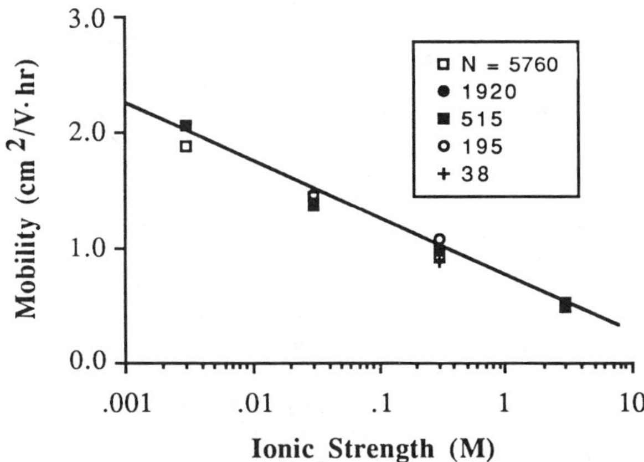

Figure 2. Ionic strength dependence of the free solution mobility for PSS, with chain length N as a parameter.

A comprehensive model of electrophoresis for a flexible coil polymer in free solution has apparently not been formulated. The effects observed in Figures 1 and 2 can be explained, at least qualitatively, by either the charged rod model of Henry (*24*) or the porous sphere model of Hermans and Fujita (*25,26*); both approaches tentatively predict the key trends just discussed, although certain parameters in the models are difficult to specify for a real system. Given the effective screening by counterions, which diminishes or even eliminates the role played by long range chain structure, the rod model appears sufficient. Refinements and more complete analyses of electrophoresis in free solution have been presented by additional authors (*22,27-30*); the more recent work has dealt mostly with issues pertaining to nonlinear electrostatic fields and the so-called "relaxation" effect associated with distortion of the counter-ion cloud during motion in an electric field. The most complete treatment appears to be the one by Stigter (*28,29*).

The key finding of the free solution study is that *the structural sensitivity observed for the mobility during electrophoresis in gels must arise from the interactions of the probe molecule with the matrix.* If there are no enthalpic interactions between the gel and the probe (i.e., in the absence of adsorption), transport is entirely controlled by configurational and/or solvent flow phenomena associated with motion of the probe polymer through the gel's pore space. These latter phenomena depend strongly on probe chain structure, an issue that will be the main topic of the remaining discussion. There is no evidence for specific chemical interaction or adsorption for systems involving PSS or DNA in agarose or poly(acrylamide) gels. Attempts at electrophoresis of more hydrophobic polyelectrolytes such as poly(ethacrylic acid) and poly(methacrylic acid) have been unsuccessful due to the appearance of weak attractive forces between gel and probe. These forces can be inferred from the broad, asymmetric bands noted after electrophoresis of narrow distribution polymer fractions. In some cases, the samples appear unable to even penetrate the gel and are then deposited on the front solvent/gel interface, which is thereby distorted and even fractured.

Electrophoresis in Gels, Chain Length Dependence. In the presence of a gel, the mobility of a flexible polyelectrolyte becomes strongly dependent on molecular weight. To simplify discussion, we will consider this dependence only at low applied fields and at high ionic strengths. The free solution conformation of PSS under these conditions is that of a relaxed flexible coil, with the radius of gyration R_g expressible as a power law in the degree of polymerization N; for the buffers selected here, the power law or excluded volume exponent ν is about 0.55-0.58 (*31*), values characteristic of a flexible coil under good solvent conditions. Recent theoretical work (*32*) suggests the value of this excluded volume exponent is preserved within a random gel; experimental confirmation of this invariance, however, has not yet been obtained and remains a goal of current investigations by several groups.

Figure 3 displays the molecular weight dependence of the mobility in a 0.6% agarose gel. (Molecular weight is expressed in terms of the degree of polymerization N since the degree of sulfonation is not known precisely for all samples.) The analogous data in a 7.0% poly(acrylamide) gel is given in Figure 4. Both figures indicate that as N increases, the mobility decreases, at first slowly and then more rapidly. As shown previously, these curves reflect at least three types of molecular transport that we have identified with *unentangled*, *weakly entangled*, and *strongly entangled* regimes; a geometrical depiction of these regimes is provided in Figure 5. The degree of entanglement is determined by the ratio of some characteristic polymer size (R_g, for example) to the mean gel mesh or pore size ξ. When this ratio is small, the probe chain is unentangled; when the ratio is very large, the probe is strongly entangled; and finally, when the ratio is order 1, the probe is weakly entangled. Regime boundaries are obviously not sharp. Experimentally, the degree of entanglement can be modified in two equivalent ways, by varying molecular weight or

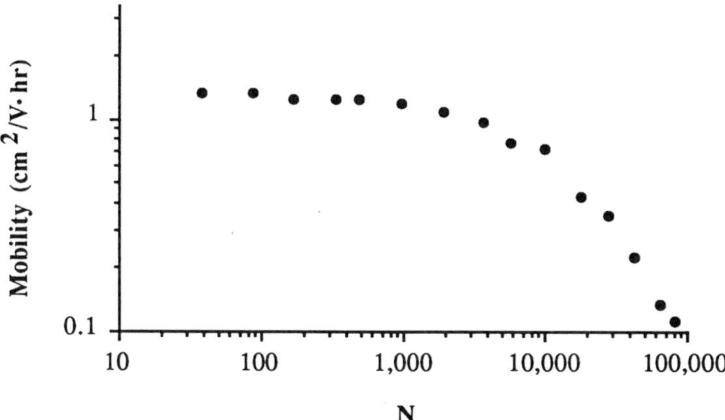

Figure 3. N dependence of PSS mobility in 0.6 wt. % agarose buffered with 0.01 M sodium phosphate dibasic. (Adapted from ref. 15.)

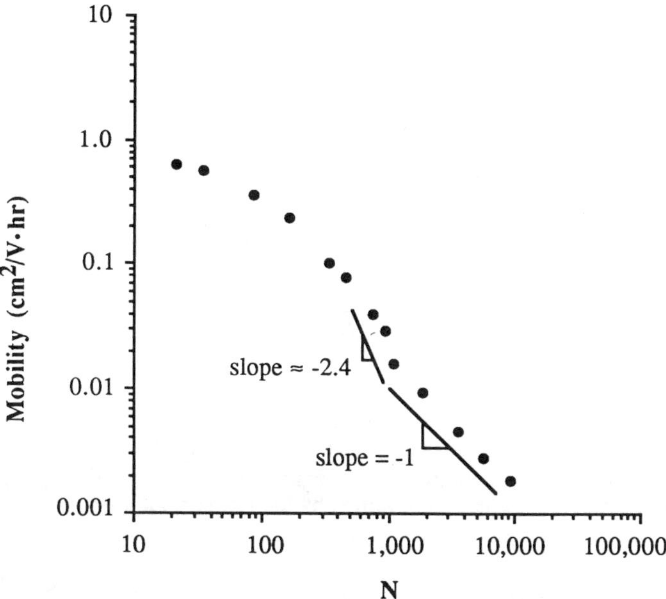

Figure 4. N dependence of PSS mobility in a 7.0 wt. % poly(acrylamide) gel buffered with 0.01 M sodium phosphate dibasic.

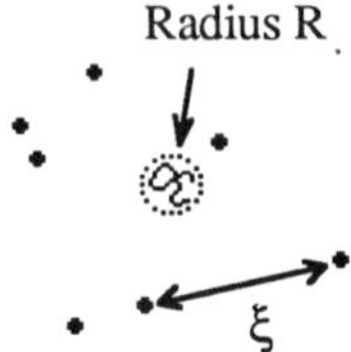

Unentangled

Weakly Entangled

Strongly Entangled

Figure 5. Depiction of the entanglement regimes for flexible chain transport in gels. The appropriate regime is determined by the ratio of polymer size R to mesh spacing ξ.

ionic strength (and thus changing polymer size) and by varying gel concentration (and thus changing ξ).

If data from the unentangled regime are extrapolated to zero N, the free solution mobility μ_o is obtained, to within an error of perhaps +/- 5%. This agreement is to be expected, as a small enough chain will have little interaction with a dilute gel network. A simple way to clearly define the unentangled regime is to restrict attention to probe/gel environments for which the probe and gel interaction is localized to a single gel site. This interaction is thus easily modeled, using a selected obstacle geometry for the gel site and the known conformation of the probe chain. Motion through a dilute environment of obstacles is essentially an excluded volume problem if hydrodynamics are ignored; a brief discussion of the importance of these hydrodynamics will be made later.

To quantitatively analyze behavior in the unentangled regime, we have modeled dilute gels as arrays of planar, cylindrical, or spherical obstacles and probe molecules as spheres, rods, or Gaussian chains (*33*); the appropriate geometries are displayed in Figure 6. Previous workers had used only the sphere depiction for the probe, and the behavior of this conformational model in an array of randomly oriented cylindrical obstacles is known as the Ogston model (*34-36*). This model, as well as all others that assume a rigid conformation for the probe, predicts that the mobility is a function of molecular size. The appropriate molecular size is not necessarily related to size measures such as the radius of gyration or the hydrodynamic radius (see refs. *33,37,38*). Figure 7 compares the dimensionless excluded volume V_{ex} for a rod-like (length=L), spherical (radius=R), and Gaussian chain (radius of gyration=R_g) probe molecule near a cylinder of radius a. The mobility μ in a sufficiently dilute array of these cylinders can be written in terms of V_{ex} and the free solution mobility μ_o,

$$\mu/\mu_o = 1 - \phi V_{ex} \tag{3}$$

so long as the volume fraction ϕ of cylinders remains small. The differences among the curves in Figure 7, particularly the "glitch" in the Gaussian chain result, indicate that polymer conformation has a striking impact on μ for molecules of similar size. Rescaling the abscissa by definition of an "effective size" cannot superimpose the curves, indicating that the mobility is not interpretable through a universal size scaling. Finally, the mobility of a series of flexible chain molecules may not be monotonic in the radius of gyration, a result for which some experimental evidence exists (*37*).

Although the unentangled regime is perhaps the easiest to model, the fractionation is not as strongly dependent on molecular structure as when the chain lengths are greater; this conclusion is immediately apparent from Figures 3 and 4. Skipping discussion of the weakly entangled regime for the moment, the chain length dependence of μ in the strongly entangled regime can be interpreted in terms of the reptation theory (*39-41*). With reptation, the gel matrix is regarded as providing sufficient lateral confinement that the probe chain advances (or recedes) only by motion at its ends. The diffusion coefficient under reptation follows the well known behavior $D \sim N^{-2}$, so from equation 2 the mobility obeys $\mu \sim N^{-1}$ (*42,43*). Figure 8 shows that this chain length scaling is well obeyed; here we are using data for DNA in agarose, but the results for PSS in agarose or poly(acrylamide) are much the same. The predicted inverse chain length dependence for μ begins at N such that the gel mesh size is about $2R_g$; the mesh size has been obtained from literature correlations based on electron microscope images of sectioned gels (*3*). At large N, it becomes extremely difficult to obtain low E data, inasmuch as fields of the order 1 V/cm can significantly deform these chains and thereby cause μ to depend on E (*36, 44,45*). Measurements may take several days if E independent data are desired; previous workers have seen the same N scaling for μ of long DNA chains under these conditions (*46,47*).

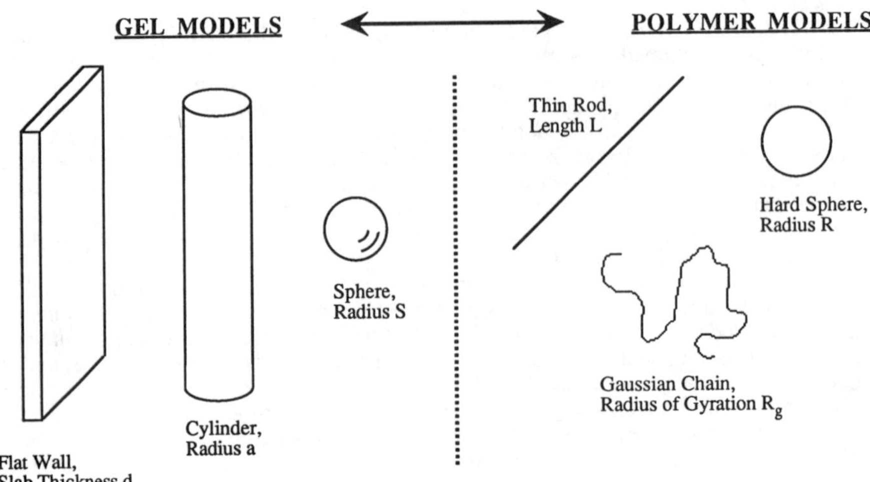

Figure 6. Gel and polymer models used to calculate the mobility for the unentangled transport regime.

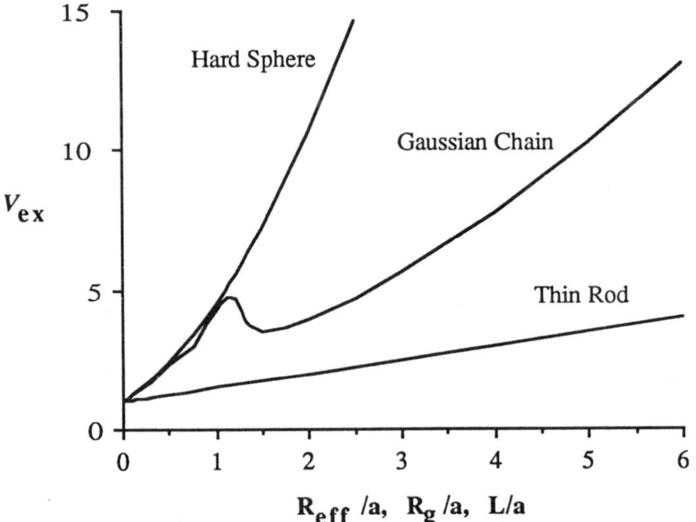

Figure 7. The mutual excluded volume V_{ex} between the different chain models presented in Figure 6 and the cylindrical gel site model. The excluded volume/cylindrical length has been made dimensionless with the cylinder cross-sectional area. R_{eff} is defined as $(\sqrt{\pi}/2)R$; this definition superimposes the Gaussian chain and hard sphere results at low V_{ex}. (Adapted from ref. 33.)

The N dependence of μ for PSS in an agarose gel, displayed in Figure 3, could be interpreted solely through a combination of the unentangled and strongly entangled transport regimes. Figure 4, which shows the analogous data for PSS in a poly(acrylamide) gel, clearly reveals an intermediate regime characterized by a stronger N dependence. Although the data on this figure are limited, the steepest region of Figure 4 corresponds to a power law relationship between μ and N with an exponent approaching -2.4, much greater than the terminal, reptation exponent of -1; use of a power law correlation is only for convenience since the slope is never constant across a wide range of N. This large intermediate exponent is incompatible with the two conventional transport mechanisms, excluded volume (unentangled) and reptation (strongly entangled), introduced earlier. We thus must consider a new mechanism, one operative over the approximate range $0.1<R_g/\xi<2$, where ξ is the gel's mean mesh spacing. Over this domain, the characteristic coil dimension is comparable to the pore spaces through which the probe molecule must past; intuitively, neither excluded volume nor reptation appears suitable to the description of this transport. Before discussing what this new transport mechanism might be, we will discuss an additional set of experiments in this entanglement regime that have greatly clarified our thoughts on this issue.

Electrophoresis in Gels, Dependence on Chain Topology. A major distinction between excluded volume controlled transport and reptation is that the former fractionates molecules based on *molecular size* while the later fractionates molecules based on *chain length*. To uncouple these parameters, the topology of the probe chain must be altered. We thus have conducted an additional series of experiments with star PSS and circular DNA. Figure 9 shows the chain length dependence of the mobility of flexible linear and circular DNA in two gels, one at 0.6% agarose and the second at 0.3% agarose (ionic strength I=0.03 M). Surprisingly, the chain topology does not appear to affect mobility in the more dilute gel, at least not within the N range studied. In the more concentrated gel, this N independence persists to some critical molecular weight, and then the mobilities for the two topologies diverge. To understand these trends, we will initially consider how the circular chain topology affects polymer conformation. Here, a simple Gaussian chain analysis will suffice; the ratio of the squared radius of gyration of a long linear polymer is exactly twice that of a circular chain containing the same number of segments. Incorporating excluded volume and electrostatic interactions more properly will alter this ratio, but not appreciably.

Applying a size perspective to the interpretation of the upper curve in Figure 9, we see that although the mobility has a sharp dependence on N, there is absolutely no dependence on molecular size. As a direct consequence of this finding, the relevant transport mechanism cannot be of the excluded volume type. At the same time, however, transport cannot be controlled by reptation since a circular molecule does not possess an "end." Next, examining linear and circular polymer data for the more concentrated gel, the divergence noted in the upper molecular weight region indicates that a circular molecule has a lower mobility than a linear molecule at equal N. Once again, this trend is inconsistent with excluded volume transport, as the smaller conformation is the one more retarded by the matrix. Perhaps a more useful comparison is to examine whether the linear polymer data across the high N regime agrees with the reptation prediction, $\mu \sim N^{-1}$. In fact, this check has already been made, with the divergent linear polymer data at high N from Figure 9 being the same data previously presented in Figure 8. As N increases, separation between ring and linear chains only appears as N becomes large enough for the onset of reptation for the linear chains. Figures 8 and 9 show that the onset of reptation with increasing N apparently occurs over a fairly narrow range of N and that a broad region of highly N dependent mobility exists for less entangled chains.

Although not discussed here, the corresponding data for star polymers shows the same trends (*48*); most importantly, the mobility is independent of topology across

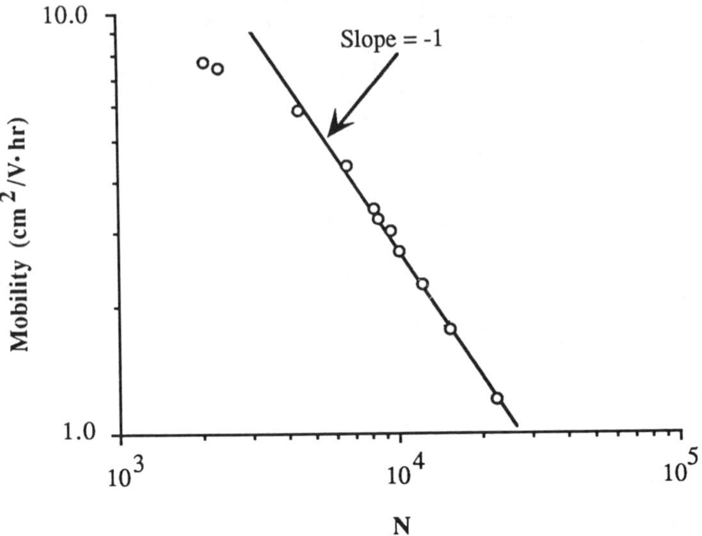

Figure 8. N dependence of DNA mobility, at large N, in a 0.6 wt. % agarose gel. The data are consistent with N scaling predicted by the reptation model.

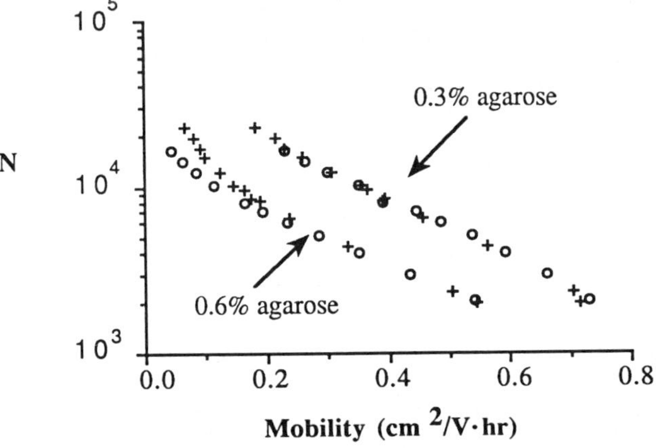

Figure 9. Comparison between the mobility of linear and circular DNA chains at two different agarose gel concentrations; the circular molecules have been made flexible by treatment with topoisomerase. The ionic strength is 0.03 M. (Adapted from ref. 48.)

a broad N range spanning $0.1<R_g/\xi<2$. At larger N, where linear polymer behavior is consistent with transport by reptation, linear chains possess greater μ than star polymers at equal N. Results were obtained for flexible PSS star polymers possessing from 3 to 12 arms ($3<f<12$) and N/f ($=N_{arm}$) varying by an order of magnitude ($540<N_{arm}<5000$).

Electrophoresis in Gels, Molecular Mechanism in the Weakly Entangled Regime. Transport in the two limiting entanglement environments appears well described by conventional theories, but a novel form of molecular transport must exist for intermediate entanglements where the polymer coil size approximately matches the gel mesh spacing. A successful description of this weakly entangled regime must account for the two preceeding observations: (i) the mobility is insensitive to chain topology and (ii) the mobility may depend on N more strongly than in either of the two bounding entanglement regimes. To our knowledge, only one model from the literature (*32,49,50*) can account for both (i) and (ii). This model, termed "entropic barriers" transport, was first formulated to explain the results of a molecular dynamics simulation of linear chain diffusion in random porous media. The basic idea of entropic barriers transport is that a real gel presents a highly inhomogeneous environment to a translating flexible molecule. To maximize configurational entropy and thereby minimize free energy, the molecule resides predominantly in the more open "voids" of the gel matrix. To achieve bulk motion, molecules must occasionally pass over "bottlenecks" that are comprised of the more highly confining regions of the matrix. At any given center-of-mass position, the chain possesses an entropy that can be calculated by counting all the equally accessible configurations. The gel thus appears to the chain as a three dimensional entropy or free energy surface; the chain occupies different locations on this surface according to the Boltzmann statistics dictated by the chain's local free energy. Under Brownian motion and the weak applied field, the chain will sporadically pass over entropic barriers, following a route marginally biased in the direction of the field; the rate of barrier passage will depend on the height and width of each barrier, which can be assessed by calculating the confinement entropy difference at each bottleneck position relative to the adjacent void. A crude depiction of this motion as well as the associated entropy and free energy functions are given in Figure 10.

Entropic barriers transport differs significantly from either of the alternative transport mechanisms discussed earlier. In contrast to the models rooted in molecular size, this new transport is regulated by confinement entropy. This result appears physically reasonable since the motion of a polymer through a highly confining network of bottlenecks ($R_g\sim\xi$) is principally governed by the configurational Brownian fluctuations that permit motion through small bottlenecks, not by the translational Brownian fluctuations that allow motion of the undistorted molecule through some fraction of sufficiently large bottlenecks. In essence, this model assumes the configurational entropy of a flexible chain dominates over the translational entropy when confinement is great. In the unentangled regime at lower confinement, the two entropies become indistinguishable. For rigid molecules, on the other hand, which possess only orientational and translational entropy, motion is dictated by the probability of proper molecular orientation to pass bottlenecks. Here, for a set of molecules of similar shape, motion is dictated by molecular size. Different molecular shapes can be directly compared via an effective molecular size parameter that is defined in a way dependent on the local structure of the matrix (*37*). There is no weakly entangled regime for rigid molecules.

Comparing entropic barriers motion to reptation, the key difference is that reptation assumes that a flexible chain molecule possesses the same configurational entropy irrespective of its location; in reptation there is no obvious thermodynamic driving force behind chain motion. Reptation can only be achieved in a heterogeneous gel when the probe chain is sufficiently large compared to fluctuations in the gel

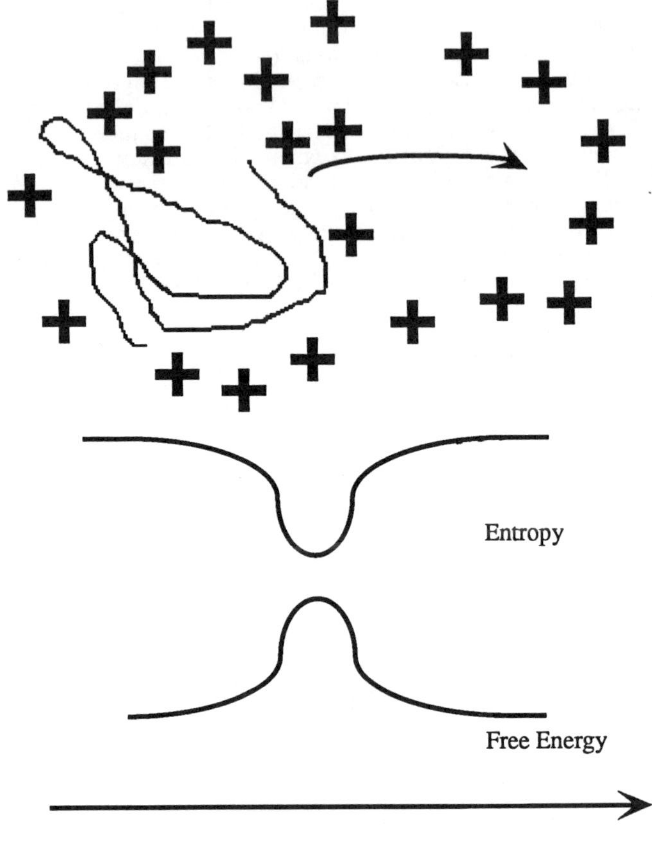

Figure 10. Pictorial representation of a typical void and bottleneck structure. For this configuration, the probe chain must be transported through the bottleneck before emerging at the second void. Corresponding changes in the confinement entropy and free energy for this process are shown below as functions of the probe's center of mass position.

environment. Using the experimental data presented here, this condition apparently occurs only when R_g exceeds about 2ξ.

Constructing a quantitative theory for entropic barriers transport appears problematical, inasmuch as the detailed features of both the probe chain and the matrix must be stipulated at the onset. For the present purposes, we need only explain how model predictions are consistent with the two striking experimental trends reported earlier for transport in the weakly entangled regime. The key variable is the probe confinement entropy, and as far as model predictions are concerned, this entropy must be known as a function of position within the gel. Since the model is rooted in thermodynamics, the specific mechanistic details of how a chain actually finds its way through constrictions is irrelevant. It would appear from the present data that in the strongly confined regime, such mechanistic details are indeed necessary to analyze chain motion.

To develop the model predictions, we employ a particularly simple geometrical model for a single constriction within the matrix; a real gel can then be regarded as an array of these constrictions. Following Figure 11, we analyze the motion of a flexible polymer through a narrow slit ($H \ll R_g$) of sufficient length to readily contain the entire probe chain ($G \gg R_g$). The open half spaces on each side of the slit represent open voids within the gel matrix. As shown by Guillot (*51*), the rate of passage of a flexible probe molecule through such a barrier is proportional to the local partition coefficient K_p, which can be written for a dilute solution

$$K_p = e^{-\Delta G_{con}/kT} = e^{\Delta S_{con}/k} \tag{4}$$

where ΔG_{con} and ΔS_{con}, the free energy and entropy of confinement of a single chain, represent the difference in these thermodynamic quantities between the slit and the free spaces on either side. In a real gel, the entire probe molecule will likely not fit entirely within most constrictions which will be too short to contain the whole chain. As shown earlier (*22*), the key thermodynamic quantities in this case are those associated with the transfer of all segments from a void into an adjacent bottleneck. Therefore, partial confinement within a finite length slit has no role in the application of equation 4 to the geometry of Figure 11.

The confinement entropies of linear and topologically complex flexible chains within a slit have been calculated by previous investigators (*38,52,53*), and this subject is extensively discussed from the standpoint of scaling theories in the text by de Gennes (*54*). Basically, a confined flexible molecule can be modeled as a necklace of blobs, each of diameter H and behaving as a hard sphere. Within a blob, the chain obeys the normal conformational statistics associated with a linear chain in free solution. Communication between blobs, however, is strongly suppressed because of confinement. Only when two blobs come into direct contact is there any communication. Thus, thermodynamic properties associated with chain confinement become linear functions of the number of blobs required to represent the chain structure. For a large enough chain, this number is linear in the degree of polymerization N. Stated more simply, when a trapped polymer is very large compared to the dimension of its local space, a small chain subsection of coil size H does not know how the chain is globally assembled. Although this argument follows a scaling approach, quantitative calculations which support this picture are readily performed for Gaussian chains in the slit geometry. Analytical calculations in more complex confinement geometries appear a formidable challenge.

Once the confinement entropy is recognized as an extensive property, i.e., linear in N, the apparently anomalous chain length and topological independence of the mobility are easily understood. With ΔS_{con} linear in N, the rate of passage through a single slit is exponential in N; it then follows that the mobility through a series of these slits, perhaps the simplest model for a gel, is also exponential in N. In a log-log format, this exponential dependence produces what appears to be N-dependent scaling exponents for μ, with the exponent increasing as N increases. The transition to reptation at large N, unfortunately, is not readily incorporated within this framework, and prediction of the terminal crossover remains an outstanding research question. In a more complex confinement geometry than the simple slit when multiple length scales are important in the description of matrix structure, the exact functional dependence of μ(N) may not be a simple exponential. The entropic barriers mechanism, however, should still be operative across some intermediate entanglement range where the gel possesses spatial heterogeneity at lengths scales comparable to the flexible probe size.

The same argument readily applies to the topological independence of μ; if the confinement entropy is explicitly linear in N, chain topology is irrelevant. Even after the chain length becomes large enough that reptation dominates for linear chains, the smaller star and circular conformations remain in the weakly entangled regime, with mobilities that remain lower than those of the equivalent linear chains. As their sizes increase further, however, transitions to new transport mechanisms are expected for topologically complex molecules (*55*); measurements corresponding to these new regimes have not yet been performed.

The N dependence of D in the weakly and strongly entangled regimes is expected to track that of μ, according to the relationship expressed in equation 2. This agreement supposes that intramolecular hydrodynamic interactions are strongly suppressed within the gel, causing both diffusion and electrophoresis phenomena to obey free-draining theories. Efficient screening of intermolecular hydrodynamic interactions by the surrounding matrix obviously requires that the matrix be of sufficient density, and the precise point of crossover to free-draining behavior with matrix density has yet to be calculated. Data has recently been reported for large apparent N exponents for D in probe chain/gel systems similar to the ones discussed here (*56-58*). These results suggest that at least some screening occurs prior to the onset of reptation. At this stage, the mobility data is somewhat more complete because of the ease of the electrophoresis experiment compared to diffusion experiments, particularly for large probe molecules. A variety of simulation techniques are also being developed to analyze diffusion and electrophoresis problems in different gel structures. One important simulation result is that the entropic barriers mechanism is absent when the gel matrix is periodic and possesses no heterogeneity at the length scale of the probe molecule (*32*). Experimental verification of this prediction waits upon the availability of a gel system with periodically placed constrictions.

Electrophoresis in Gels, Role of Chain Charge Density. Variations of the charge density within a chain population have been ignored to this point, and it has implicitly been assumed that μ is a function only of chain length N, at fixed ionic strength I and field strength E. Intuitively, one expects that if two chains possess equal length but different total charge Q, μ for the more highly charged chain would be greatest. If this suggestion were correct, interpretation of mobility data for synthetic samples would be difficult, inasmuch as synthetically produced polyelectrolytes inevitably exhibit poorly characterized, but often appreciable, polydispersity in charge density. Natural polymers, on the other hand, are either uniform in native charge density (DNA) or readily modified to produce a structure with uniform charge density (SDS-protein complexes).

To assess the significance of charge density variation, we have prepared a series of hydrolyzed poly(acrylamide)s from a single unhydrolyzed poly(acrylamide) parent of weight average molecular weight 720,000. Increasing levels of hydrolysis raise the density of carboxyl groups but leave the chain length unchanged; when the hydrolysis occurs under basic conditions, as was done here, the charge is believed to be uniformly distributed along the backbone. These samples remain water-soluble at all degrees of hydrolysis, but significant conformational rearrangements are probable as the charge density is increased, all other conditions held fixed. To prevent these conformational changes from hindering data interpretation, electrophoresis experiments are performed only in the weakly and strongly entangled regimes, where conformational variations are not likely to substantially perturb the mobility. Finally, to assure that carboxyl groups are equally ionized across the set of samples, the solvent for the electrophoresis experiment is adjusted to pH 7.5.

Figure 12 shows how the degree of hydrolysis affects the mobility in a 0.6% agarose gel with I = 0.03 M [tris(hydroxymethyl)aminomethane buffer]. At low charge densities, a strong dependence on charge density is apparent, but above 30-40% substitution, μ becomes independent of this variable. The practical consequences of these results are that the electrophoretic mobilities of highly charged polyelectrolytes - those with charge approaching magnitude e per repeat unit - are insensitive to small variations in the actual charge density. When the charge group substitution falls below 30-40%, on the other hand, uncoupling of the chain length and charge density effects in the mobility will be difficult. The PSS samples employed throughout experiments discussed in the previous sections are all characterized by sulfonation of at least 85% of the styrenic repeat units; the mobility appears completely insensitive to the slight variations in the sulfonation level (+/-10%) that have been detected by C^{13} NMR measurements. The trends revealed in Figure 12 are consistent with the phenomenon known as "counterion condensation" (*59,60*). A more complete discussion of experimental results, as well as theories for the functional dependence of μ on charge density, will be presented in a later publication. A quantitative experimental study of these effects in the context of the free solution mobility can be found in a contribution by Whitlock (*21*).

Conclusions

This study was initiated by a need to more completely understand electrophoretic transport mechanisms if enhancements to the quality and scope of the method were to be implemented for synthetic polymers. The most significant result is there are at least three distinct transport mechanisms at low field associated with the unentangled, weakly entangled, and strongly entangled regimes. We have speculated that each form of transport arises solely from *configurational* effects, with the appropriate transport mechanism dictated by the ratio of the polymer size to the gel's mesh spacing. Although no unequivocal proof can be offered here, there is no compelling reason to evaluate explicitly the *hydrodynamics* associated with this transport. The conventional models, such as excluded volume transport (the Ogston model) and reptation, also ignore or circumvent hydrodynamic analysis. In all cases, structure-sensitive separations are postulated to arise from configurational effects, and the hydrodynamics are lumped into a *segmental* friction coefficient which is inversely proportional to the mobility. Following equation 2, this friction coefficient can depend on the density of charge, the ionic strength, and (probably) the cross-sectional dimension of the chain backbone.

This neglect of hydrodynamic detail is supported by the present data, with the free solution mobility independent of chain length and the mobility across the weakly entangled regime independent of topology. It is difficult to envisage a realistic, solely hydrodynamic model which would display these behaviors because correspondence with the data would then imply that hydrodynamic interaction of the chain with the matrix sites must correlate with N independent of chain structure, which appears physically implausible. Although weak probe/matrix hydrodynamic interactions must occur, especially in more dilute gels, we believe these interactions are too weak to be isolated from the strong, approximately exponential N dependence that arises from configurational effects that dominate across the unentangled and weakly entangled regimes. In the strongly entangled regime, we are certain that the matrix totally screens long range intramolecular hydrodynamic interactions.

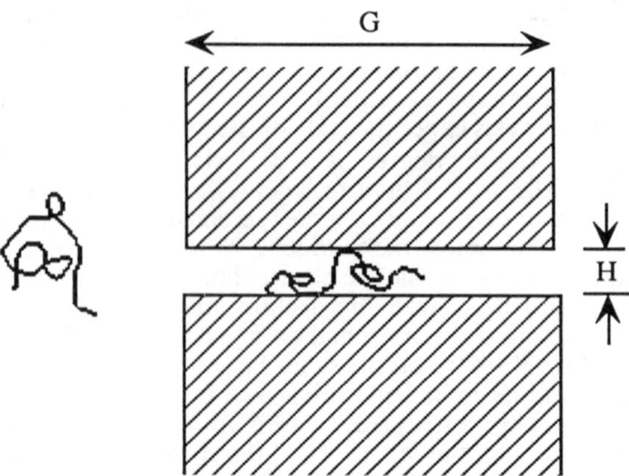

Figure 11. Simple geometrical model for a bottleneck connecting two large voids. The channel is much longer and wider than the chain, but its height is much less.

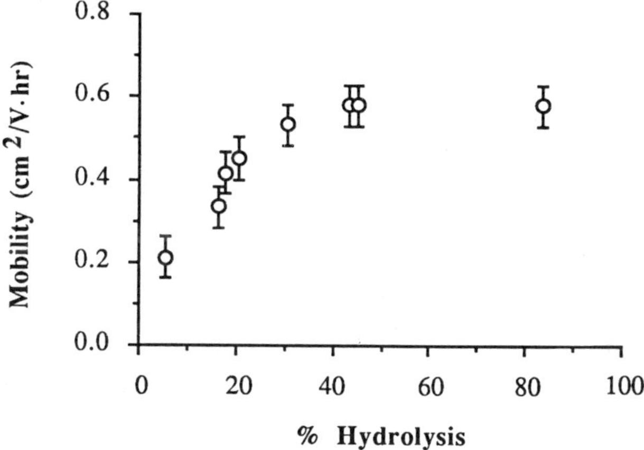

Figure 12. Mobility of hydrolyzed poly(acylamide) as a function of the degree of base-catalyzed hydrolysis for a 0.6 wt. % agarose gel at I = 0.03 M.

Electrophoretic measurements are rather simple to perform, and all the measurements reported here were conducted on equipment that can be assembled for less than $2,000. Given the connection between diffusion coefficient and mobility, strong incentives exist to study probe diffusion using electrophoretic methods. The connection between diffusion coefficient and mobility, however, will be strengthened when duplicate measurements are performed on a single probe/gel system. Such measurements will better pinpoint when the matrix is fully capable of screening intrachain hydrodynamic interactions in the diffusion experiment. Our speculation is that the screening becomes highly efficient even as the matrix density increases within the unentangled regime. Thus, the diffusion coefficient and mobility information are presumed to reflect the same physics. If this speculation is incorrect, still another regime transition is expected as the gel's mesh size is decreased.

Provided that both a sharp molecular weight fractionation is established and that the mobility is insensitive to charge density, gel electrophoresis can readily be employed for empirical molecular weight measurement of unknown samples using analytical standards for calibration. In fact, within our own laboratory we routinely measure poly(styrene) molecular weights by performing electrophoresis on the sulfonated analogs. In this way, we can sometimes measure polydispersities for anionically polymerized samples at the level 1.01 +/- 0.005; the same separation quality extends from the smallest to the highest molecular weight PSS samples examined ($2x10^3$ < molecular weight < $2x10^7$). The resolution is better than with SEC, and the experimental procedures are more easily conducted. Applications to additional polyelectrolyte species await the rather straightforward development of more general band staining procedures and a better understanding of the polymer/gel binding problem. Also, there is not yet an adequate theory for predicting the segmental friction coefficient associated with electrophoresis, so the calibration standards must be of the same polymer species as the unknown. Although not presented here, the list of synthetic polymer types we have examined, sometimes only qualitatively, now contains 10 entries. Even in the absence of appropriate standards, useful general information is normally discernable by evaluating the relative locations of stained bands by eye. Most of our current efforts are directed toward development of improved gel matrices to optimize the separation quality and further expand this list.

Acknowledgments

This work has been supported on a sustained basis by the National Science Foundation through the University of Massachusetts Materials Research Laboratory.

Literature Cited

1. Andrews, A. T. *Electrophoresis: Theory, Techniques, and Biochemical and Clinical Applications*; Oxford University Press: New York, 1986.
2. Schwartz, D.C.; Cantor, C.R. *Cell* **1984**, *37*, 67.
3. Attwood, T. K.; Nelmes, B. J., Sellen, D. B. *Biopolymers* **1988**, *27*, 201.
4. Serwer, P.; Hayes, S. J. *Anal. Biochem.* **1986**, *158*, 72.
5. Ruchel, R.; Brager, M. D. *Anal. Biochem.* **1975**, *68*, 415.
6. Giddings, J. C. In *Advances in Chromatography* ; Giddings, J. C.; Grushka, E.; Cazes, J.; Brown, P. R., Eds.; Marcel Dekker: Washington, DC,1982; Vol 20, pp 217-254.
7. Muller, G.; Yonnet, C. *Makromol. Chem. Rapid Commun.* **1984**, *5*,197.
8. Fawcett, J. S.; Morris, C. J. O. R. *Separation Sci.* **1966**, *1*, 9.
9. Dunker, A. K.; Rueckert, R. R. *J. Biophys. Chem.* **1969**, *244*, 5074.
10. Chrambach, A.; Rodbard, D. *Science* **1971**, *172*, 440.
11. Stellwagen, N. C. *Biopolymers* **1985**, *24*, 2243.
12. Edmondson, S. P.; Gray, D. M. *Biopolymers* **1984**, *23*, 2725.
13. Hervet, H.; Bean, C. P. *Biopolymers* **1987**, *26*, 727.
14. Chen, J.L.; Morawetz, H. *Macromolecules* **1982**, *15*,1185.
15. Smisek, D. L.; Hoagland, D. A. *Macromolecules* **1989**, *22*, 2270.
16. Vink, H. *Macromol. Chem.* **1981**, *182*, 279.
17. Olivera, B. M.; Baine, P.; Davidson, N. *Biopolymers* **1964**, *2*, 245.
18. Nagasawa, M.; Noda, I.; Takahashi, T.; Shimamoto, N. *J. Phys. Chem.* **1972**, *76*, 2286.
19. Noda, I.; Nagasawa, M.; Ota, M. *J. Am. Chem. Soc.* **1964**, *86*, 5075.
20. Meullenet, J. P.; Schmitt, A.; Drifford, M. *J. Phys. Chem.* **1979**, *83*, 1924.
21. Whitlock, L. R. In *New Directions in Electrophoretic Methods*; Jorgenson, J. W.; Phillips, M.; Eds.; ACS Symposium series 335; American Chemical Society: Washington, DC, 1987.
22. Manning, G. S. *J. Phys. Chem.* **1981**, *85*, 1506.

23. Russel, W.B.; Saville, D.A.; Schowalter, W.R. *Colloidal Dispersions*; Cambridge Univ. Press; New York, 1989.
24. Henry, D. C. *Proc. Roy Soc.* **1931**, *A133*, 106.
25. Hermans, J. J.; Fujita, H. *Proc. K. Ned. Akad. Wet., Ser. B:Phys. Sci.* **1955**, *B58*, 182.
26. Hermans, J. J. *J. Polym. Sci.* **1955**, *18*, 527.
27. Overbeek, J. Th. G.; Stigter, D. *Recl. Trav. Chim. Pays-Bas* **1956**, *75*, 543.
28. Stigter, D. *J. Phys. Chem.* **1978**, *82*, 1417.
29. Stigter, D. *J. Phys. Chem.* **1978**, *82*, 1424.
30. Sherwood, J.D. *J. Chem. Soc., Faraday Trans. 2* **1982**, *78*, 1091.
31. Wang, L.; Bloomfield, V. A. *Macromolecules* **1990**, *23*, 194.
32. Muthukumar, M.; Baumgartner, A. *Macromolecules* **1989**, *22*, 1937.
33. Arvanitidou, E.; Hoagland, D. A.; Smisek, D. L. *Biopolymers* **1991**, *31*, 435.
34. Rodbard, D.; Chrambach, A. *Proc. Natl. Acad. Sci. U.S.A.* **1970**, *65*, 970.
35. Rodbard, D.; Chrambach, A. *Anal. Biochem.* **1971**, *40*, 95.
36. Slater, G. W.; Rousseau, J.; Noolandi, J.; Turmel, C.; Lalande, M. *Biopolymers* **1988**, *27*, 509.
37. Giddings, J. C.; Kucera, E.; Russell, C. P.; Myers, M. N. *J. Phys. Chem.* **1964**, *72*, 4397.
38. Cassasa, E. F. *Polymer Lett.* **1967**, *5*, 773.
39. De Gennes, P. G. *J. Chem. Phys.* **1971**, *55*, 572.
40. De Gennes, P. G. *J. Chem. Phys.* **1980**, *72*, 4756.
41. Doi, M.; Edwards, S.F. *J. Chem. Soc. Farad. Trans. 2* **1979**, *74*, 1789.
42. Lerman, L.S.; Frisch, H. L. *Biopolymers* **1982**, *21*, 995.
43. Lumpkin, O.J.; Zimm, B.H. *Biopolymers* **1982**, *21*, 2315.
44. Lumpkin, O.J.; Dejardin, P.; Zimm, B.H. *Biopolymers* **1985**, *24*, 1573.
45. Slater, G.W.; Noolandi, J. *Biopolymers* **1986**, *25*, 431.
46. Stellwagen, N.C. *Biopolymers* **1985**, *24*, 2243.
47. Hervet, H.; Bean, C.P. *Biopolymers* **1987**, *26*, 727.
48. Smisek, D. L.; Hoagland, D. A. *Science* **1990**, *248*, 1221.
49. Baumgartner, A.; Muthukumar, M. *J. Chem. Phys.* **1987**, *87*, 3082.
50. Muthukumar, M.; Baumgartner, A. *Macromolecules* **1989**, *22*,1941.
51. Guillot, G.; Leger, L.; Rondelez, F. *Macromolecules* **1985**, *18*, 2531.
52. Casassa, E. F.; Tagami, Y. *Macromolecules* **1969**, *2*, 14.
53. Halperin, A.; Alexander, S. *Macromolecules* **1987**, *20*, 1146.
54. De Gennes, P. G. *Scaling Concepts in Polymer Physics*; Cornell University Press: Ithaca, New York, 1979.
55. Rubinstein, M. *Phys. Rev. Lett.* **1986**, *57*, 3023.
56. Wheeler, L. M.; Lodge, T. P. *Macromolecules* **1989**, *22*, 3399.
57. Kim, H.; Chang, J. M.; Yohanan, M.; Wang, L.; Yu, H. *Macromolecules* **1986**, *19*, 2737.
58. Nemoto, N.; Kishine, M.; Inoue, T.; Osaki, K. *Macromolecules* **1990**, *23*, 659.
59. Manning, G. S. *J. Chem. Phys.* **1969**, *51*, 924.
60. Manning, G. S. *Ann. Rev. Phys. Chem.* **1972**, *23*, 117.

RECEIVED August 19, 1991

Chapter 13

Moving Ion Exchange Fronts in Polyelectrolyte Gels

Stevin H. Gehrke[1], Gaurav Agrawal[2,3], and Ming-Chien Yang[1,4]

[1]Department of Chemical Engineering, University of Cincinnati, Cincinnati, OH 45221–0171
[2]Department of Materials Science and Engineering, University of Cincinnati, Cincinnati, OH 45221–0012

The classic shrinking core model can be used to correlate the movement of a sharp front that develops when poly(acrylamide-*co*-sodium methacrylate) gels equilibrated at pH 4 are immersed in 0.010 N aqueous sodium hydroxide. At this front, observable as a sharp refractive index gradient, the gel is converted from the nonionic hydrogen form to the ionized sodium form as the hydroxide reacts with the carboxylic acid proton. The boundary film mass transfer coefficients and the diffusion coefficients obtained from a least squares fit of the model to the data are physically reasonable. However, the diffusion coefficients obtained are significantly less than independently estimated values.

In many applications of polyelectrolyte gels, the gels are converted from one ionic form to another during use; in other words, ion exchange occurs. The rate at which this exchange occurs is often critical to the success of the application. In most cases, the mechanism of ion exchange must be inferred by measurement of changes in the bulk solution surrounding the gel. However, an optically sharp front can develop within the polyelectrolyte gel. Ion exchange occurs at this front, so measurement of the rate at which the front advances toward the center of the gel allows a means of studying ion exchange kinetics by observing changes within the gel itself rather than in the solution.

In this chapter, applications for polyelectrolyte gels are briefly reviewed with an emphasis upon gels which respond to changes in their environment. The development of moving ion exchange fronts within gels is explained and past work on the measurement and modeling of this phenomenon in ion exchange resins is reviewed. Then a shrinking core model is developed and applied to a pH-responsive polyelectrolyte gel. The success of this model in describing the movement of ion exchange fronts is evaluated.

[3]Current address: Department of Materials Science and Engineering, University of Illinois at Urbana–Champaign, Urbana, IL 61801
[4]Current address: Fiber and Polymer Engineering Program, National Taiwan Institute of Technology, 43 Keelung Road, Section 4, Taipei, Taiwan, 10772, Republic of China

0097–6156/92/0480–0211$07.75/0
© 1992 American Chemical Society

Applications of Ion Exchange Resins and Polyelectrolyte Gels

The difference between a polyelectrolyte gel and an ion exchange resin is primarily a difference in terminology. However, ion exchange resins generally have high exchange capacities and are rather rigid, polymer-dense materials whose volumes do not change significantly as a result of the ion exchange. In contrast, polyelectrolyte gels may have rather low concentrations of ionogenic (ionizable) groups and are usually highly compressible materials that swell and shrink substantially during ion exchange.

Ion exchange resins have a long history of diverse technological application, especially for water treatment and purification (*1*). These resins can be effectively used to recover dilute metal ions from solution as well as separate complex mixtures of high molecular weight species like proteins (*2*). Ion exchange resins can also be loaded with doses of ionic drugs; taken internally, the resins release the drugs in a controlled fashion (*3*). The most common ion exchange resins are derivatized styrene-divinylbenzene copolymers; however, ion exchange resins are made from many different polymers.

Covalently crosslinked polyacrylamide copolymer gels are the subjects of this report. Polyacrylamide gels are important chromatographic supports, especially for gel filtration chromatography, and are also the separation media most commonly used for gel electrophoresis. The inclusion of ionizable comonomers like acrylic acid or sodium methacrylate in the gel formulation makes the gel a polyelectrolyte; hydrolysis of polyacrylamide gel in base will also create a polyelectrolyte gel. Ionizable polyacrylamide gels have been widely studied due to the large changes in their swelling degrees as a function of solution composition. These gels are highly swollen at high pH values due to the osmotic swelling pressure of the counterions associated with the ionized acidic groups, but have much reduced swelling degrees at low pH values where the acidic groups are in the unionized hydrogen form. This change in swollen volume from a highly swollen state to a much less swollen state has been shown to occur abruptly with infinitesimal changes in acetone concentration in solution or electric field. Such a volume transition has been identified as a phase transition.

Applications of Responsive Polyelectrolyte Gels. Polymer gels whose properties, most notably their solvent-swollen volumes, respond to specific environmental stimuli have been termed "actuated," "stimuli-sensitive" and "responsive" gels. Most of these gels are polyelectrolytes. The stimuli which have been demonstrated to induce these changes in the gel properties are startlingly diverse, and include pH, ionic strength, electric field, solvent composition, light intensity, pressure and temperature. Perhaps no other class of materials can be made to respond to so many different signals. The properties which can change reversibly include:

- swelling degree (ratio of swollen to dry mass) - from less than 1.2 to over 100;
- swelling/shrinking kinetics - time constants less than 1s to over 10^5s;
- solute permeability - impermeable to highly permeable;
- clarity - completely opaque to completely transparent;
- mechanical properties - stiff to flexible.

These changes can occur gradually over a broad range of stimulus values or suddenly and discontinuously at a specific stimulus level.

The unique properties of responsive gels have resulted in substantial applications research, especially since the early 1980's. However, research in the field of responsive polymer gels was initiated in 1949 by Katchalsky (*4*). He investigated aqueous solutions of polymeric acids and found that the chains could be made to reversibly stretch and contract with a change in the degree of ionization of the polymer. These polymers were then crosslinked to form a three-dimensional network or gel which reversibly swelled and deswelled in response to pH. He and his coworkers designed a number of elegant machines which were driven by changes in pH and ionic strength (*5,6*).

The responsive gel applications being developed today can be loosely grouped as mechanical devices, controlled solute delivery devices, and chemical separation mechanisms. As mechanical devices, Suzuki and Tatara have both determined that the generated power density of gel actuators is on the order of 0.1 W/cm^3, about that of skeletal muscles (*7,8*). Urry *et al.* measured a responsive gel made of an ionizable polypentapeptide to pick up weights 1000 times its dry weight (*9*). Shiga *et al.* recently published an interesting demonstration of the potential for responsive gels, showing that an electrically responsive polyelectrolyte gel can lift and carry a quail's egg (*10*). De Rossi and coworkers have mathematically modeled the dimensional changes of electrically responsive polyelectrolyte gels (*11-13*). Osada *et al.* described the use of polyelectrolyte gels as switching devices based on the repetitive oscillations of electric current in the gels when a constant electrical potential was applied (*14,15*). In addition, since a similar phenomenon is observed in the electric response of human nerves, these gels might be used as artificial nerves.

However, the most active area of application development for responsive gels is in the area of controlled delivery of solutes, especially of drugs to the body. The possibilities in this area have recently been reviewed by Gehrke and Lee, Hoffman, and Urry (*16-18*). These devices can function in several different ways. In response to an appropriate stimulus, gels can be used to expand like a piston to pump drug from a reservoir (*19*), contract around a reservoir, squeezing solution from it (*17,18*), or alter the permeability of a gel membrane, turning drug release on and off (*17,20*). Many different types of systems have been proposed to do these things; see, for example, the work by the research groups of Hoffman, Horbett, Kim, Ratner or Siegel (*21-32*).

The controllable permeability variations of responsive gels has also been used to achieve a variety of size- or charge-selective separations. For example, Gehrke *et al.* have used pH-sensitive gels to concentrate solutions of macromolecules and ionic solutes, and temperature sensitive gels to dewater coal slurries (*33,34*). This process, originally proposed by Cussler *et al.* (*35*), has also been used by other groups for dewatering various sludges and slurries (*36*). Osada's (*37-39*) group has developed chemical valves to control the permeability of water and macromolecules through electrically responsive membranes, as has Grodzinsky's group (*40-41*).

Most of the applications proposed for responsive hydrogels depend critically upon the kinetics of the volume change. The rate limiting steps for the kinetics could be either the stimulus rate itself or the rate of network response. Gehrke and Cussler showed that swelling kinetics are often limited by ion exchange rates, but that shrinking tends to be dominated by the network response rate (*42*). Grimshaw *et al.* have developed an elegant model for treating the kinetics of electrically and pH-responsive gels, which quantified such observations. Network response-limited volume change can be characterized using either the equations of motion (*43*) or Fick's law (*42,44*), except when the phase transition is crossed, where certain regular deviations from these theories are observed. According to Akhtar and Gehrke, the rate of network response depends primarily upon the concentration of polymer in the network; this dependence has been interpreted using scaling concepts developed originally for polymer solutions (*44,45*). However, research groups have recently reported that gels of a heterogeneous, microporous structure may shrink and swell in response to changes in temperature roughly 1000 times faster than did previously reported homogeneous gels (*46-49*). It appears that an interconnected porous structure which allows water to be expelled from the pores by a convective process rather than a diffusive one results in this enhanced response rate. The response rate of polyelectrolyte gels made by these techniques would likely be limited by ion exchange kinetics.

Kinetics of Ion Exchange in Polyelectrolyte Gels

Since the ion exchange reactions are very fast, the rate limiting steps of ion exchange are generally either mass transfer of ions through the boundary film surrounding the gel particle or diffusion of the ions within the gel particle. Intraparticle diffusion tends to be rate-limiting for gels with low degrees of swelling and low capacity contacted with well-stirred, concentrated solutions. In contrast, film diffusion tends to be rate-limiting for small diameter particles of highly swollen, high capacity gels contacted with unstirred, dilute solutions. The polyelectrolyte gels studied here are highly swollen, of low capacity, large size, contacted with unstirred solutions of intermediate concentration. Such conditions are likely to be representative of a number of polyelectrolyte gel drug delivery systems. In oral drug delivery using polyelectrolyte gels, for example, stirring is likely to be minimal so that significant film-limited effects may arise. Such conditions may also persist in systems where polyelectrolyte gels are expected to function as mechano-chemical actuators, swelling and shrinking in response to small changes in their environmental pH. Thus, both film and particle diffusion are expected to influence the rate of ion exchange in a variety of polyelectrolyte gel systems.

The type and concentration of fixed ionogenic groups are generally the most important properties of polyelectrolyte gels, regardless of application. For ion exchange resins, these groups determine the nature and number of ions captured and also the types of ions excluded from the gel by the Donnan effect *(1)*. For ion-exchange drug delivery systems, the groups determine the drug dosage and release conditions *(3)*. For pH-sensitive gels, the groups determine the shape of the swelling degree *vs.* pH curve. For all of the different applications of polyelectrolyte gels, the rate at which ion exchange occurs will influence and sometimes determine the effectiveness of the gels. Most of the kinetic studies on ion exchange systems focus on the rate of ion uptake or release measured in terms of concentration changes in the external solution, since it is difficult to determine the concentration profiles within the gel as a function of time. However, for certain types of ion exchange resins or gels, a sharp front develops that can be observed to move toward the center of the gel as the ion exchange process takes place. This boundary indicates a region of a sudden concentration change, as it separates an inner core containing the counterion (that is, the ion with a charge opposite of the fixed gel groups) present within the gel at time zero, from an outer shell containing the exchanged ion which originated from the solution. Thus, the measurement and modeling of the rate of movement of these fronts provides a means of testing ion exchange kinetic theory that focuses upon the interior of the gel rather than upon changes in the bulk solution. In the next section the reasons for the development of moving fronts in certain ion exchange systems are examined qualitatively; later, a mathematical model is developed to describe their motion. This chapter concludes with experimental verification of the model for responsive polyelectrolyte gels, in which such fronts had not been previously observed or modeled.

Development of Moving Ion Exchange Fronts. Conventional ion exchange processes simply involve the displacement of one counterion associated with an ionogenic group in the gel with a different counterion. A common example is the exchange of Na^{+} and Ca^{++} in water softeners. However, some ion exchange processes involve the consumption of counterions in a chemical reaction. Helfferich has shown that such reactive ion exchange systems behave quite differently from systems in which all counterions maintain their identity *(50-52)*.

There are four basic ion exchange reaction types, with examples of each given in Figure 1:

1) Type 0 is the conventional ion exchange process, where all ions maintain their identities - the counterion originally in the solution simply trades places with the counterion originally on the gel. There are two possible ion exchange rate-limiting steps in this system: diffusion of the ions within the gel ("particle diffusion") and diffusion of the ions across the boundary layer of unstirred solution surrounding the gel ("film diffusion").
2) The Type I ion exchange reaction consumes a freely diffusible counterion from within the ion exchange gel. For example, this situation occurs when a gel with strongly acidic ionogenic groups is immersed in a basic solution. In such strong-acid gels, the protons are not strongly associated with the acidic groups and thus the gel is ionized even before the protons are consumed upon reaction with the base. As in ordinary Type 0 ion exchange, this exchange can be either film or particle diffusion controlled.
3) Type II differs from ordinary ion exchange (Type 0) in that the H^+ counterion from the solution associates strongly with the ionic gel group, neutralizing its charge. This is the case when weakly acidic gels in the ionized form (e.g., sodium form) are immersed in an acid bath. For pH-sensitive responsive gels, Type II exchange corresponds to the shrinking process. Usually only particle diffusion is expected for Type II ion exchange systems. A moving boundary can develop in this system, as described below.
4) Type III ion exchange consumes the counterion released from the gel, as in Type I, but differs in that the counterions initially associated with the gel are bound with the gel's ionic groups. Thus the gel is virtually uncharged prior to the ion exchange reaction, unlike the initially ionized Type I systems. This behavior is seen for initially unionized (hydrogen form), weakly acidic gels immersed in a basic solution. For pH-sensitive responsive gels, Type III exchange corresponds to the swelling process. Type III shows different mechanisms of particle diffusion, depending upon the solution composition. At high counterion concentrations a moving boundary can develop. Film diffusion usually is not rate-limiting in Type III ion exchange systems, but it would be similar to Type I film diffusion.

Let us examine the reasons for the development of moving fronts in Type II ion exchange systems and at high counterion concentrations in Type III systems. For weakly acidic polyelectrolyte gels, Type II ion exchange occurs when the highly-swollen ionized (e.g., sodium form) gel is immersed in acid, neutralizing the gel and causing it to shrink. However, the reverse of this process, immersion of the unionized hydrogen from of the gel in a basic solution, is Type III ion exchange.

The process involved in Type II ion exchange is shown in Figure 2. As H^+ enters the gel, it quickly associates with the weak-acid (e.g., carboxylic) groups, eliminating their charge. This association results in the development of a nonionic shell surrounding an ionized core. Development of the nonionic shell helps speed up the process, since H^+ and Cl^- (or other anion) can diffuse into the nonionic shell together and up to the reaction front. In contrast, in Type 0 ion exchange, the coion (here, Cl^-) is largely excluded from the gel due to the Donnan potential, so counterions can enter the particle only as quickly as others leave. Despite this increased rate of intraparticle diffusion, rate limitation by the boundary film is unlikely, simply because the diffusion of the acid in the film will still be faster than diffusion of the acid in the gel (in a sense, the unionized shell is an extension of the boundary film). Most particle diffusion processes are independent of solution concentration since the ion concentration at the gel surface is equal to the equilibrium ion capacity, regardless of the solution concentration. Then the driving force for the flux within the gel - the concentration difference between the surface and the center of the gel - is nearly independent of the solution concentration. However, for the Type II system the front concentration is

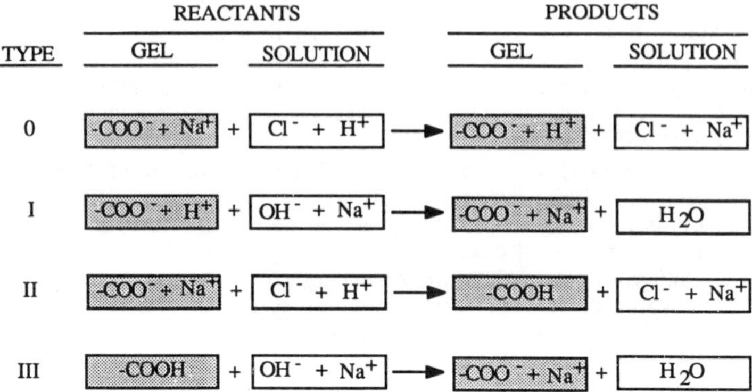

Figure 1. Four basic types of ion exchange reactions.

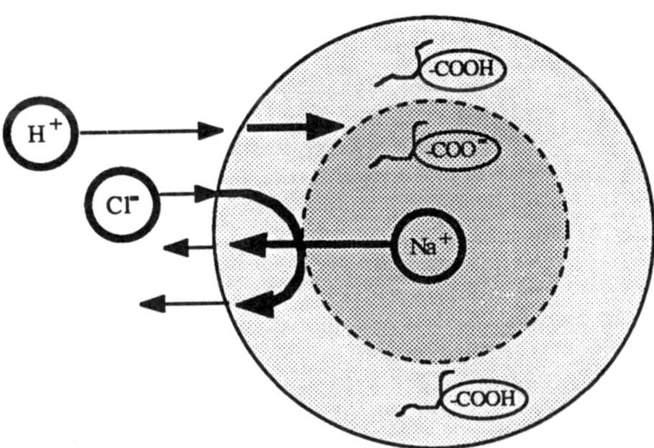

Figure 2. The Type II ion exchange reaction, as observed during the shrinking of pH-sensitive gels. Particle diffusion is the rate-controlling step shown; the ions and steps which control the rates are indicated in bold-face.(Adapted from ref. 51).

always about zero while the surface concentration is about the solution concentration. Thus, the particle diffusion rate will be proportional to the solution concentration. Since three mobile ions are involved in the ion exchange process, theoretical prediction of the exchange rate would be complex *(50,51)*. However, the progression of the moving front could be analyzed with the shrinking core model developed later in this chapter.

Type III ion exchange occurs when a neutral (hydrogen form) gel is ionized by an invading solute. As illustrated in Figures 3a, 3b and 3c, Type III ion exchange has three different potential rate-controlling steps: film diffusion plus two different particle diffusion steps which may develop depending upon the solution concentration. Figure 3a illustrates the low concentration particle diffusion process. Initially, a base like NaOH can diffuse freely into the nonionic gel, but as it ionizes groups on the gel, a Donnan potential quickly builds up, inhibiting any further diffusion of NaOH into the gel. Exchange then becomes quite slow, as it is limited by the rate at which neutral acid groups dissociate, freeing H^+ ions to exchange with Na^+. The protons do not react with the hydroxides until the protons reach the solution. A moving boundary will not develop in such a system.

However, a different exchange process is observed if the concentration of NaOH in solution is high (or if the solution has a high ionic strength). Under these conditions the Donnan potential is greatly reduced, so NaOH can readily diffuse into the gel, as shown in Figure 3b. Because the hydroxide will quickly react with the protons associated with the weak-acid groups, a moving boundary develops, separating an ionized shell from the neutral core. This problem is simpler than that of the Type II problem since there are no mobile ions in the core and because the hydroxide ions are consumed at the front. Because the OH^- concentration is much less than the Na^+ concentration in the shell, hydroxide diffusion should be the dominant variable controlling the exchange rate (*50, 51*).

Helfferich solved this problem under simplified conditions, obtaining the usual relations for a diffusion-controlled process: t/r^2 scaling and short-time $\sqrt{t}$ dependence [50]. His analysis also showed that the rate should be directly related to the solution concentration, not independent of it, as for ordinary ion exchange. He estimated that the high concentration mechanism is operative above solution concentrations of 10^{-2} M, while the low concentration mechanism appears below 10^{-3} M. These predictions are tested later in this chapter.

Because Helfferich expected both of these processes to be slow, he did not believe that film diffusion would be important for Type III ion exchange. However, Gehrke and Cussler found that the low concentration process could be significantly influenced by film limitations *(42)*. The importance of film resistance in the high concentration region is examined later in the results section of this chapter. In Type III film diffusion, H^+ from the gel diffuses to the solution/ gel interface where it meets the sodium hydroxide which has diffused through the film to the interface, as shown in Figure 3c. The H^+ and OH^- immediately react at the surface and the sodium ion diffuses into the gel to replace the hydrogen ion. Thus, the rate controlling step for film diffusion in Type III ion exchange is the diffusion of Na^+ and OH^- together through the film, instead of the interdiffusion of the two counterions through the film in opposite directions as in conventional Type 0 ion exchange.

Observations and Modeling of Moving Ion Exchange Fronts. The moving ion exchange fronts observed in polyelectrolyte gels are reminiscent of the shrinking core models used to analyze fluid-solid reactions in chemical reaction engineering. Several authors have adapted this approach to apply to a number of ion exchange systems (*53-63*). The systems that have been successfully treated with this type of model include:

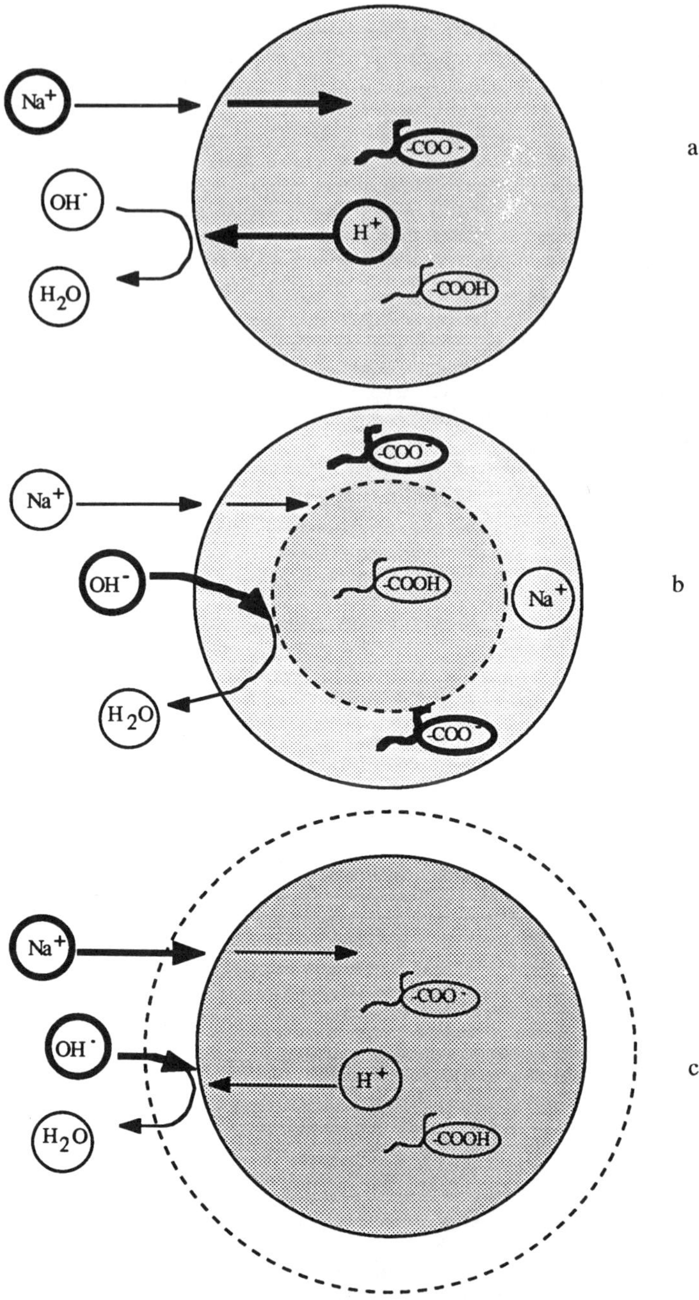

Figure 3. The Type III ion exchange reaction as observed during the swelling of pH-sensitive gels. The rate-controlling step shown is (a) the particle diffusion mechanism that occurs at low NaOH concentrations, (b) the particle diffusion mechanism at high NaOH concentrations, and (c) film diffusion. Ions and steps that control the rates are indicated in boldface. (Adapted from ref. 51.)

1) Acid elution of cupric ammine complex from a strong-acid cation-exchange resin (Dowex 50WX8 resin, a sulfonated copolymer of styrene and 8% divinylbenzene) (*53,54*).
2) Adsorption of bovine serum albumin by a weak-base anion exchange resin (Sephadex A50, a crosslinked dextran gel functionalized with tertiary amine groups) (*55*).
3) Various metal ion and proton exchange systems in weak (carboxylic) acid cation exchangers (Dowex CCR-2 (*56*), Amberlite IRC resins (*57-59*), Lewatit CNP resins (*58-59*).
4) Uptake of aqueous acid and base by γ-alumina pellets (*62*).
5) Various weakly basic anion exchangers (*63*).

In the literature diffusion limited theories and fronts have received the most attention, since ion exchange processes most commonly operate under these conditions. For example, Selim and Seagrave solved the unsteady state version of the moving boundary problem in the absence of film limitations (*64,65*). Other researchers have developed shrinking core models similar to the one developed in the next section, but under conditions where the boundary film could be neglected (*56-59*). Their experimental work was therefore carried out under conditions where particle diffusion was rate controlling. Pinto and Graham incorporated film resistance by estimating the film thickness and assuming diffusive transport through it (*55*). Dana and Wheelock developed a shrinking core model which included both an intraparticle diffusion coefficient and a film mass transfer coefficient (*53*). These researchers also carried out experiments which tested the theory under a variety of mass-transfer limiting conditions; specifically, the acid elution of cupric ammine complex from a strong-acid cation-exchange resin.

These researchers were able to correlate their observed moving boundary problems well with their theories. These theories are quite similar in nature; their primary differences lie in the treatment of the various dissociation equilibria. For example, the heuristic arguments of Helfferich outlined in the previous section neglect the diffusion of H^+ in basic solutions and the diffusion of OH^- in acidic solutions (*50*). Nativ *et al.* assume that the exchange reactions are irreversible (*57*). Höll and Sontheimer consider the effect of weak acid dissociation equilibria on the diffusion of hydrogen ions without considering the dissociation reaction of water (*58*). Kataoka and Yoshida also neglect water dissociation while using the general Nernst-Planck diffusion equations (*56*). Lewnard *et al.* have developed the most thorough model of the moving boundary ion exchange process, considering both film and particle diffusion plus the coupling of the acid and base fluxes through electroneutrality and dissociation equilibrium constraints (*62*). Their formulation provides a generally valid description of H^+ and OH^- diffusion within polyelectrolytes at all values of pH.

Mathematical Modeling of Moving Ion Exchange Fronts

The system modeled in this chapter is that of a carboxylic group-containing gel in the undissociated hydrogen form immersed in a solution of sodium hydroxide at a sufficient pH that the moving front postulated by Helfferich develops (high concentration Type III ion exchange) (*50-51*). The ion exchange reaction is:

$$RCOOH + Na^+ + OH^- \rightarrow RCOO^- + Na^+ + H_2O$$

Because the hydroxide reacts with the hydrogen ion to form a neutral species, counter-diffusion of the cation being exchanged is not required, as in ordinary Type 0 ion exchange. Instead, the sodium ion diffuses along with hydroxide to the exchange site, thus maintaining electroneutrality at all points within the resin. Because the hydroxides are consumed, leaving a large number of sodium ions present within the shell, Helfferich postulated that the rate-limiting species would actually be the hydroxide anion because its concentration within the gel will be much less than sodium. This idea is evaluated in the experimental results section.

A variety of models have been developed to correlate the moving fronts observed, making different types of approximations as discussed earlier. Here we follow the basic approach of Dana and Wheelock (*53*), but for cylinders rather than spheres. This shrinking core model is based upon work in chemical reaction engineering (e.g., Levenspiel (*66*)). The work explicitly considers both mass transfer resistance in the boundary film surrounding the gel and diffusive resistance within the gel.

The major assumptions behind this shrinking-core model include:

1) The reaction rate at the exchange site is fast in comparison to diffusion;
2) The gel dimensions are constant;
3) The gel properties (density, concentration of exchange sites) are independent of position;
4) The diffusivities of ions are constants;
5) Mass transfer of ions to the front is much faster than the movement of the front (this allows a pseudo-steady state approximation to be made);
6) Mass transfer within the gel occurs only by diffusion and the only concentration gradients which exist are perpendicular to the gel surface.

The development of this model begins with the material balance on the hydroxide ions within a gel cylinder of radius R and diameter d (overbars indicate gel phase concentration):

$$\frac{\partial \overline{C}}{\partial t} = D \left\{ \frac{1}{r} \frac{\partial}{\partial r} \left(r \frac{\partial \overline{C}}{\partial r} \right) \right\} \tag{1}$$

The boundary conditions associated with this system are zero concentration at the leading edge of the moving front (r_c) and continuity of flux at the solution/ gel interface:

$$\overline{C}|_{r = r_c} = 0 \tag{2}$$

$$D\left(\frac{\partial \overline{C}}{\partial r}\right)\Big|_{r = R} = k\left(C_o - \frac{\overline{C_s}}{\lambda}\right) \tag{3}$$

where:

k = mass transfer coefficient;
D = diffusion coefficient inside the gel;
C_o = bulk solution concentration;
$\overline{C_s}$ = surface concentration of the gel;
C_i = solution concentration at the gel/solution interface;
λ = distribution coefficient at the gel/solution interface;
= $\overline{C_s}/C_i$;
≈ 1 for hydroxide ions.

If the pseudo-steady state approximation is made, $\partial\overline{C}/\partial t \approx 0$. Under this condition, Equation 1 can be integrated and the boundary conditions applied to yield the following equation:

$$\overline{C} = \frac{kR}{D}\left(C_o - \frac{\overline{C_s}}{\lambda}\right) \ln\left(\frac{r}{r_c}\right) \tag{4}$$

The surface concentration can be eliminated from this equation by noting that $\overline{C}_s = \overline{C}|_{r=R}$, yielding an equation for the concentration profile of sodium ions within the shell:

$$\overline{C} = \frac{C_o \lambda \ln(\frac{r}{r_c})}{\frac{\lambda}{2\,Sh} + \ln(\frac{r}{r_c})} \tag{5}$$

where: Sh = Sherwood Number
= kd/D

At this point it is necessary to determine the position of the front boundary r_c at an arbitrary time t_c. If the pseudo-steady state approximation holds, the rate of hydrogen ion consumption will equal the rate at which hydroxide ions reach the interface. The rate of hydrogen ion consumption is:

$$-\frac{dN_{H^+}}{dt} = \overline{C_g} \cdot 2\pi r_c L \frac{dr_c}{dt} \tag{6}$$

where: $\overline{C_g}$ = concentration of carboxylic groups in the gel;
L = cylinder length.

The rate at which hydroxide ions reach the front is:

$$\frac{dN_{OH^-}}{dt} = (2\pi r_c L)\ D(\frac{d\overline{C}}{dr})|_{r=r_c} \tag{7}$$

Using the pseudo-steady state approximation, equations 6 and 7 can be combined to yield:

$$-\overline{C_g}\,\frac{dr_c}{dt} = D\frac{d\overline{C}}{dr}|_{r=r_c} \tag{8}$$

At this point the concentration gradient at the front is determined by differentiating equation 5 to yield:

$$\frac{d\overline{C}}{dr}|_{r=r_c} = \frac{\lambda C_o}{\frac{\lambda r_c}{2Sh} + r_c \ln(\frac{R}{r_c})} \tag{9}$$

Substitution of equations 9 into 8, followed by integration from r = R at t = 0 to $r = r_c$ at $t = t_c$, yields:

$$t_c = \frac{\overline{C_g}\ R^2}{4C_o}\left[\frac{2(1-\xi^2)}{kR} + \frac{1 + 2\xi^2 \ln\xi - \xi^2}{\lambda D}\right] \tag{10}$$

where: $\xi = r_c/R$.

From this equation the total time T required for the moving front to vanish along the axis of the cylinder ($\xi = 0$) is:

$$T = \frac{\overline{C_g}\, R^2}{4C_o}\left[\frac{2}{kR} + \frac{1}{\lambda D}\right] \tag{11}$$

Division of equation 10 by equation 11 yields a dimensionless expression which relates the fractional front penetration ξ to the fraction of front lifetime τ:

$$\tau = \frac{4(1 - \xi^2) + (\frac{Sh}{\lambda})(1 + 2\xi^2 \ln\xi - \xi^2)}{4 + \frac{Sh}{\lambda}} \tag{12}$$

The analogous expression for spheres given by Dana and Wheelock can be written as *(53)*:

$$\tau = \frac{4(1 - \xi^3) + (\frac{Sh}{\lambda})(1 - 3\xi^2 + 2\xi^3)}{4 + \frac{Sh}{\lambda}} \tag{13}$$

Equations 12 and 13 are plotted in Figures 4a and 4b, respectively, for values of Sh/λ ranging from 0 (film diffusion control) to $+\infty$ (intraparticle diffusion control). The results of Selim and Seagrave for the unsteady state solution to the particle-diffusion limited case match the $Sh/\lambda \rightarrow +\infty$ curves in Figure 4a and 4b, indicating the accuracy of the pseudo-steady state approximation *(64,65)*.

Experimental Procedures

The polyelectrolyte gels studied in this work were copolymers of acrylamide and sodium methacrylate, crosslinked by N,N'-methylenebisacrylamide. Adapting terminology from the gel electrophoresis literature, the compositions of the gels were described in terms of %T (total monomer), %C (crosslinking ratio), and %Ch (molar charge substitution), defined as follows:

%T = 100 × total mass of monomers / mass of water;
%C = 100 × mass of crosslinking agent / total mass of monomers;
%Ch = 100 × Moles of ionizable monomer (charge) / total moles of monomer.

All gels used here were synthesized with a composition of 16% T and 4% C, with either 0.5%, 1.0%, 2.4%, or 5.0% Ch.

Sample Preparation. The following procedure was used to synthesize these gels. The appropriate amounts of acrylamide (J. T. Baker, "Baker-analyzed" grade), sodium methacrylate (Polysciences, reagent grade) and N,N'-methylenebisacrylamide (BioRad, electrophoresis grade) were dissolved in distilled, deionized water. This solution, along with freshly made solutions of the initiators ammonium persulfate and sodium metabisulfite (each solution was 0.0030 g initiator in 20 ml of distilled, deionized water), was then put under 25 in. Hg vacuum in a vacuum oven for 90 minutes to be certain that all dissolved oxygen was removed from the solution. Oxygen is a potent inhibitor of the free radical polymerization reactions used to make the gel. Degassing was used to remove oxygen instead of nitrogen sparging because it reduced problems with gas bubble formation within the gel during polymerization.

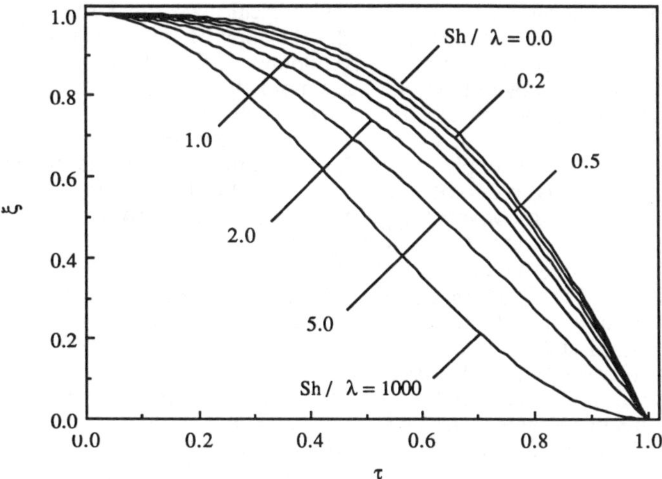

Figure 4a. Movement of ion exchange fronts within gel cylinders according to equation 12. ζ and τ are dimensionless position and time, respectively.

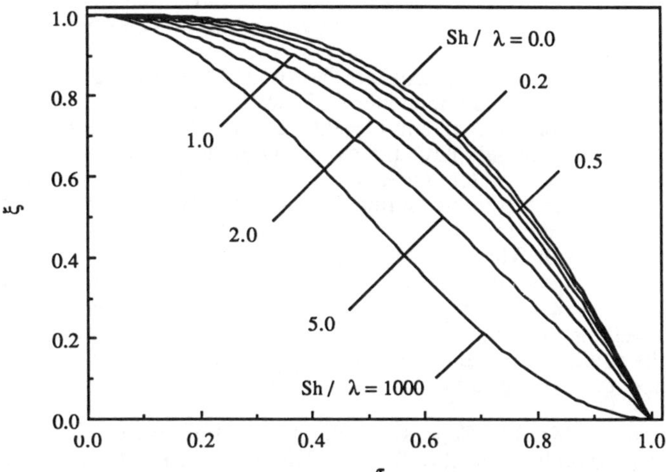

Figure 4b. Movement of ion exchange fronts within gel spheres according to equation 13. ζ and τ are dimensionless position and time, respectively.

After degassing, the solutions were transferred to a glove box with a nitrogen atmosphere. To 50 g of monomer solution, 3.5 ml of each initiator solution was added by pipette. The mixture was stirred without vortex formation on a magnetic stirrer for 30 s; then the solution was poured into cylindrical glass tubes sealed at the bottom. After filling all of the cylindrical molds, the tops of the molds were sealed with Parafilm. Gelation occurred within 5 - 10 minutes. After 24 h or more, the cylindrical pieces of gel were removed from the glass molds by gently cracking the glass with a hammer and washing the cracked glass away with distilled water.

To prepare the gels for the experiments, they were conditioned in a series of HCl solutions. There were a couple reasons for taking this step. One is simply to leach out of the gel any residual monomers, initiators, or other low molecular weight species. Also, it was desired to convert the gel from the sodium form to the hydrogen form. The conditioning scheme involved soaking the gels in a large volume of pH 2 solution and exchanging this solution after a number of hours for a fresh pH solution. This low pH facilitated the H^+ - Na^+ ion exchange. After 24 h the gels were then immersed in a pH 4 HCl solution; this solution was exchanged periodically over a couple days until the pH stabilized at 4.0. The ends of the gel cylinders were cut off, leaving samples with length/ diameter ratios of approximately 15.

The sorption capacity and water contents of the gels were determined from measurements of the gels' mass and volumes. The mass was measured by weighing after careful blotting of surface moisture with filter paper; the relative standard deviation of this method was 0.6%. The volume was determined by measurement of the linear dimensions with a caliper; the relative standard deviation of this method was 2.2%. Since the volume measurements were both more difficult and less precise than the mass measurements, for most gels only mass measurements were made; these were converted to volume via the measured density of 1.02 ± 0.02 g/cm^3 (the density was the same within experimental error for all gels under all conditions observed in this report). The polymer weight and volume were found by air-drying the samples at ambient conditions for a month, followed by weighing to determine mass and liquid displacement to determine volume.

The results of these measurements are shown in Table I (in the gel sample notation, the number indicates %Ch and the letter indicates the sample diameter at pH 4.0: A = 0.18 cm; B = 0.28 cm; C = 0.42 cm; D = 0.64 cm; E = 0.86 cm; F = 1.00 cm). In pH 4.0 HCl solution all the gels have the same swelling degrees (swollen volume /dry volume) and equilibrium water contents regardless of composition. In contrast, in pH 9.2 NaOH solution (in the high-swelling plateau of the swelling vs. pH curve for these gels), the swelling degrees and water contents increase with the level of ionic comonomer in the gel. This swelling behavior demonstrates that all of the gels are in the unionized hydrogen form at pH 4.0, but in the ionized sodium form at pH 9.2. At pH 12, hydrolysis of the acrylamide groups to acrylic acid leads to a continuously increasing swelling ratio with time. While hydrolysis makes it impossible to determine a value of the equilibrium swelling degree at pH 12, it occurs at such a slow rate that it does not affect the moving fronts.

Measurement of Moving Fronts. When a gel equilibrated with the pH 4.0 HCl solution is immersed in a pH 12 (0.01 N) NaOH solution, a distinct front is seen within the gel, and this front gradually moves toward the center of the gel. The front is not immediately noticed since it is simply a sharp refractive index gradient, but once observed it is noted to be quite distinct and well-defined throughout its lifetime. To determine if this front was associated with a sharp pH gradient, gel samples were first equilibrated with a pH 4.0 solution containing a universal indicator dye mixture (Fisher Scientific), and then immersed in the pH 12 solution. Indeed, a sharp color change at the front was observed, indicating that the front separated an acidic core from a basic shell. In contrast, when the gel was immersed in pH 9.2 solution, no front was

Table I. Equilibrium Data for Gel Samples

GEL SAMPLE	*PRIOR TO SWELLING* *solution pH 4.0*			*AFTER SWELLING* *solution pH 9.2*			
	EWC% [1]	$Q_{4.0}$ [2]	L/R [3]	EWC% [1]	$Q_{9.2}$ [2]	$Q_{9.2}/Q_{4.0}$	L/R [3]
0.5D	86.8	7.58	30.0	90.7	10.72	1.38	31.1
0.5C	86.9	7.66	33.6	90.4	10.44	1.36	33.9
0.5B	86.8	7.56	39.3	90.4	10.45	1.38	40.0
0.5A	86.7	7.52	51.8	90.5	9.48	1.26	54.3
1.0D	86.4	7.36	25.8	91.3	11.46	1.56	26.0
1.0C	86.4	7.36	30.0	91.2	11.33	1.54	30.4
1.0B	86.4	7.37	32.6	91.0	11.13	1.51	32.5
1.0A	86.6	7.49	25.9	90.8	10.85	1.45	26.3
2.4F	86.4	7.38	26.8	93.4	15.26	2.07	26.9
2.4E	----	----	29.4	----	----	2.08	29.5
2.4D	86.5	7.43	32.4	93.5	15.40	2.07	32.1
5.0D	86.2	7.23	30.7	95.8	23.58	3.3[4]	31.1
5.0C	----	----	32.8	----	----	3.36	33.4
5.0B	86.2	7.24	32.5	95.9	24.32	3.36	32.3

[1]equilibrium water content of the gel
= 100 × (swollen volume - dry volume)/swollen volume, ±0.2
[2]swelling ratio
= swollen volume / dry volume, ±0.05
[3]length : radius ratio for the cylindrical gel samples, ±0.05.
[4]gel sample not fully equilibrated with the solution.

observed and the color gradient was gradual from the surface of the gel to the core. The moving front was seen in sodium hydroxide solutions as dilute as 0.005 N (pH 11.7). Figure 5 shows photographs of this front as it progresses toward the center of the cylinder with time.

Measurements of the front position with time were made by two methods: an optical comparator and a microscope with a calibrated eyepiece. In both cases the gels were taken from the pH 4.0 HCl solution and immersed directly into a large excess of 0.010 N NaOH solution. The diameters of the cylinder and the internal core delineated by the moving front were measured at appropriate time intervals until the fronts met at the center of the gel. The solution was not deliberately stirred, but because the gels are almost neutrally buoyant, they tended to drift from their original positions. The internal core outlined by the front generally had a very consistent diameter along its length, but in some samples variations of ± 0.2 mm along the length were observed. Thus it was necessary to mark the gels with a razor nick to enable measurement of the core diameter at the same point over time. Although the gel was swelling throughout the time the front was present, the rate of diameter increase was small in comparison to the rate of front movement. Generally, the cylinder diameter increased about 10% between the time of sample immersion and disappearance of the central core.

The first means of front measurement, used with the 2.4% Ch gels only, used a Mitutoyo Series #183 Optical Comparator with a floodlight for illumination. With this device the front and the gel diameter could be measured to ± 0.1 mm. To make a measurement, an end of the gel was held against the reticle and the dimensions were

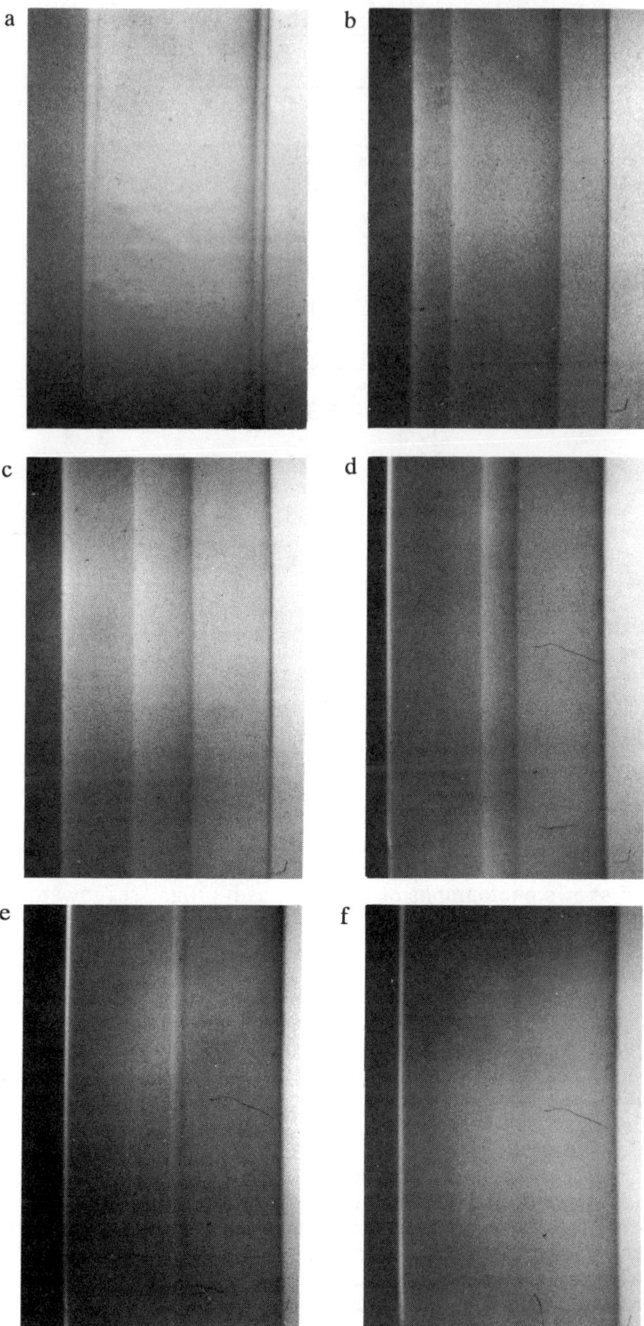

Figure 5. Progression of the moving front within gel sample 5.0C. Photographs are taken 9.8, 77.8, 144.3, 178.5, 187.1 and 190.1 minutes after immersion in pH 12 NaOH solution (gel was initially equilibrated with pH 4 HCl solution).

noted by comparison against the calibrated grid. This method worked well on larger diameter samples (>4mm) until the final stages of the front movement, since the front also moves into the gel from the ends.

For all the other gels, front movement was measured with a microscope (American Optical StereoStar 580T) fitted with 10X eyepieces and a calibrated eyepiece reticle good to ±0.005 in. Illumination provided by two high-intensity fiber optic light sources was adjusted to bring out the front most clearly for measurement. The photographs shown in Figure 5 were taken with this microscopic system.

Results and Discussion

The measurements of the front position as a function of time for gel cylinders of different dimensions and compositions are given in Figures 6a-6d. Because the gel is swelling as the front moves into the gel, the front position can be defined either relative to the original gel diameter or relative to the actual diameter at the time of front measurement. Since the gel diameter increases only on the order of 10% by the time the fronts vanish at the center, this correction is not significant. In these figures front position is defined relative to the original dimensions of the sample. It is apparent that for a given gel composition, the fronts move at about the same velocity, independent of the cylinder diameter, until the axis of a cylinder is approached. At this point the rate increases sharply. Furthermore, the velocity of the front decreases with an increasing concentration of ionic groups on the gel.

Using the theory described in the previous section, a diffusion coefficient and a mass transfer coefficient can be extracted from each curve if equation 10 is rewritten as:

$$\left\{\frac{2C_o}{R\overline{C}_g}\,\frac{t_c}{(1-\xi^2)}\right\} = \frac{1}{k} + \frac{1}{D}\left\{\frac{R}{2\lambda}\left(\frac{1-2\xi^2\ln\xi-\xi^2}{1-\xi^2}\right)\right\} \tag{14}$$

This equation has the form of y = mx + b, so a plot of the term in brackets on the left hand side of equation 14 against the term in brackets on the right yields a straight line with a slope of 1/D and an intercept of 1/k.

The results of this analysis are given in Tables IIa - IIe. Averages of the diffusion coefficients are included, since these coefficients should be independent of sample geometry (though they are expected to depend upon gel charge; see equation 17 below). However, the mass transfer coefficient is expected to be a function of the cylinder diameter (see equation 19 below). Figures 7a - 7d replot the data from Figures 6a-6d with the theoretical curves generated using the best-fit coefficients from each trial. For the sake of clarity values of penetration depth in Figures 6a-6d have been normalized with respect to the original gel diameter in Figures 7a-7d. The fit between theory and experiment is seen to be excellent. The question at this point is whether the values of D and k obtained from the curve fit of the theory to the data are physically reasonable or if they simply function as adjustable parameters. If they are reasonable, this theory could be used to make predictions of front position with time using independently obtained values of D and k.

The diffusion coefficients will be examined first. These values should be independent of the dimensions of the sample, although they may depend upon the charge content of the gels (see equation 17 below). Within each group of gel compositions, the relative standard deviation of the values of D obtained from the curve fit ranged from 3% for the 1%Ch gels to 32% for the 0.5% Ch gels. The relative standard deviation in the values of D obtained for all the trials is 26% (however, note that D is expected to be a function of gel composition, as discussed below). Thus the internal consistency of the results is good.

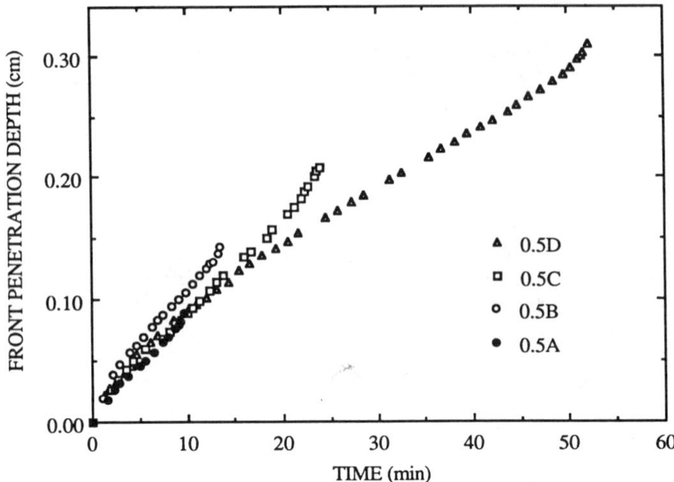

Figure 6a. Front position within 16% T, 4%C, 0.5% Ch poly(acrylamide-*co*-sodium acrylate) gel as a function of time for gel cylinders with different diameters.

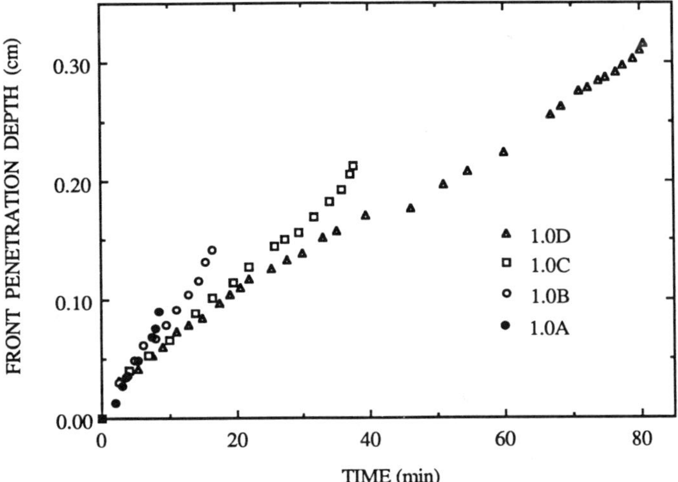

Figure 6b. Front position within 16% T, 4%C, 1.0% Ch polyacrylamide-*co*-sodium acrylate gel as a function of time for gel cylinders with different diameters.

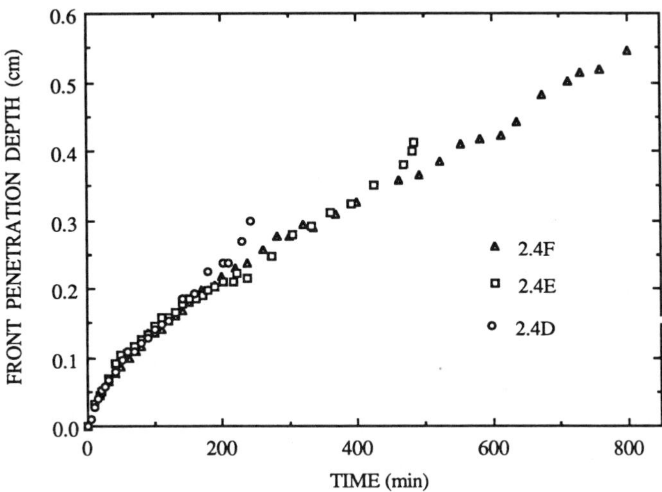

Figure 6c. Front position within 16% T, 4%C, 2.4% Ch polyacrylamide-*co*-sodium acrylate gel as a function of time for gel cylinders with different diameters.

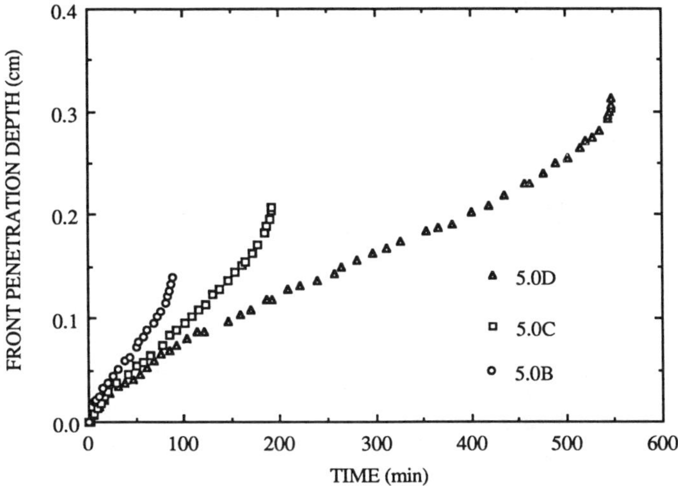

Figure 6d. Front position within 16% T, 4%C, 5.0% Ch polyacrylamide-*co*-sodium acrylate gel as a function of time for gel cylinders with different diameters.

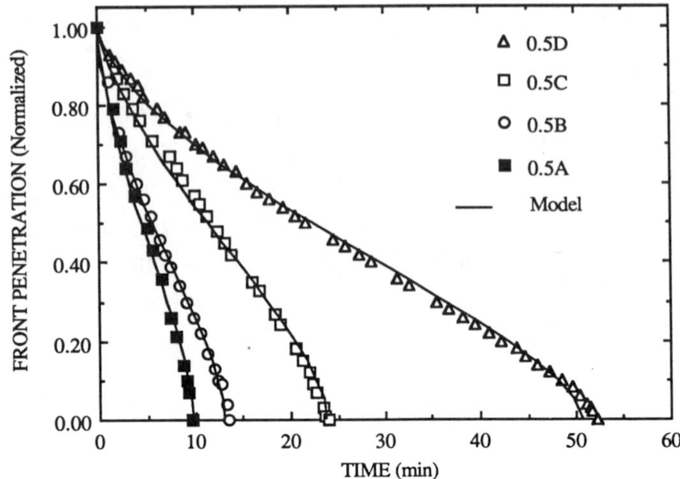

Figure 7a. Fit of the shrinking core model to the data of Figure 6a using the values of k and D obtained via linear regression using equation 14.

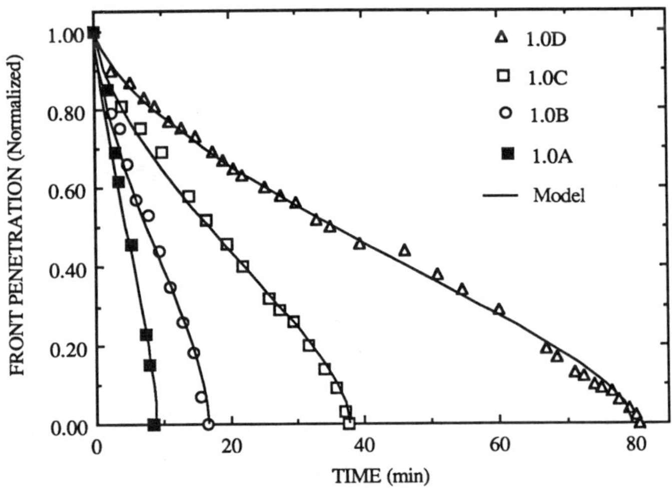

Figure 7b. Fit of the shrinking core model to the data of Figure 6b using the values of k and D obtained via linear regression using equation 14.

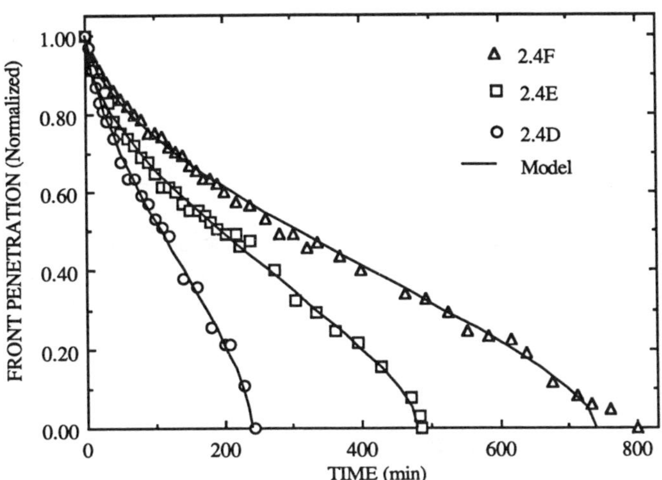

Figure 7c. Fit of the shrinking core model to the data of Figure 6c using the values of k and D obtained via linear regression using equation 14.

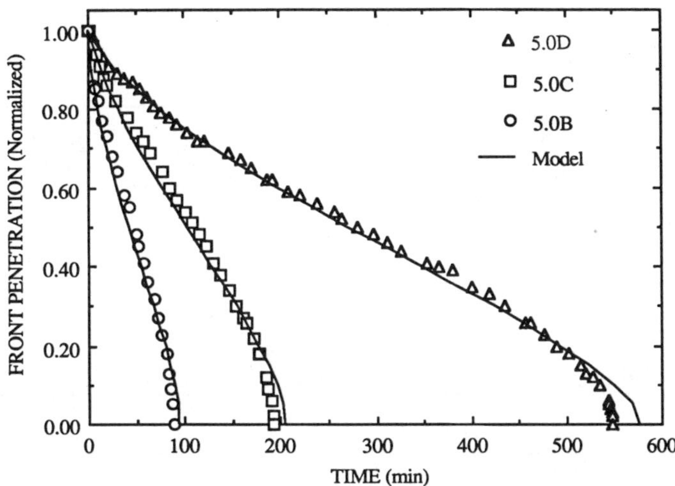

Figure 7d. Fit of the shrinking core model to the data of Figure 6d using the values of k and D obtained via linear regression using equation 14.

Table IIa. Results of Moving Front Analysis
Gel Charge Conc. = 0.0089 N; Solution NaOH Conc.: 0.00987 N

Sample Number	*d* *cm*	*D* $10^{-6}cm^2/s$	*k* $10^{-4}cm/s$	*Sh* = *kd/D*	*Re*	*v* *cm/s*
0.5D	0.618	0.79	4.2	33	45.7	0.70
0.5C	0.414	0.86	3.0	14	9.0	0.21
0.5B	0.288	0.65	9.4	42	65.5	2.16
0.5A	0.180	0.37	3.8	18	8.7	0.46
Average (excluding v >1cm/s):		0.67				0.46
Standard Deviation:		0.22				0.25
Relative Standard Dev.:		32%				54%

Table IIb. Results of Moving Front Analysis
Gel Charge Conc. = 0.018 N; Solution NaOH Conc.: 0.00988 N

Sample Number	*d* *cm*	*D* $10^{-6}cm^2/s$	*k* $10^{-4}cm/s$	*Sh* = *kd/D*	*Re*	*v* *cm/s*
1.0D	0.632	1.1	4.9	28	41.6	0.63
1.0C	0.424	1.0	6.3	26	32.9	0.74
1.0B	0.284	1.1	6.9	17	15.7	0.53
1.0A	0.180	1.1	4.2	7.1	2.3	0.12
Average (excluding v >1cm/s):		1.1				0.50
Standard Deviation:		0.035				0.27
Relative Standard Dev.:		3.2%				53%

Table IIc. Results of Moving Front Analysis
Gel Charge Conc. = 0.042 N; Solution NaOH Conc.: 0.00990 N

Sample Number	*d* *cm*	*D* $10^{-6}cm^2/s$	*k* $10^{-4}cm/s$	*Sh* = *kd/D*	*Re*	*v* *cm/s*
2.4F	1.09	8.8	7.4	92	387.1	3.37
2.4E	0.826	1.1	2.6	20	20.0	0.23
2.4D	0.600	7.6	2.1	17	11.1	0.18
Average (excluding v >1cm/s):		9.1				0.20
Standard Deviation:		1.6				0.04
Relative Standard Dev.:		18%				19%

Table IId. Results of Moving Front Analysis
Gel Charge Conc. = 0.087 N; Solution NaOH Conc.: 0.00992 N

Sample Number	*d* *cm*	*D* $10^{-6}cm^2/s$	*k* $10^{-4}cm/s$	*Sh* = *kd/D*	*Re*	*v* *cm/s*
5.0D	0.624	7.8	2.9	23	22.4	0.34
5.0C	0.416	6.4	15	98	359.7	8.22
5.0B	0.280	7.4	5.7	21.6	18.7	0.63
Average (excluding v >1cm/s):		7.2				0.49
Standard Deviation:		0.72				0.21
Relative Standard Dev.:		10%				42%

Table IIe. Average of All Trials, Excluding v > 1 cm/s

	D 10^{-6} *cm²/s*	*k* 10^{-4} *cm/s*	*Sh* = *kd/D*	*Re*	*v* *cm/s*
Average Value (excluding v >1cm/s):	8.8	4.2	20.5	20.7	0.43
Standard Deviation:	2.3	1.6	7.0	14.0	0.23
Relative Std. Dev.:	26%	37%	34%	67%	52%

The next question is whether the diffusion coefficients obtained are reasonable values. The values of D for ions and solutes within the gel can be estimated using two different approaches. The relationship given by Mackie is [1]:

$$D_g = D_s\left(\frac{\varepsilon}{2-\varepsilon}\right)^2 \tag{15}$$

where: D_g = diffusion coefficient in the gel;
D_s = diffusion coefficient in the solution;
ε = void fraction (water volume fraction).

The relationship given by Wheeler is simply [1]:

$$D_g = D_s \frac{\varepsilon}{2} \tag{16}$$

Table III lists the diffusion coefficients of the different ions in free solution. However, the diffusion coefficients of the sodium and hydroxide ions are coupled due to electroneutrality and zero electric current constraints (*1*). Within the shell, behind the moving front, the concentration of the sodium ions is greater than the hydroxide concentration by an amount equal to the charge density of the gel. The diffusion coefficient for the sodium hydroxide within the shell can be calculated from the following equation, where the subscripts A and C indicate the anion and the cation, respectively (*1*):

$$D_{AC} = \frac{D_A D_C\,(z_A^2 C_A + z_C^2 C_C)}{z_A^2 C_A D_A + z_C^2 C_C D_C} \tag{17}$$

where: z = valence of the ion.

This equation predicts that the diffusion coefficient of the more dilute ion will dominate the diffusion rate of the ion pair within the gel. As a result, Helfferich expected that the diffusion coefficient of the hydroxide ion would control the movement of the front observed under these conditions, since sodium ions are in excess in the hydrogen-ion depleted shell surrounding the unexchanged core. For diffusion of ions within a nonionic gel, the ion concentrations are in stoichiometric proportion, and equation 19 reduces to:

$$D_{AC} = \frac{D_A D_C\,(|z_A| + |z_C|)}{|z_A| D_A + |z_C| D_C} \tag{18}$$

Assuming that the concentration of the hydroxide ion in the shell equals the solution concentration and that the sodium ion concentration in the shell equals the solution concentration plus the gel charge density (C_g), the NaOH diffusion coefficient within the gel can be estimated from the free solution values for Na^+ and OH^- by combining

Table III. Comparison of Measured and Estimated Diffusion Coefficients

Species	*Environment*	*Diffusion Coefficient (Experimental)*	*Diffusion Coefficient (Mackie Estimate)*	*Diffusion Coefficient (Wheeler Estimate)*
Na^+	Solution	1.3 §		
	Nonionic Gel		0.76	0.56
OH^-	Solution	5.3 §		
	Nonionic Gel		3.1	2.3
NaOH	Solution	2.1 §		
	Nonionic Gel		1.2	0.91
	0.5% Ch Gel	0.67	1.5	1.1
	1.0% Ch Gel	1.1	1.7	1.3
	2.4% Ch Gel	0.91	2.1	1.5
	5.0% Ch Gel	0.72	2.4	1.8

§ Taken from (*68*)
All diffusion coefficients have units of 10^{-5} cm^2/s
Void fraction ε is 0.865 for all gels

equation 17 with either equation 15 or 16. Table III compares these predictions with the values of D obtained from the moving boundary measurement. The diffusion coefficients obtained from the shrinking core model fall between the values of the sodium and the hydroxide ionic diffusion coefficients predicted within the gel. However, the measured values of D are quite a bit less than the estimated values for NaOH, since equation 17 predicts that the OH^- diffusion coefficient should dominate the diffusion of the ion pair. The significance of this discrepancy is unclear, since the estimates themselves are rough. Both the Mackie and Wheeler models for estimating the diffusion coefficients of solutes within gels are rather simplistic and may not be accurate. Furthermore, since the actual concentrations of ions within the gel are unknown, equation 17 can be applied only as an approximation. Thus the final conclusion is that the shrinking core model yields reasonable, internally consistent values of diffusion coefficients, although it would be difficult to accurately predict the movement of the moving boundary based on estimates of D within the gel based upon the free solution ionic diffusion coefficients.

Evaluating the physical reasonableness of the mass transfer coefficients is also difficult. To begin with, the mass transfer coefficient is simply an empirical parameter, unlike the diffusion coefficient. It also depends upon the cylinder dimension. But the biggest problem is that it depends upon the flow field of the solvent past the gel. Since this was not well-defined for these experiments, nor would it be in practical applications for polyelectrolyte gels, k will be difficult to estimate, and in fact probably varied over the course of the experiment. However, each gel was treated similarly during the experiments.

Many expressions are available in the literature to correlate heat transfer coefficients for flow perpendicular to a cylinder; using the Colburn analogy between heat and mass transfer, these expressions can be extended to correlate mass transfer coefficients as well. Considering the ill- defined flow field, it does not seem necessary to choose anything more complex than the following expression, valid for Reynolds Numbers between 1 and 4000 (*67*):

$$Sh = 0.43 + 0.532\ Re^{0.5}\ Sc^{0.31} \tag{19}$$

where: Sh = Sherwood Number
= kd/D
Re = Reynolds Number
= dv/ν
Sc = Schmidt Number
= ν/D
ν = kinematic viscosity
d = cylinder diameter
v = velocity of fluid approaching the cylinder at right angles to its axis.

Since the flow velocity is unknown, the Sherwood number cannot be calculated directly from this expression. Thus the approach taken here is to solve for the Reynolds number using the values of D and k obtained from the model for the moving front to calculate the Sherwood and Schmidt Numbers. The average flow velocity can then be derived from the Reynolds number. This value should be small and nearly the same for for all samples since since the solution was not deliberately stirred in any trial.

From Tables IIa-IIe, the Reynolds numbers average 20.7, which is within the range of applicability for equation 19. Three essentially random trials came out with velocities drastically higher than the average; if these points are excluded, the average flow velocity is 0.43 cm/s with a relative standard deviation of 52%. This internal consistency is excellent, considering the lack of control of the flow field. Furthermore, this velocity is of an appropriate order of magnitude. Equation 19 also predicts that the mass transfer coefficient should vary with $1/\sqrt{d}$. However, this dependence is not clearly seen since the variation in d between the different samples is of the same order of magnitude as the deviation in the estimated flow velocities.

Conclusions

The pseudo-steady state shrinking core model described in this paper can effectively describe the moving front which develops during the Type III ion exchange process examined in this work. As a simple curve fit, the model describes the movement of the front toward the center of the gel cylinder very well. Furthermore, the values of the diffusion and mass transfer coefficients obtained from the model are physically reasonable, although they differ somewhat from independent estimates of their values.

Acknowledgements

This work was partially supported by Grant CBT-8809271 from the National Science Foundation.

Literature Cited

1 Helfferich, F. *Ion Exchange*, McGraw-Hill: New York, NY, 1962.
2 Scopes, R. *Protein Purification*, 2nd edn., 1988.
3 Schacht, E.H. In *Recent Advances in Drug Delivery Systems* ; Anderson, J.M.; Kim, S.W. Eds., Plenum Press, New York, NY, 1984, pp. 259-278.
4 Katchalsky, A. *Experientia*, **1949**, 5, p. 319.
5 Sussmann, M. V.; Katchalsky, A. *Science*, **1970**, 167, p. 45.
6 Katchalsky, A.; Sussmann, M. V.; Oplatka, A.; Steinberg, I. *US Patent No. 3321908* ,1967.
7 Suzuki, M. *Kobunshi Ronbunshu*, **1989**, 46, p. 603.

8 Tatara, Y. *Adv. Robotics*, **1987**, 2, pp. 69.
9 Urry, D. W.; Harris, R. D.; Prasad, K. U. *J. Am. Chem. Soc.*, **1988**, 110, p. 3303.
10 Shiga, T.; Hirose, Y.; Okada, A.; Kurauchi, T. *Polym. Prepr.*, **1990**, 30, 1, p. 310.
11 De Rossi, D. E.; Parrini, P.; Chiarelli, P.; Buzzigoli, G. *Trans. Am. Soc. Artif. Intern. Organs*, **1985**, 31, p. 60.
12 De Rossi, D. E.; Chiarelli, P.; Buzzigoli, G.; Domenici, C.; Lazzeri, L. *Trans. Am. Soc. Artif. Intern. Organs*, **1986**, 32, p. 157.
13 Chiarelli, P.; De Rossi, D. *Progr. Colloid Polym. Sci.,* **1988**, 78, p. 4.
14 Osada, Y.; Umezawa, K.; Yamauchi, A. *Makromol. Chem.*, **1988**, 189, p. 597.
15 Umezawa, K.; Osada, Y. *Chem. Lett.*, **1987**, p. 1795.
16 Gehrke, S. H.; Lee, P. I. in *Specialized Drug Delivery Systems, Manufacturing and Production Technology*, Tyle, P. ed.; Marcel Dekker, New York, NY, 1990, pp. 333-392.
17 Hoffman, A.S. *J. Controlled Release*, **1987**, 6, p. 297.
18 Urry, D. W. *Polym. Mat. Eng. Sci*, **1990**, 63, p. 329.
19 Cornejo-Bravo, J. M.; Siegel, R. A. *Proc. Int. Symp. Controlled Release of Bioactive Materials*, **1990**, 17, p. 174.
20 Okano, T.; Bae, Y. H; Jacobs, H.; Kim, S. W. *J. Controlled Release*, **1990**, 11, p. 255.
21 Horbett, T. A.; Kost, J.; Ratner, B. D. in *Polymers as Biomaterials*, Shalaby, S. W.; Hoffman, A. S.; Ratner, B. D.; Horbett, T. A., Eds.; Plenum Press, 1984, pp. 167-179.
22 Mukae, K.; Bae, Y. H.; Okano, T.; Kim, S.W. *Polym. J.*, **1990**, 22, p. 250.
23 Hoffman, A. S.; Afrassiabi, A.; Dong, L. C. *J. Controlled Release*, **1986**, 4, p. 213.
24 Siegel, R. A.; Firestone, B. A. *Macromolecules*, **1988**, 21, p. 3254.
25 Firestone, B. A.; Siegel, R. A. *Polym. Commn.*, **1988**, 29, p. 204.
26 Afrassiabi, A.; Hoffman, A. S.; Cadwell, L. A. *J. Membr. Sci.*, **1987**, 33, p. 191.
27 Mukae, K.; Bae, Y. H.; Okano, T.; Kim, S.W. *Polym. J.*, **1990**, 22, p. 206.
28 Bae, Y. H.; Okano, T.; Kim, S. W.; *J. Controlled Release*, **1989**, 9, p. 271.
29 Okano, T.; Bae, Y. H.; Jacobs, H.; Kim, S. W. *J. Contolled Release*, **1990**, 11, p. 255.
30 Horbett, T. A.; Ratner, B. D.; Kost, J.; Singh, M. In *Recent Advances In Drug Delivery Systems*, Anderson, J. M.; Kim, S. W., Eds.; Plenum Press, 1984, pp. 209-220.
31 Horbett, T. A.; Ratner, B. D. *Proc. Pharm Tech Conference'88*, East Rutherford, NJ, **1988**, p. 356.
32 Kost, J.; Horbett, T. A.; Ratner, B. D. *J. Biomed. Mater. Res.*, **1985**, 19, p. 1117.
33 Lyu, H. L.; Gehrke, S. H. *Extended Abstracts, Annual AIChE Meeting*, San Francisco, **1989**, paper 98j.
34 Gehrke S. H.; Andrews G. P.; Cussler E. L. *Chem. Eng. Sci.*, **1986**, 41, p. 2153.
35 Cussler, E. L.; Stokar, M. R.; Varberg, J. E. *AIChE J.*, **1984**, 30, p. 578.
36 Huang, X.; Akehata, T.; Unno, H.; Hirasa, O. *Biotech. Bioeng.*, **1989**, 34, p. 102.
37 Osada, Y.; Takeuchi, Y. *J. Polym. Sci. Poly. Lett. Ed.*, **1981**, 19, p. 303.
38 Osada, Y.; Takeuchi, Y. *Polym. J.*, **1983**, 15, p. 279.
39 Osada, Y.; Hasebe, M. *Chem. Lett.*, **1985**, p.1285.

40 Grimshaw, P. E.; Grodzinsky, A. J.; Yarmush, M. L.; Yarmush, D. M. *Chem. Eng. Sci.*, **1989,** 44, p. 827.

41 Grimshaw, P.E.; Nussbaum, J.H.; Grodzinsky, A. J.; Yarmush, M.L.*J. Chem. Phys.*, **1990,** 93, p. 4462.

42 Gehrke, S. H.; Cussler, E. L. *Chem. Eng. Sci.*, **1989,** 44, p. 559.

43 Tanaka, T.; Fillmore, D. J. *J. Chem. Phys.*, **1979,** 70, p. 1214.

44 Gehrke, S. H.; Akhtar, M. K. *Extended Abstracts, 33rd IUPAC Int. Symp. on Macromolecules*, Montreal, Canada, July 8-13, 1990.

45 Akhtar, M. K. Volume Change Kinetics In Near Critical Gels, M.S. Thesis, University of Cincinnati, 1990.

46 Huang, X.; Unno, H.; Akehata, T.; Hirasa, O. *J. Chem. Eng. Jpn.*, **1979,** 21, p. 10.

47 Huang, X., Unno, H., Akehata, T., Hirasa, O., *J. Chem. Eng. Jpn.*, **1987,** 20, p. 123.

48 Kabra, B. G.; Akhtar, M. K.; Gehrke, S. H. *Polymer*, accepted, **1991**.

49 Kabra, B. G.; Gehrke, S. H. *Polymer Communications.*, in press, **1991**.

50 Helfferich, F. *J. Phys. Chem.*, **1965,** 69, pp. 1178-1187.

51 Helfferich, F. In *Ion Exchange Kinetics;*, Marinsky, J. A., ed; Ion Exchange Kinetics - A Series of Advances, Marcel Dekker Inc., New York, NY, 1966, Vol. I, pp. 65-100.

52 Helfferich, F. In *Ion Exchange Kinetics - Evolution of a Theory;* Liberti , L; Helfferich, F., Eds; Mass Transfer and Kinetics of Ion Exchange NATO ASI Series E: Applied Sciences, Martinus Nishoff Publishers, The Hague, 1983, Vol. 71, pp. 157-179.

53 Dana, P.R.; Wheelock, T.D. *Ind. Eng. Chem., Fundam.*, **1974,** 13, p. 20.

54 Selim, M.S.;Seagrave, R.C. *Ind. Eng. Chem., Fundam.*, **1973,** 12, p. 14.

55 Pinto, N.G.; Graham, E.E *Reactive Polymers*, **1987,** 5, p. 49.

56 Kataoka, T; Yoshida, H *Can. J. Chem. Eng.*, **1981,** 59, p. 475.

57 Nativ, M.; Goldstein, M.; Schmuckler, G. *J. Inorg. Nucl. Chem.*, **1975,** 37, p. 1951.

58 Höll, W.; Sontheimer, H. *Chem. Eng. Sci.*, **1977,** 32, p.755.

59 Höll, W. *Reactive Polymers*, **1984,** 2, p. 93.

60 Höll, W.; Geiselhart, G. *Desalination*, **1978,** 25, p. 217.

61 Bolto, B.A.; Warner, R.E. *Desalination*, **1970,** 8, p. 21.

62 Lewnard, J.J.; Peterson, E.E.; Radke, C.J. *J. Chem. Soc., Faraday Trans.*, **1984**, 1, p.3927.

63 Höll, W.; Kirch, R. *Desalination*, **1978,** 26, p.163.

64 Selim, M.S.;Seagrave, R.C. *Ind. Eng. Chem., Fundam.*, **1973,**12, p. 1.

65 Selim, M.S.;Seagrave, R.C. *Ind. Eng. Chem., Fundam.*, **1973,**12, p. 9.

66 Levenspiel, O. *Chemical Reaction Engineering,* J. Wiley and Sons, New York, NY, 1972, 2nd Ed., pp. 361-373.

67 Treybal, R.E. *Mass Transfer Operations,* McGraw Hill, New York, NY, 1980, 3rd Ed., pp. 69-76.

68 Cussler, E.L. *Diffusion*, Cambridge University Press, Cambridge, 1984, p. 147.

RECEIVED August 19, 1991

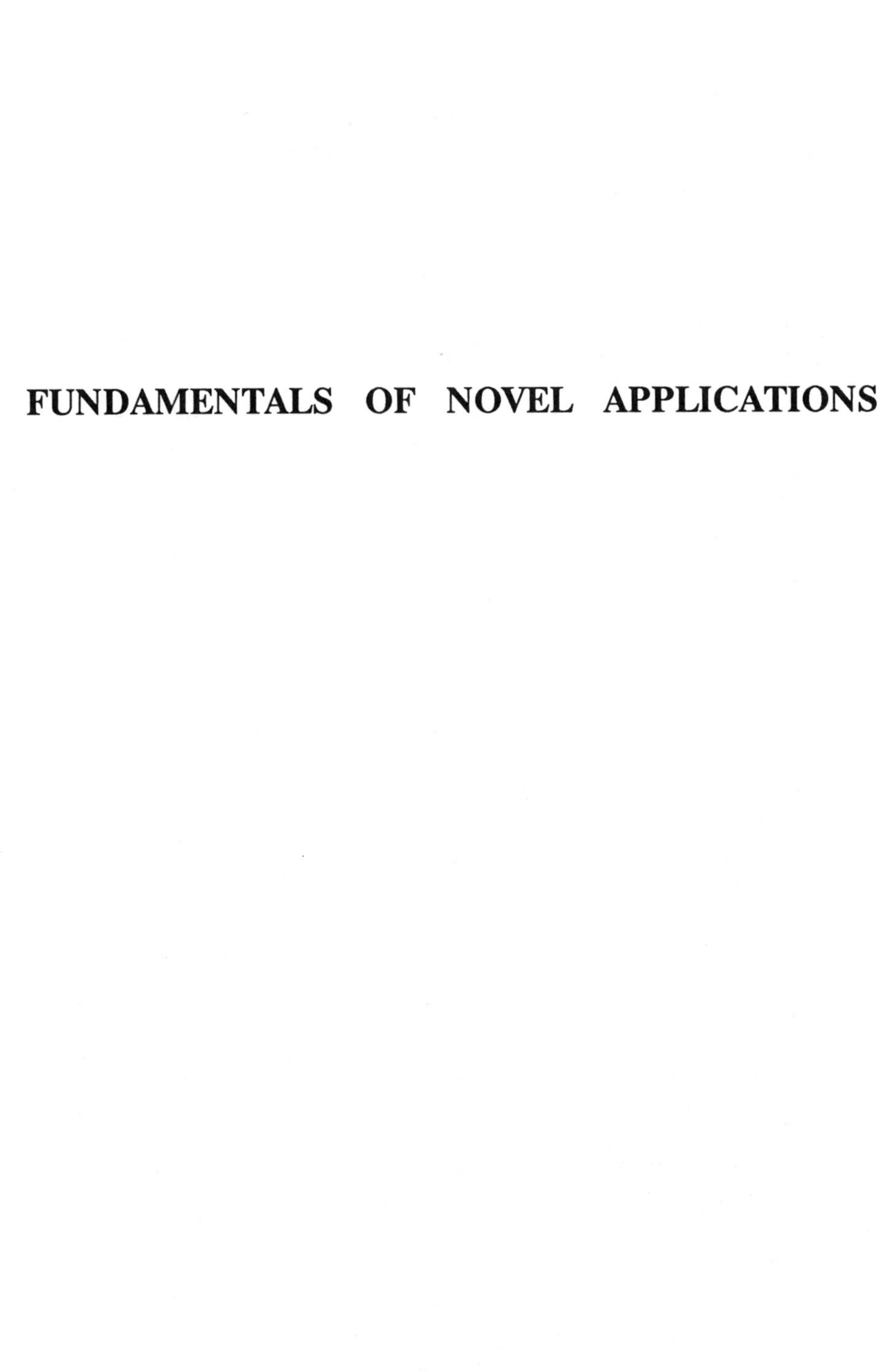

FUNDAMENTALS OF NOVEL APPLICATIONS

Chapter 14

Development of Electrochemical Sensors for Hydrogen, Oxygen, and Water Using Perfluorosulfonic Acid Membranes

David R. Morris, Xiadong Sun, and Lietai Yang

Department of Chemical Engineering, University of New Brunswick, P.O. Box 4400, Fredericton, New Brunswick E3B 5A3, Canada

The development of solid state potentiometric sensors for hydrogen and oxygen utilizing Nafion polymer electrolyte is described. The reference electrode comprises a mixture of Fe^{II} and Fe^{III} sulphate hydrates. For pure hydrogen at 1 atm. pressure, E_H^O = + 0.68V; for pure oxygen at 1 atm. pressure, E_O^O = - 0.36V. For hydrogen inert gas mixtures in the pressure range $\sim 10^{-4} < p_H \leq 1$ atm, and for oxygen-inert gas mixtures in the pressure range $\sim 10^{-2} < p_O \leq 1$ atm, Nernstian response is observed. For mixtures of hydrogen, oxygen and an inert gas, a mixed potential is observed which is a linear function of the ratio of the pressures of hydrogen and oxygen over the range $\sim 0.25 < p_H/p_O < \sim 10.0$.

The sensor is also responsive to hydrogen introduced into steel e.g. by a corrosion reaction. The elapsed time between the initiation of the corrosion reaction on the inner surface of a steel vessel and detection of hydrogen by the sensor at the outer surface permits estimation of the diffusion coefficient of hydrogen in steel ~ 2×10^{-6} cm^2/s at ambient temperature.

Data on water sorption by Nafion, the diffusion coefficient of water in Nafion and the electrical conductivity of Nafion are presented. Nafion

0097–6156/92/0480–0240$06.00/0
© 1992 American Chemical Society

absorbs ~ 14 mols H_2O/mol H^+ when equilibrated with water at unit activity. The diffusion coefficient of water is a strong function of the water content. The electrical conductivity is also a strong function of the water content. This latter dependency is exploited in the development of an amperometric sensor for water vapour in gases.

The availability of perfluoro sulphonic acid membranes (e.g. Nafion, Du Pont Co. Inc.) has enabled the development of relatively simple solid state electrochemical sensors for the monitoring of hydrogen, oxygen and water in various environments. The results of initial development work have been described by Morris, Kumar and Fray *(1)* ; this work was primarily concerned with the hydrogen sensor. In this paper the results of further developmental work on the hydrogen sensor, for improved long-term stability and reproducibility, and development work on the oxygen and water sensors, are described. In addition, measurements of water sorption by Nafion and some transport properties are presented.

Sensors for Hydrogen and Oxygen

The sensors for hydrogen and oxygen are of the potentiometric type and may be represented schematically as the sequence of phases,

Gas, $H_2O_{(g)}$, Pt | Nafion | reference mixture, C, stainless steel.
I II

The test gas is humidified by saturation with water at ~20°C; Pt represents platinum black applied to the surface of the Nafion from a slurry in Nafion solution. A sectioned drawing of the sensor is shown in Figure 1. *(2)* . The reference mixture comprises Fe^{II} and Fe^{III} sulphate hydrates with Nafion powder. A stainless steel ram with a graphite disc is used to compress the mixture against the Nafion and serves as an electrical terminal. The graphite disc is inserted to prevent corrosion of the ram by the reference material.

The equilibria at the interfaces I and II of the sensor are postulated to be,

I Hydrogen gas, $H_3O^+_{(Naf.)} + e^-_{(Pt)} === \frac{1}{2} H_{2(g)} + H_2O_{(g)}$

I Oxygen gas, $O_{2(g)} + e^-_{(Pt)} === O^-_{2}(Pt)$

II Reference mixture, $Fe^{3+}_{(s)} + e^-_{(c)} === Fe^{2+}_{(s)}$

The Nernst equation leads to the following expressions for the sensor voltage E_H or E_O for fixed gas humidity, and fixed activities of the components of the reference mixture,

Hydrogen-inert gas mixtures: $E_H = E^O_H + (RT/2F)\,\ell n\, p_H$ (1)

Oxygen-inert gas mixtures: $E_O = E^O_O - (RT/F)\,\ell n\, p_O$ (2)

where E^O_H and E^O_O are constants, p_H and p_O are the ratios of the hydrogen pressure and oxygen pressure to the standard atmospheric pressure in the mixtures with inert gas, and R, T and F have their usual meaning. Sensors similar in principle have been described by Miura et al. *(3)* .

Calibration of Sensors; Mixtures with an Inert Gas. The sensors were calibrated by mounting in a small stainless steel vessel with the gas electrode I pressed against stainless steel gauze. Eight such vessels were connected in series with continuous gas flow. The hydrogen (or deuterium) partial pressures in the gas mixtures were in the range $1.0 \geq p_H \geq 1 \times 10^{-4}$ atm.; the oxygen partial pressures were in the range $1.0 \geq p_O \geq 0.01$ atm. The inert gas components of the mixtures were usually nitrogen or carbon dioxide. Voltage was measured with a Keithley high impedance electrometer.

Typical calibration charts for the hydrogen and oxygen sensor are shown in Figures 2 and 3. The regression equations and correlation coefficients r are,

$E_H = 0.6808 + 0.0297 \log p_H$, r = 0.996 (3)

$E_O = -0.3564 - 0.0599 \log p_O$, r = 0.999 (4)

The gradients, $\Delta E/\Delta \log p$, agree well with the predicted values from Equations 1 and 2 with T ~ 298K. Responsiveness to change of hydrogen pressure is essentially instantaneous; responsiveness to change of oxygen pressure is less rapid. The

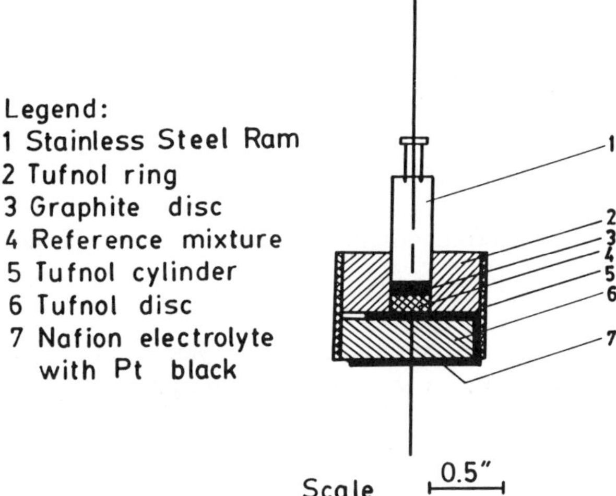

Figure 1. Section of Electrochemical Sensor.

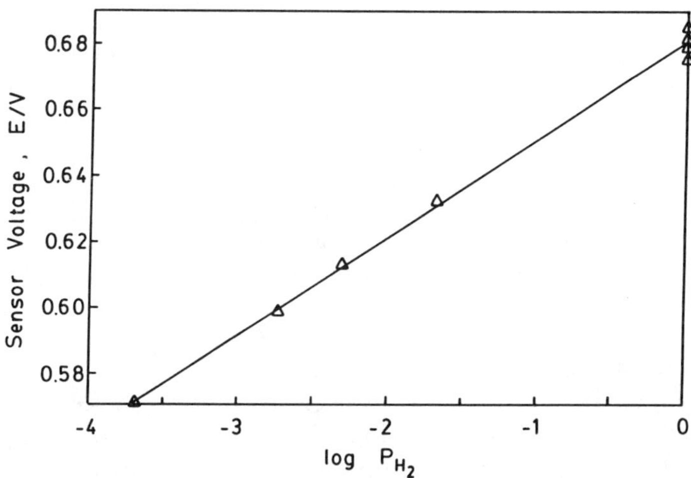

Figure 2. Nernstian Response of Sensor to Hydrogen Pressure. Humidified hydrogen–inert gas mixtures at ambient temperature.

slow response to change of oxygen pressure is attributed to relatively slow adsorption/desorption kinetics of the O_2^- ion on platinum. Standard deviation of the voltage of a sensor operating under apparently constant conditions for long term tests (10^3 to 10^4h) is in the order 1mV. Variability of voltage between sensors is in the order 10mV.

Reactive Gas Mixtures. The sensor voltage attained with a reactive gas mixture (e.g. hydrogen + oxygen + inert gas) is interpreted as a mixed potential from the simultaneous anodic ($\frac{1}{2}H_2 + H_2O \rightarrow H_3O^+ + e^-$) and cathodic ($O_2 + e^- \rightarrow O_2^-$) reactions occurring, and kinetic parameters are paramount. The results of typical measurements of sensor voltage E_{HO} with various reactive gas mixtures are presented in Figure 4. The empirical regression line equation is,

$$E_{HO} = 0.291 + 0.252 \log (H/O), \; r = 0.994 \qquad (5)$$

where H/O represents the ratio of partial pressures of hydrogen and oxygen in the mixture. Relatively few tests have been conducted with reactive mixtures, and it is uncertain whether a sensor would operate satisfactorily over extended periods of time.

Sensing of Hydrogen in Steel. It is well known that hydrogen in steel causes embrittlement of the metal, blistering and failure, particularly in a "sour" environment of sulphides. In a corrosive situation sulphides promote the entry of atomic hydrogen into the steel by suppressing the formation of hydrogen gas molecules. The sensor described appears to have application for monitoring hydrogen in steel.

Responsiveness of the sensor to hydrogen in steel has been demonstrated by clamping the sensor to the side of a small steel vessel on which a flat surface had been milled; the thickness of the steel was ~1.8 mm. A smear of silicone grease was applied between the surfaces to exclude air *(1)*. The sensor voltage, E, measured between the stainless steel ram, and the steel vessel was recorded as a function of time. The results of a typical experiment are presented in Figure 5. The following notes summarise the observations:

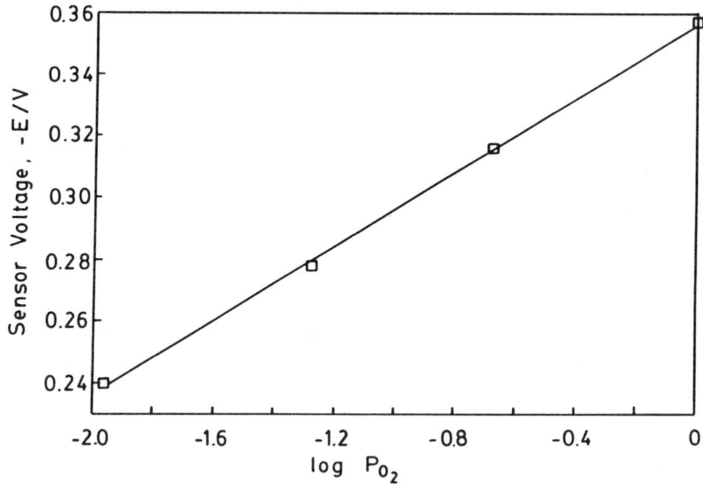

Figure 3. Nernstian Response of Sensor to Oxygen Pressure. Humidified oxygen–inert gas mixtures at ambient temperature.

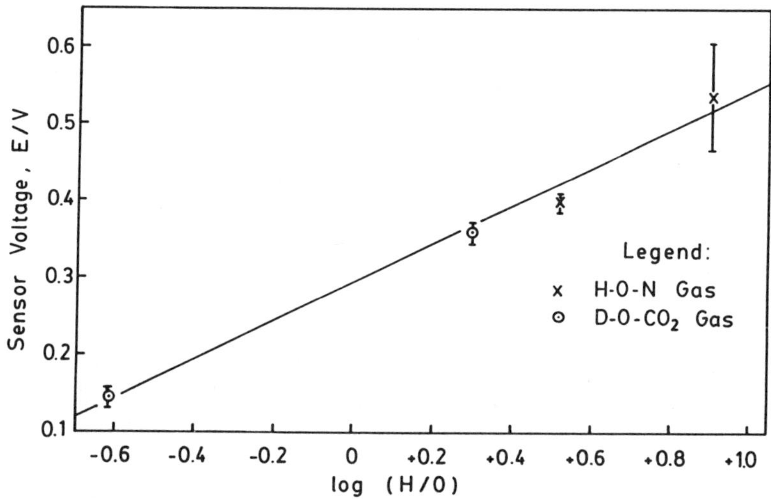

Figure 4. Response of Sensor to Hydrogen–Oxygen Pressure Ratio. Humidified hydrogen–oxygen–inert gas mixtures at ambient temperature.

1. Sensor mounted on outer surface of steel vessel. Sensor voltage stabilised at ~0.55V after approximately 30 h. The equivalent hydrogen pressure was ~ 4 X 10^{-5} atm, Equation 3.
2. Hydrochloric acid solution was introduced to the steel vessel at time 78 h. Sensor voltage rose rapidly to ~ 0.70V, equivalent to a hydrogen pressure of ~ 4 atm.
3. A small quantity of zinc sulphide was added to the acid solution at time 119 h. The sensor voltage rose rapidly to ~0.72V. The equivalent hydrogen pressure was ~ 30 atm.
4. Thereafter, the sensor voltage declined as the acid was consumed and removed from the steel vessel at time 160 h.
5. In order to verify the interchangeability of sensors, the sensor was replaced at time 305 h with a second sensor. The voltage generated by the second sensor fluctuated initially as indicated by the dashed line.
6. Fresh hydrochloric acid was introduced to the steel vessel at time 353 h. The sensor voltage rose rapidly as previously observed to ~ 0.72V.
7. Thereafter, the sensor voltage declined following consumption of the acid and removal of the acid from the steel vessel at time 380 h.

The above observations are attributed to the diffusion of hydrogen introduced into the steel by the cathodic reaction,

$$H_3O^+_{(aq)} + e^- \rightarrow H_{(ads)} + H_2O \qquad (6)$$

where $H_{(ads)}$ represents a hydrogen atom adsorbed on the surface of the steel. The adsorbed hydrogen may then form molecular hydrogen gas,

$$2H_{(ads)} \rightarrow H_{2(g)} \qquad (7)$$

or enter the steel as interstitial atoms,

$$H_{(ads)} \rightarrow H_{(Fe)} \qquad (8)$$

and diffuse through the steel. The diffusing H atoms may escape to the atmosphere at the external surface or at faults in the metal structure to form hydrogen gas. The latter phenomenon can result in the formation of hydrogen blisters as noted earlier *(4)* . The response of the sensor following the addition of zinc sulphide to the solution is also in accord with previous observations. *(4)* .

A portion of a response chart similar to that presented in Figure 5 is shown in Fig. 6. The observed break-through time permitted an estimate of the diffusion coefficient *(5)* for hydrogen in steel as ~ 2 X 10^{-6} cm^2/s at ~ 25°C in agreement with published data *(6)* .

Some recent experiments with zirconium into which hydrogen had been introduced by corrosion reactions with water suggest that the sensor may have application to this system also.

Water Sorption and Transport Properties of Nafion.

Water sorption isotherms for Nafion H 1100 were determined at 25°, 50° and 100°C with a Cahn microbalance and at 25°C by equilibration with the vapor phase of a solution of lithium chloride of known concentration *(7)* .

A small piece of Nafion (~ 0.03 g) was boiled in hydrochloric acid solution and water, suspended in the microbalance and heated to ~150°C under vacuum (~10^{-5} mm Hg) to a constant mass. Variation of the temperature(±10°C) caused a small (~ 0.7%) variation in the final mass; some darkening of the Nafion was observed. Water vapor was admitted to the system after cooling to 25°C and the mass of the Nafion recorded as a function of time. Upon attainment of a constant mass, additional water vapor was admitted and measurements continued. Sorption measurements were followed by desorption measurements at the temperatures noted. Sample preparation for the isopiestic measurements was identical; the specimen was periodically weighed until a constant mass was achieved. The equivalent weight of the Nafion was found to be 1097.7 ± 5.0 by elution with sodium chloride solution followed by titration *(7)* .

The results of the water sorption experiments are presented in Figure 7 as a plot of the mol ratio N (≡ mols H_2O/mol H^+) as a function of the water activity in the gas phase, a_{H_2O} (≡ partial pressure water / saturated vapor pressure). It is evident that all data form a single curve; at a_{H_2O} = 1, N ~14.0 mols H_2O/mol H^+ corresponding to a mass fraction ~0.23 water in the Nafion. These data are in close agreement with those of Pushpa et al. *(7)* . Water sorption was accompanied by swelling

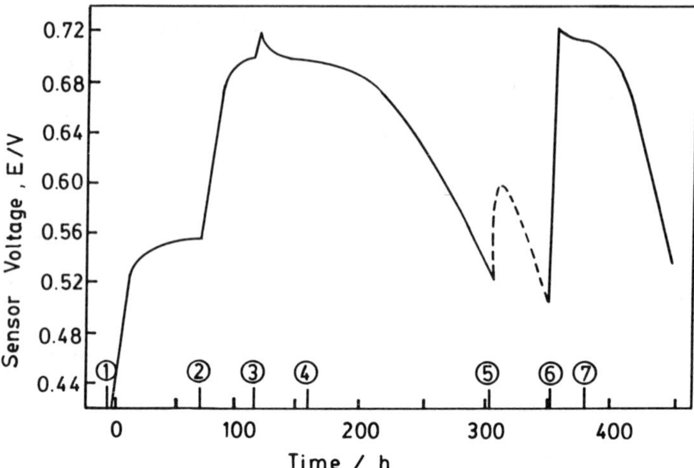

Figure 5. Response of Sensor to Hydrogen in Steel. Numerals on diagram refer to text.

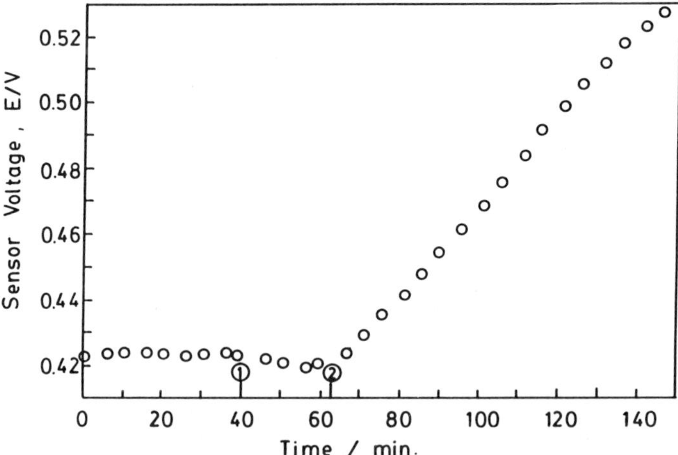

Figure 6. Breakthrough Time for Response of Sensor to Hydrogen in Steel. 1, Acid solution introduced to steel vessel; and 2, sensor voltage increasing.

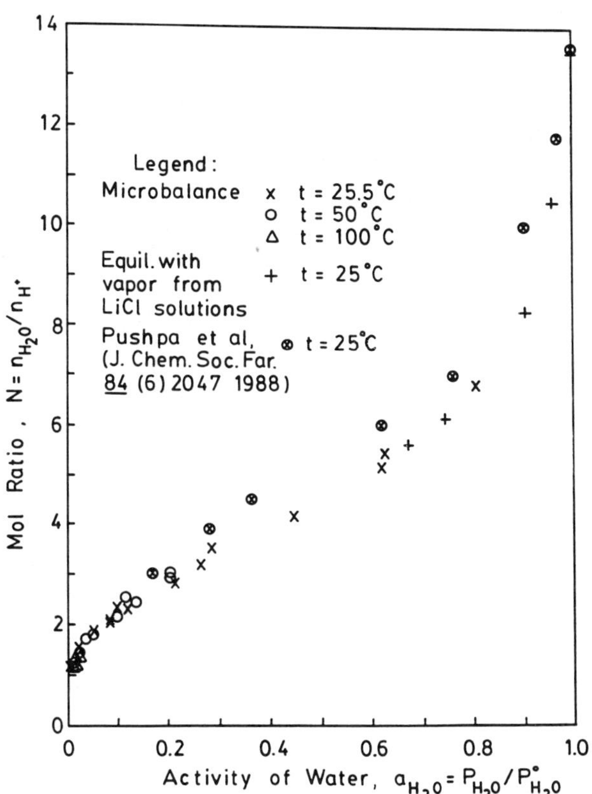

Figure 7. Sorption of Water by Nafion. Measurements with microbalance and equilibration with LiCl–H_2O solutions.

of the Nafion; the specific volume of the material increased from 0.49 cm^3/g for material dried at 130°C under vacuum, to 0.55 cm^3/g for material equilibrated with pure water.

Water sorption data obtained from the microbalance permitted evaluation of the average diffusion coefficient for water in Nafion over the range of the concentration increment (or decrement). The diffusion coefficient D was determined from the data using the expression *(8)* ,

$$M_t/M_\infty = (4/\pi^{1/2})(Dt/\ell^2)^{1/2} \tag{9}$$

where M_t and M_∞ are the mass of water at time t and mass at equilibrium and ℓ is the thickness of the material. Plots of M_t/M_∞ vs $t^{1/2}$ showed good linearity to $M_t/M_\infty \sim 0.5$, for both sorption and desorption.

Diffusion coefficients thus determined are presented in Figure 8 as a function of the average mol ratio N for the particular increment (or decrement). It is seen that the diffusion coefficient is a strong function of the water content, and that data obtained from sorption experiments differ from those obtained from desorption at a given value of N. These differences are attributed to dimensional changes of the Nafion accompanying sorption and desorption. Furthermore, it is seen that D increases with increase of water content to $N \sim 4.0$ followed by decreasing D for $N > 4.0$.

Data obtained at 100°C are least reliable due to the small mass change of the Nafion and the restricted experimental range. In particular at the lowest water activity levels, oscillation of the microbalance obscured the measured mass change of the specimen. At the water activity level $a_{H_2O} \sim 0.02$, the activation energy for water diffusion has been estimated as ~ 3.46 kcal/mol.

Electrical Conductivity of Nafion. The electrical conductivity of Nafion was measured with the following schematic arrangement,

Air, H_2O, Pt | Nafion | Pt, H_2O, Air

A Nafion disc, to each side of which Pt black had been applied, was held between platinum gauze and clamped with plexiglass flanges, provided with gas access ports. The unit was mounted in a small vessel provided with connections to an air supply of known humidity, and located in a constant temperature oven. The electrical resistance of the Nafion was measured with an AC conductivity bridge or by use of an AC power supply in conjunction with an ammeter. The electrical conductivity of the Nafion is primarily a function of the mol ratio N as shown in Fig. 9. The regression equation is,

$$\log \sigma = 6.48 \log N - 5.08; \; r = 0.996 \qquad (10)$$

It is evident that the electrical conductivity is a strong function of the water content of the Nafion; the Grotthuss mechanism of conduction is suggested to account for this dependency.

Amperometric Sensor for water. The strong dependency of the electrical conductivity of Nafion on the water concentration indicates possible application as a water sensor. The responsiveness of the sensor, constructed in a similar manner to that described in the previous section, to changes in the humidity of the air stream at ambient temperature is shown in Figure 10. It is evident that the device shows good responsiveness and reproducibility.

Conclusions

1. Electrochemical sensors of the potentiometric type for hydrogen and oxygen in inert gas mixtures utilizing Nafion polymer electrolyte show a Nernstian response with good long-term voltage reproducibility and stability.
2. With a reactive gas mixture, e.g., hydrogen and oxygen with an inert gas, the sensor voltage is interpreted as a mixed potential.
3. The sensors have application for monitoring of hydrogen in metals, e.g., H in steel and zirconium.
4. The equilibrium water concentration in Nafion is determined by the activity of water in the gas phase.
5. The diffusion coefficient of water in Nafion is a strong function of the water content.
6. The electrical conductivity of Nafion is primarily a function of the water content.
7. An amperometric sensor for water utilizing Nafion has been developed. This unit showed good reproducibility and responsiveness to changes in gas humidity.

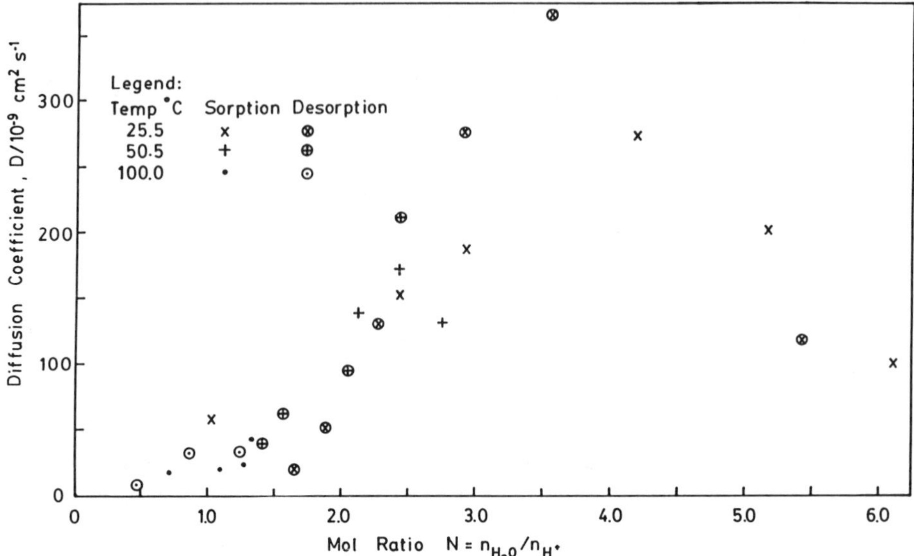

Figure 8. Diffusion Coefficient of Water in Nafion as Function of Temperature and Water Content.

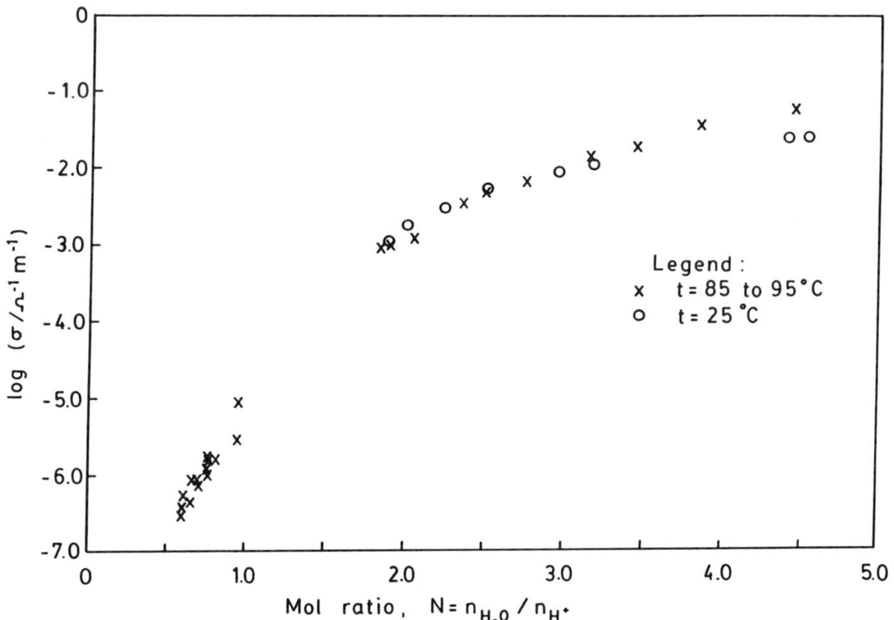

Figure 9. Electrical Conductivity of Nafion as Function of Water Content.

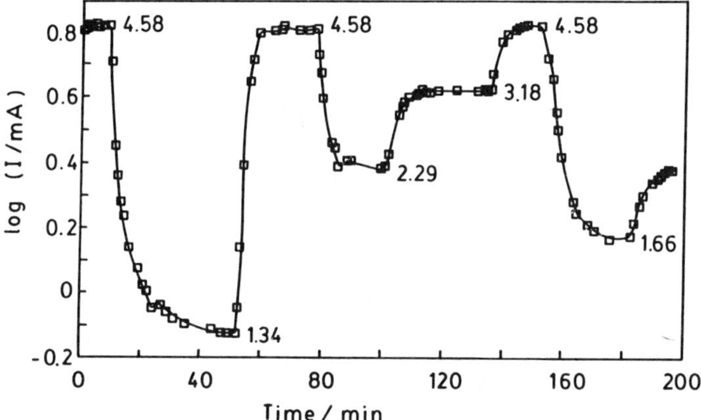

Figure 10. Response of Amperometric Humidity Sensor to Humidity Change of Equilibration Gas. Numerals represent water partial pressure, mm Hg in air.

Literature Cited

1. Morris, D. R., Kumar, R. V.; Fray, D. J. *Ironmaking and Steelmaking*, **1989**, *vol. 16*, pp. 429-434.
2. Fray, D. J.; Morris, D. R., U.S. Patent No. 4879005 Nov. 7, **1989**.
3. Miura, N.; Kato, H.; Yamazoe, N.; Sieyama, T. in *Chemical Sensors*; Editors, T. Seiyama et al; *Analytical Chemistry, Symposium Series vol. 17* , pp. 233-238, Elsevier **1983**.
4. Fontana, M. G.; Greene, N.D. *Corrosion Engineering* 2nd ed. McGraw-Hill, New York, **1978**.
5. Crank, J.; *The Mathematics of Diffusion* , Oxford University Press, Oxford, **1964**, pp. 47-49.
6. Ellerbrock, H.G.; Vibrans, G.; Stuwe, H.P. *Acta Metallurgica* , **1972**, vol. 20, pp. 53-60.
7. Pushpa, K.K.; Nandan, D.; Iyer, R.M., *J. Chem. Soc.* , Faraday Trans. I, **1988**, *vol. 84* , pp. 2047-2056.
8. see ref. 5, pp. 247-248.

RECEIVED August 19, 1991

Chapter 15

Polyvinylpyrrolidone-Grafted Silica Resin

Promising Packing Material for Size-Exclusion Chromatography of Water-Soluble Polymers

Yoram Cohen and Patricia Eisenberg

Department of Chemical Engineering, University of California, Los Angeles, CA 90024

General approaches to the modification of silica resins for aqueous size exclusion chromatography (ASEC) are reviewed with emphasis on surface modification by polymers. A new approach to the modification of silica SEC resins via graft polymerization with vinylpyrrolidone is described. The resulting VP-silica resins was found to be useful for ASEC of water soluble polymers. The ability of the VP surface layer to swell in various solvents, while the silica support maintains the mechanical integrity of the resin, provides for an additional degree of freedom in developing ASEC applications.

Size exclusion chromatography is an analytical technique in which molecules in a mobile phase are separated according to their effective size; thus, retention is principally controlled by the pore size distribution of the packing material. Size exclusion chromatography has been used to characterize water-soluble polymers since the 1950's (*1*). When size exclusion chromatography (SEC) was first introduced, it was performed exclusively using crosslinked polymeric resins. For example, dextran gels were used to characterize biopolymers in aqueous media and macroreticular polystyrene resins have been used to separate synthetic polymers in organic solvents (*2*). The early polymeric-type gel resins were restricted to low solvent flow rates to avoid gel compaction due to high operating pressures, with the consequence of long analysis times. More recently, a variety of highly cross-linked resins that can be used at pressures as high as 3000 psi have been introduced. As an alternative to polymeric resins, rigid porous silica-based packings of controlled pore size were developed. These porous silica beads are incompressible up to pressures of about 400 atm (6000 psi). However, problems of polymer adsorption and denaturation of proteins on silica packings have severely limited their application.

0097–6156/92/0480–0254$06.00/0
© 1992 American Chemical Society

When silica-based packings are used, particularly with aqueous mobile phases, band distortion due to secondary retention can occur. Secondary retention can be attributed to ion exclusion, ion exchange or non-specific surface adsorption (*3*). **Ion exclusion** occurs when the pH of the mobile phase is greater than about 4 and thus the silica surface carries a negative charge due to the ionization of the surface silanol groups. Consequently, negatively charged solute molecules are repelled from the surface and thus may not enter the pores of the packing material. Solute molecules which are excluded from the pores will experience shorter retention times and hence the molecular weight obtained from the universal calibration curve will be higher than the true value. Secondary retention due to **ion exchange** is experienced when the pH of the mobile phase is higher than 4 and the polymer molecules being analyzed are cationic species. Given the above conditions, solute molecules can be attracted by the surface leading to an increase in the solute's retention and consequently the determined molecular weight will be lower than the true value. Higher retention times can also be caused by non-specific solute adsorption as well as adsorption due to hydrogen-bonding with the silica silanol groups. In addition to undesired retention of the injected samples, adsorption onto the silica surface results in low solute recovery.

To overcome the above problems associated with silica resins, interaction between the silica and the polymer solute to be separated must be reduced or eliminated. Various procedures for minimizing polymer-silica interactions have been proposed. These procedures include modification of eluent composition by adding salt to the mobile phase or increasing buffer concentration, or changing the pH of the mobile phase so neither the surface nor the sample are ionized. These procedures have been partially successful, but they require tedious method development time, and reproducibility at long operating times can be difficult.

The breakthrough in controlling solute-silica interaction was achieved by chemical modification of the silica surface. These modifications include silica surface silylation, adsorption of polymers, polymer adsorption followed by crosslinking of the adsorbed layer, and surface-grafted polymers (*1*). In this chapter modified silica packing materials are briefly reviewed with a focus on gel-like polymer surface layers that have shown promise as effective modifiers of silica SEC resins.

Silylated-Silica Resins. The feasibility of using various silylated silica surfaces in size exclusion chromatography of water soluble polymers was reported by numerous investigators (*4-10*). For example, Engelhardt and Mathes (*4,5*) demonstrated that a bonded amine phase was suitable for exclusion chromatography of polyvinylpyrrolidone (PVP) and dextran, but it partially retained polyethylene glycol (PEG).

In another study the use of a polar diol bonded phase was demonstrated by Herman and Field (*11*). The diol-silica resins, however, retained up to 75% and 100% of polyethylene oxide (PEO) and PVP samples, respectively. It was found that upon addition of an organic solvent to the aqueous mobile phase (50% methanol in water for PEO, and 40% acetonitrile in 0.01 M KH_2PO_4, pH=2.1 for PVP), the recoveries of both samples increased to 100%. Herman and Field *(11)* suggested that since PVP may behave as a weakly acidic polyelectrolyte, the adsorptive retention due to the diol columns using water as the mobile phase are due to hydrophobic interactions with the diol phase and not due to adsorption on the residual silanols. The above study and the recent work of Parnas et al. (*12*) demonstrated that surface chemistry of the modified silica clearly has an important factor in controlling surface-polymer interactions.

Adsorbed-Polymer Silica Resins. A number of investigators have proposed that porous silica packing modified with an adsorbed water soluble polymer layer can be utilized for SEC of water soluble polymers and proteins (*13-15*).

The early study of Hiatt, et al. (*13*) demonstrated that porous glass conditioned with an adsorbed layer of PEO with high molecular weight (100,000 and 200,000) prevents the adsorption of rabies virus. Since PEO has a strong affinity for complex formation by hydrogen bonding, the displacement by the rabies virus was effectively blocked. In another study Hawk et al. (*16*) investigated the performance of a resin made by adsorbing polyethylene glycol (PEG, MW=20,000) onto controlled-pore glass minimized the adsorption of protein and enzymes. Adsorption of proteins and enzymes was apparent with the above resin, and thus phosphate buffer mobile phases were used to reduce polymer solute adsorption onto the modified silica packing. In a landmark study Letot, et al. (*17*) investigated the aqueous SEC of PEG, PEO, PVP, and Dextran using a porous silica resin modified with an adsorbed PVP (MW= 10,000) surface layer. Their study revealed that injected PVP samples showed a delayed elution. Letot et al. (*17*) suggested that dipolar interactions between adsorbed PVP and PVP chains in solution was responsible for the above behavior.

The studies with adsorbed polymer layers demonstrated that surface modification with polymers has merit. Unfortunately, long term stability with such silica resins with a gel-like polymer surface layer is poor, due to the eventual attrition of the polymer layer.

Cross-linked adsorbed polymer-silica Resins. Polymers adsorbed onto silica can be stabilized by cross-linking the adsorbed polymer and thereby creating a networked coating. In this approach a polymer layer is first adsorbed onto the silica surface, and it is subsequently crosslinked using an appropriate crosslinking agent (*18-22*). For example, using the above method Köhler (*19*) immobilized polyvinylpyrrolidone (molecular weight in the range of 10,000-360,000) on small and large-pore silica (5 μm particle size and 10 nm pore diameter and 7 μm particle size, with 30 nm pore diameter) and evaluated the performance for protein analysis. In one illustrative example a PVP-silica resin (MW=24,000) was used with a protein mixture consisting of transferrin (MW=80,000), hemoglobin (MW=68,000), ribonuclease A (MW=13,700) and lysozyme (MW=14,400). Recovery studies revealed that all the investigated proteins were eventually quantitatively eluted, with the exception of conalbumin (MW=86,000) for which sample recovery was in the range of 80%

to 85%. More recently, Santarelli et al. (*20*) developed a SEC silica resin (40-100 μm particle size, 30 and 125 nm pore size) prepared by coating the silica resin with 2-dietylaminoethyl (DEAE)-dextran in which a small fraction of dextran units contained the DEAE groups. In order to avoid attrition of the coated polymer, the DEAE-dextran layer was cross-linked. Most of the proteins investigated, using the DEAE-dextran-silica resin were eluted successfully, except for cytochrome c, which was significantly retained by the modified silica. The authors attributed this abnormal behavior to a low ionic strength of the mobile phase and to the small size of this protein. They concluded that some silanol functions inside the pores of the modified silica were not completely masked by the DEAE-dextran coating.

Polymer Grafted-Silica Resins. Silica SEC resins can also be modified by a direct covalent bonding of polymer chains onto the silica surface. Polymer chains may be grafted onto the surface of a solid or by graft polymerization or polymer grafting. Graft polymerization involves the growth of polymer chains from surface active sites by step or chain polymerization reaction mechanism. In contrast, polymer grafting involves the chemical bonding of live polymer chains to the support surface. The polymer grafting approach allows for the attachment of monodispersed polymers. However, since polymer molecules must diffuse to the solid surface, diffusional limitations and steric hindrance effects severely reduce the degree of surface coverage and graft yield obtained by the polymer grafting method (*23*). In contrast, in graft polymerization diffusion limitations and steric hindrance effects in porous supports are diminished due to the smaller size of the monomer molecules. Thus, a higher surface concentration and a more uniform surface coverage by the grafted polymerized chains are possible. Recently, a PVP-grafted silica resin prepared by Krasilnikov and Borisova (*21*) was used to purify different viruses without the loss of biological activity. Their study with the PVP-grafted column revealed that a tris-HCl buffer mobile phase (pH=7) could be used for the complete elution of the viruses. In contrast, a much higher pH was required for complete recovery of viruses on unconditioned silica resins.

The interest in PVP-Silica resins has increased in recent years due to the unique potential of PVP in enabling the separation of proteins and water soluble polymers. Polyvinylpyrrolidone, which is of particular interest in this work, is a nontoxic and biocompatible polymer which is soluble in aqueous as well as many organic solvents (*24*). The following sections focus on a description of a novel PVP-Silica resin synthesized by graft polymerization.

Experimental

Chemical Modification of Porous Silica by Graft Polymerization. The synthesis of the PVP-grafted silica resin involves two steps: a) a silylation step, which provides active sites on the silica surface for a graft polymerization reaction with the monomer; and b) a graft polymerization step in which polymer chains grow from active sites of the silica surface, by a free-radical polymerization reaction. The silica support resin used in this study was

Nucleosil 1000-10, with an average particle diameter of 10 μm, average pore size of 1000 Å, and a surface area of 50 m^2/g. The process of graft polymerization for which a U.S. Patent was recently issued to the University of California is briefly described below.

Silylation. The silylation procedure involves the exposure of the silica substrate to an excess of organosilane dissolved in either an anhydrous solvent or an aqueous solvent (*25*). Prior to silylation the silica particles are washed with dilute hydrochloric acid to remove trace contaminants (*26*). Subsequent washing of the silica resin with deionized water also results in the hydrolysis of surface siloxane groups to form silanols groups, which provide the reactive sites for silane bonding (*27*). In the present process the silylated silica particles were dried at 150°C under vacuum to remove surface water and avoid the formation of polysilanes which may limit the development of a uniform surface coverage (*28*). High drying temperature should be avoided in order to eliminate the condensation of the surface silanol groups to form siloxane bonds.

The silylation procedure employed in this work followed the method of Chaimberg et al. (*25*) (see Figure 1). Accordingly, a predetermined amount of silica was placed in a round-bottom flask into which a 10% solution of vinyltriethoxysilane in xylene was added. The reaction was performed for 5 hours, and the condenser was kept at a temperature above the boiling point of the displaced alcohol (ethanol in this case), but below the boiling point of the silane solution. The silylated particles were washed several times with xylene and allowed to cure overnight at 50°C under vacuum (*29*).

Since the subsequent graft-polymerization reaction was performed in an aqueous solution, the remaining ethoxy groups on the surface grafted triethoxyvinyl silane were first converted to hydroxyl groups in order to obtain hydrophilic silica particles which can be effectively dispersed in the aqueous reaction mixture. This latter step was carried out by hydrolysis of the ethoxy groups (Figure 1b) in water at a pH of 9.5 (*30*).

The amount of surface-bonded vinylsilane was determined by thermogravimetric analysis (TGS-2, Perkin Elmer, Norwalk, CT) of the silica particles. The silica particles were first held at 100°C until no weight loss was observed for a period of 10 minutes. This procedure was performed in order to eliminate the nonchemically bonded water. Subsequently, the weight loss upon heating of the silica particles in an air atmosphere, at 25°C/min from 100 to 700°C, was recorded. The upper limit of 700°C was sufficient to remove the bonded silane from the silica surface. The amount of surface-bonded vinylsilane was then determined by subtracting the weight loss of the virgin silica from the weight loss of the silylated silica sample. The results for two different resins synthesized in this work are given in Table I.

Graft Polymerization. The graft polymerization reaction of vinylpyrrolidone onto vinylsilane modified porous silica particles was performed in an aqueous slurry batch reactor (*31*) (Figure 2) under a nitrogen atmosphere. All the experiments were conducted at 70°C. The polymerization

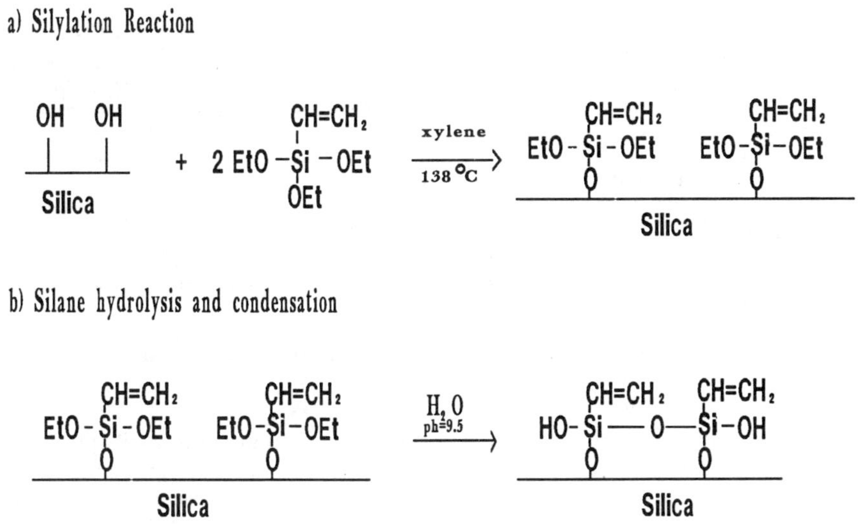

a) Silylation Reaction

OH OH | | Silica + 2 EtO–Si(CH=CH$_2$)(OEt)–OEt $\xrightarrow[138\,^{\circ}C]{xylene}$ EtO–Si(CH=CH$_2$)(O)–OEt EtO–Si(CH=CH$_2$)(O)–OEt Silica

b) Silane hydrolysis and condensation

EtO–Si(CH=CH$_2$)(O)–OEt EtO–Si(CH=CH$_2$)(O)–OEt Silica $\xrightarrow[ph=9.5]{H_2O}$ HO–Si(CH=CH$_2$)(O)—O—Si(CH=CH$_2$)(O)–OH Silica

Figure 1. Silylation procedure using organic solvents.

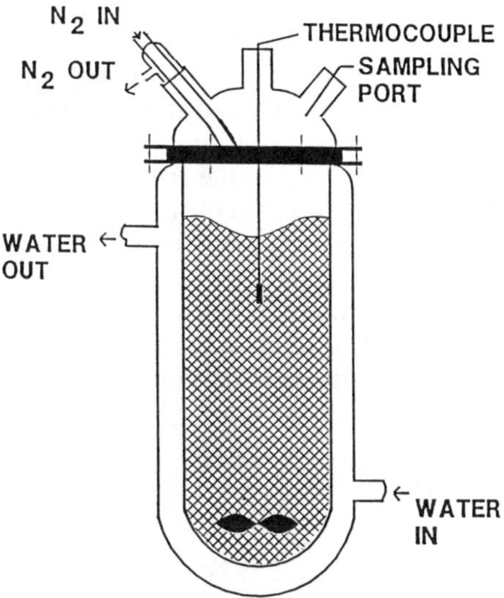

Figure 2. Graft polymerization reactor.

TABLE I. Vinyl Group Concentration and Polymer Graft Yield

Type of column	Silica Type	Vinyl Groups Concentration (μmol/m^2)	Polymer Graft Yield (mg/m^2)
PVP-30 (30% VP[a])	Nucleosil 1000-10	2.4	1.10
PVP-10 (10% VP)	Nucleosil 1000-10	3.2	1.05
Adsorbed PVP-silica	Nucleosil 1000-10	-	0.75[b]

[a]VP - vinylpyrrolidone
[b]Resin prepared by adsorbing PVP (MW = 10,000) at the adsorption plateau region.

procedure consisted of first placing a predetermined amount of silylated silica particles into the reaction vessel along with a 200 ml solution of vinylpyrrolidone monomer and deionized water. The reaction mixture was heated until the solution reached the desired temperature. Subsequently, 0.4 ml of ammonium hydroxide (58%) was added, and the grafting reaction was then initiated with 1.0 ml of hydrogen peroxide (30%). Ammonium hydroxide acts as a buffer for the reaction mixture, and thus it prevents the formation of undesirable acetaldehyde by-product under acidic conditions (*32*). In addition, ammonium hydroxide has a strong activating effect on the polymerization reaction which shortens the latent period and increases the rate of reaction. The nitrogen atmosphere was necessary in order to eliminate the atmospheric oxygen which is known to lead to an increase in the latent period of polymerization, as well as a decrease in the rate of polymerization.

The conversion of the monomer in the reaction mixture was determined by UV analysis. Vinylpyrrolidone monomer and polyvinylpyrrolidone both contain carbonyl groups which absorb UV radiation at 196 nm. The vinyl groups which absorb at 233 nm are only present in the monomer. Therefore, by taking the ratio of the absorbance at 233 and 196 nm, one obtains the ratio of monomer to polymer concentration and thus the conversion is given by:

$$conversion = \left[\frac{(Abs\ 233/196)_{t=0} - (Abs\ 233/196)_{t}}{(Abs\ 233/196)_{t=0}}\right]$$

The concentration of grafted polymer in the reaction mixture was negligible compared to the amount of homopolymer. Thus, a correction to account for the grafted polymer was not needed in the calculation of the conversion as given above. The conversion versus time results for the PVP-grafted silica resins; for resins PVP-30 and PVP-10 prepared using an initial monomer concentration of 30% and 10% (v/v) respectively are depicted in Figure 3.

The initial rate of homopolymer formation was found to increase with increasing monomer concentration (Figure 3). These results are in qualitative agreement with the previous homopolymerization study of Kline (*32*).

The amount of grafted polyvinylpyrrolidone in the silylated silica particles was also determined by a thermogravimetric analysis as described previously. The graft yield, defined as the mass (mg) of grafted polymer per square meter of silylated silica, as a function of the reaction time is shown in Figure 4 for the synthesis at initial monomer concentration of 10% (v/v). The final graft yields for resins PVP-30 and PVP-10, prepared following the procedure described above, are given in Table I.

Size Exclusion Chromatography with PVP-Silica Resins. The PVP-Silica resins were packed into stainless steel columns (25 cm length, 0.46 cm I.D with 2 μm frits, Beckman Instruments, Fullerton, CA) by a slurry technique using an Isco pump Model 3250 (Isco Inc., Lincoln, Nebraska). The columns were packed at a temperature of 40°C and a maximum flow rate of 0.6 ml/min.

Aqueous size exclusion chromatography (SEC) analysis of PEG, PEO and PVP using untreated, adsorbed PVP-Silica and grafted PVP-Silica porous silica resins was performed using a high pressure liquid chromatographic system. The stability of the columns was evaluated through the use of a sample recovery analysis. The HPLC system consisted of a ternary gradient system (Isco Inc, Lincoln, Nebraska) connected to an autoinjector (ISIS model, Isco Inc, Lincoln, Nebraska) with a sample loop injector (Valco C6W, Houston, Texas) and a metering pump (FMI RHOCTC Oyster Bay, New York). A Hewlett-Packard series 1050 multiple wavelength UV detector (Hewlett-Packard, Richmond, California) was used as the concentration detector.

Results and Discussion

Unmodified Silica resin. The initial experiments with the freshly packed virgin NU-1000 silica column demonstrated total retention of low molecular weight PEG and PEO polymer standards (MW=4,250 - 100,000). The injection of higher molecular weight polymers yielded low sample recovery consistent with the finding of Hawk (*16*) and Letot (*17*) for SEC of PEG and PVP using silica resins. Upon continued injections with PVP, the peak areas gradually increased. This behavior indicated that as the adsorption capacity was reached, the fraction of the injected polymer sample that could adsorb decreased. These results suggested that a fully coated PVP adsorbed layer is required for complete elution of injected polymer standards.

Adsorbed PVP-Silica resin. A second SEC column was prepared in which low molecular weight PVP was used to develop an adsorbed PVP layer that would allow better coating within the silica resin pores. The adsorbed PVP-Silica resin was prepared by eluting a 2% PVP solution (MW 10,000) through the column at 0.5 ml/min for 15 hours. At the conclusion of the coating period, the PVP solution was replaced with deionized water, which was eluted until a stable baseline at the detector was observed. Upon completion of the

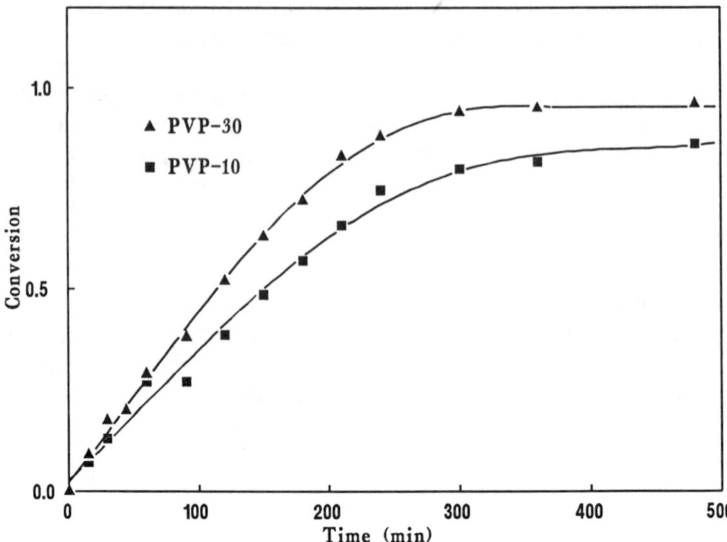

Figure 3. Vinylpyrrolidone conversion for two different initial monomer concentrations (T=70°C).

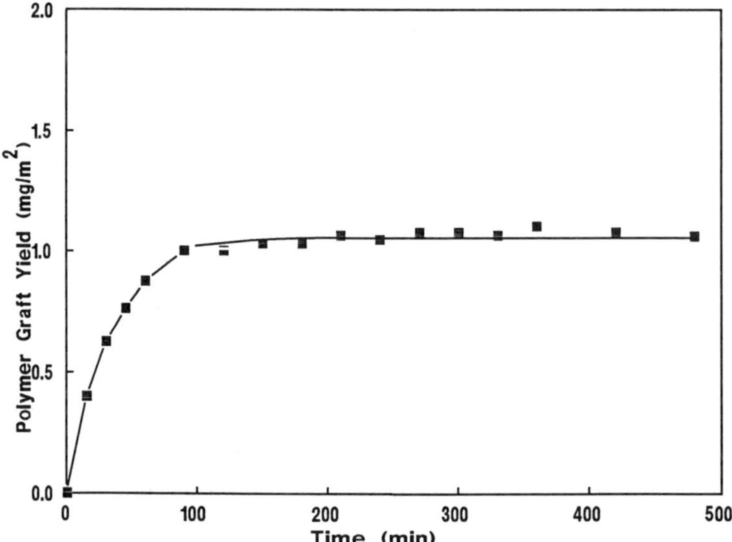

Figure 4. Polyvinylpyrrolidone graft yield as a function of reaction time. Initial vinylpyrrolidone monomer concentration of 10% (T=70°C).

adsorption procedure, the chromatographic peak areas were found to be significantly higher relative to the untreated silica resin. PEG and PEO polymer standards (Mw/Mn<1.12) were used to characterize the calibration curve with typical operating pressures in the range of 350-400 psi at a flow rate of 1.0 ml/min. The total permeation limit of the PVP-10K column was approximately 3.25 ml which corresponds to a molecular weight of approximately 10,000. The upper exclusion limit was not found due the unavailability of sufficiently high molecular weight standards.

In addition to the PEG and PEO standards, broad molecular weight distribution samples of dextran and PVP were injected for comparison of elution volumes. The lower molecular weight PVP samples showed a delayed retention which is in agreement with the results of Letot et al. (*17*). Although the adsorbed PVP-10K column demonstrated reasonable resolution, degradation of the adsorbed polymer layer occurred with continued operation. In order to demonstrate this degradation effect, timed hourly injections were made with three PEO standards of molecular weights 23,000, 105,000, and 400,000 over a period of four days. The variability of the normalized peak height of the chromatograms with time of operation (Figure 5) reveals complex adsorption/desorption effects that demonstrate the instability of the adsorbed PVP layer. The above results suggest that, while the original idea of Letot et al. (*17*) of using a PVP surface layer is appealing, such a polymer layer should be covalently bonded to the silica resins to impart long-term stability.

Grafted PVP-Silica Resin. The PVP-10 and PVP-30 grafted PVP-Silica resins prepared in this work had a PVP surface layer with an average molecular weight of 7,000 and 13,500, respectively. The molecular weight was estimated based on the analysis of Chaimberg (*33*). Since the grafted PVP layer consists of terminally anchored chains, it can have a significant effect on the permeability of the packed SEC column. In this study, the permeability of the PVP-10 and PVP-30 SEC columns was evaluated for four different mobile phases. The mobile phases were deionized water (at 25°C and 28°C) which is a good solvent for PVP, water:acetone (33.2:66.8 v/v) at 25°C and aqueous 0.55 M sodium sulfate solution at 28°C. The two latter mobile phases are theta solvents for PVP (*34*). The experimental procedure consisted of evaluating the flow rate-pressure drop curves for the SEC columns and subsequently determining the permeability from Darcy's law for flow through packed beds (*35*):

$$Q = k\left[\frac{A\,\Delta P}{\mu\,L}\right]$$

where Q is the flow rate (cm^3/sec), A is the cross sectional area of the column (cm^2), ΔP is the pressure drop (Pa), μ denotes the viscosity of the mobile phase (g/(cm sec)), L length of the column (cm), and k is the permeability (cm^2). The permeabilities calculated from the slope of the pressure drop-flow rate curve (see Figure 6 for PVP-30) for each mobile phase and for each column are listed in Table II. The permeability for water at both temperatures was

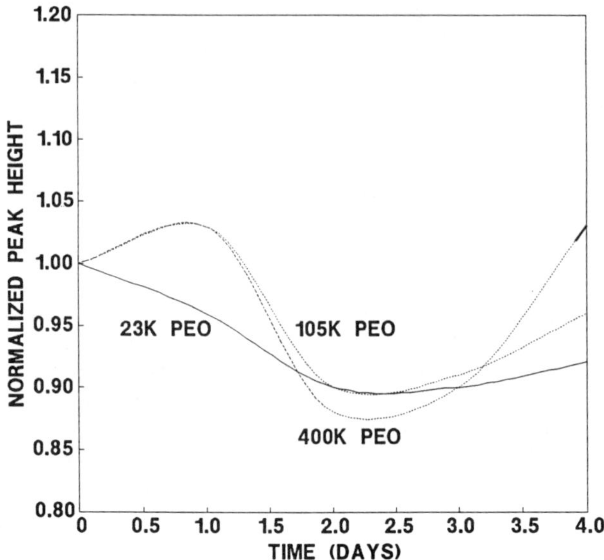

Figure 5. Illustration of column stability for adsorbed PVP-silica resin for different injected molecular weight samples of PEO. (Peak height is normalized with respect to the initial peak height.)

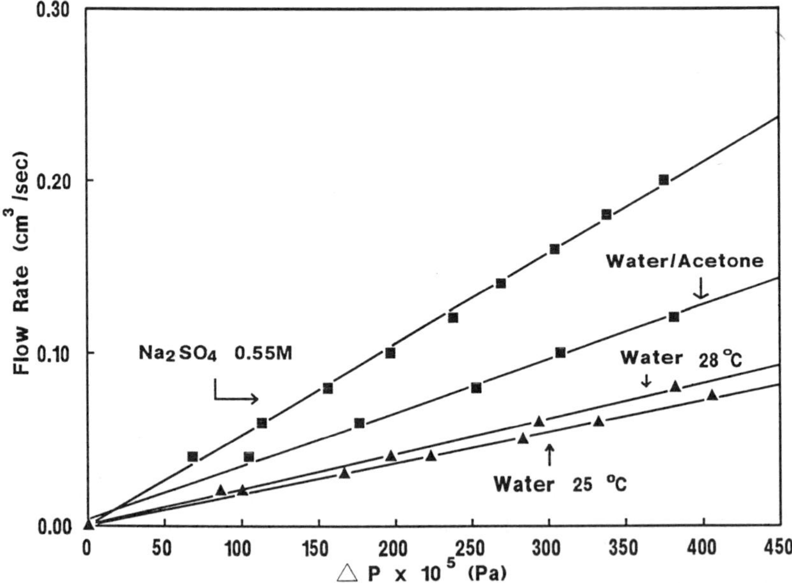

Figure 6. Flow curves for PVP-silica packed columns (PVP-30) for different solvents.

TABLE II. Permeability of PVP-Grafted Silica Columns

Mobile Phase	k (cm^2) x 10^{-12}	
	PVP-30	PVP-10
Water (25°C)	4.09	13.05
Water:Acetone (33.2:66.8 v/v) (25°C)	8.63	15.70
Water (28°C)	4.43	14.22
Sodium Sulfate 0.55 M (28°C)	13.67	18.55

lower than for the two theta solvents (water:acetone 33.2:66.8 v/v and sodium sulfate 0.55M). This trend is consistent with the fact that in a good solvent the grafted polymer chains are expected to have an expanded conformation, which results in a greater resistance to the flow and thus a lower permeability. In a theta solvent the polymer chains adopt a more compact conformation, and thus the resistance to flow is lower and in turn the permeability of the column is higher relative to a good solvent. The above implied change in the surface conformation with solvent power suggests that an additional degree of freedom may be available in designing polymer-silica matrix resins for SEC applications.

The performance of the PVP-10 and PVP-30 SEC columns is illustrated by the calibration curves plotted in Figure 7. Both columns demonstrate good size exclusion behavior with well defined total permeation and exclusion limits. The selective permeation or fractionation ranges for the PVP-30 and PVP-10 columns were 500-200,000 and 1000-500,000, respectively. The calibration curves for the PVP samples for both PVP-grafted columns are qualitatively similar, in distinct from the calibration curve for the PVP-adsorbed column. It is plausible that effects other than dipolar interactions as suggested by Letot (*17*) are operating with the adsorbed-PVP column. A comparison of the two PVP-grafted columns and the PVP-adsorbed column is also shown in Figure 7. It is quite clear that the PVP-grafted Silica columns demonstrate good molecular weight resolution in size exclusion chromatography for a large range of molecular weights.

Percent sample recovery studies were performed for PEG, PEO, dextran, and PVP polymer samples to determine the adsorption retention for the adsorbed PVP-silica and grafted-PVP silica resins. The adsorbed PVP-Silica resin displayed significant adsorption of the injected polymer solutes for the initial injections of the solute. The percent recovery increased after multiple injections once the resin's surface was fully covered with polymeric solute. The adsorbed PVP-Silica resin, however, was unstable as discussed previously (see Figure 5). In contrast, the grafted PVP-Silica resins displayed minimal interaction with the eluting polymer solute with an average recovery

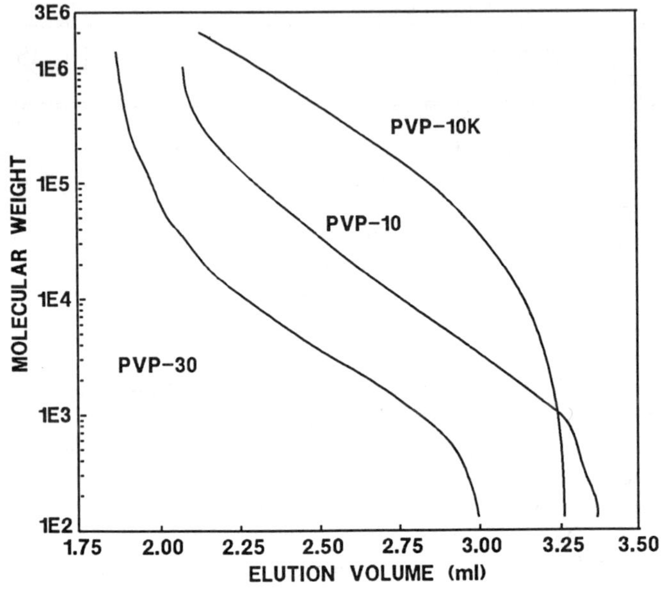

Figure 7. Molecular weight versus elution volume calibration curve for adsorbed PVP-silica (PVP-10K) and PVP grafted-silica (PVP-10 and PVP-30).

of the injected solute of 93%, 97% and 98% for PEO, dextran and PVP, respectively. Given that the estimated error in the analytical technique is about ±5%, it can be concluded that the performance of the grafted PVP-Silica SEC resin is excellent.

Finally, it should be noted that the grafted PVP-Silica resins were not optimized for high performance SEC analysis. Work is currently in progress to optimize the degree of grafted polymer surface density and molecular weight to produce SEC resins that will have a high resolution and high permeability to allow operation at high flow rates.

Acknowledgments

This work was funded in part by the National Science Foundation Grant CBT-8416719, the National Science Foundation Engineering Research Center on Hazardous Substance Control Grant CDR-86-22184, the United States Geological Survey Department of the Interior under Award No. 14-08-0001-G1315, and the Academic Senate of the University of California, Los Angeles.

Literature Cited

1. *Development in Polymer Characterization - 5*; Dawkins, J. V., Ed.; Chapter 4; Elsevier Applied Science: London, 1986.
2. *Aqueous Size-Exclusion Chromatography*; Dubin, P. L., Ed.; Chapter 8; Elsevier Science: New York, N.Y, 1988; Vol. 40.
3. Dolan, J. W.; *LC.GC.* **1990**, *8*, 291.
4. Engelhardt, H.; Mathes, D.; *J. Chromatogr.* **1977**, *142*, 311.
5. Engelhardt, H.; Mathes, D.; *J. Chromatogr.* **1979**, *185*, 305.
6. Becker, N.; Unger, K. K.; *Chromatographia* **1979**, *12*, 539.
7. Roumeliotis, P.; Unger, K. K.; *J. Chromatogr.* **1981**, *218*, 535.
8. Kiselev, A. V.; Khoklova, T. D.; Nikitin, Y. S.; *Chromatographia* **1984**, *18*, 69.
9. Feibush, B.; Cohen, M. J.; Karger, B. L.; *J. Chromatogr.* **1983**, *282*, 3.
10. Miller, N. T.; Feibush, B.; Karger, B. L.; *J. Chromatogr.* **1985**, *316*, 519.
11. Herman, D. P.; Field, L. R.; *J. Chromatogr. Sci.* **1981**, *19*, 470.
12. Parnas, R. S.; Chaimberg, M.; Taepaisitphongse, V.; Cohen, Y.; *J. Colloid. Interface. Sci.* **1989**, *129*, 441.
13. Hiatt, C. W.; Shelokov, A.; Rosenthal, E. J.; Galimore, J. M.; *J. Chromatogr.* **1971**, *56*, 362.
14. Shechter, I.; *Anal. Biochem.* **1974**, *58*, 30.
15. Darling, T.; Albert, J.; Russell, P.; Albert, D. M.; Reid, T. W.; *J. Chromatogr.* **1977**, *131*, 383.
16. Hawk, G. L.; Cameron, J. A.; Dufault, L. B.; *Preparative Biochem.* , **1972**, *2*, 193.
17. Letot, L.; Lesec, J.; Quivoron, C.; *J. Liq. Chromatogr.* **1981**, *4*, 1311.
18. Alpert, A.; Regnier, F. E.; *J. Chromatogr.* **1979**, *185*, 375.
19. Köhler, J.; *Chromatographia* **1986**, *21*, 573.

20. Santarelli, X.; Muller, D.; Jozefonvicz, J.; *J. Chromatogr.* **1988**, *443*, 55.
21. Krasilnikov, I.; Borisova, V.; *J. Chromatogr.* **1988**, *446*, 211.
22. Zhou, F. L.; Muller, D.; Santarelli, X.; Jozefonvicz, J.; *J. Chromatogr.* **1989**, *476*, 195.
23. Papier, E.; Nguyen, V. T.; *Polym. Letters.* **1972** , *10*, 167.
24. Blecher, L.; Lorenz, D. H.; Lowd, H. L.; Wood, A. S.; Wyman D. P. In *Handbook of Water-Soluble Gums and Resins*; Davison, R. L., Ed.; "Polyvinylpyrrolidone", Chapter 21; McGraw-Hill: New York, N.Y.,1980.
25. Chaimberg, M.; Cohen, Y.; *J.Colloid Interface. Sci.* **1990**, *134*, 576.
26. Eastmond, G.C.; Mucciariello, G.; *Polymer*, **1984**, *23*, 164.
27. Iler, R.K.; *The Chemistry of Silica*; John Wiley and Sons: New York, N.Y., 1979.
28. Majors, R.E.; Hopper, M.J.; *J.Chromatogr.* **1974**, *120*, 321.
29. Urban , M.W.; Koenig, J.L.; *Appl.Spectrosc.* **1986**, *40*, 513.
30. *Silylated Surfaces*; Pleuddemann, D.E.; Collins, W.T., Eds.; Gordon and Breach: New York, N.Y., 1979.
31. Chaimberg, M.; Parnas, R.; Cohen, Y.; *J. Appl. Polym. Sci* **1989**, *137*, 2921.
32. Kline, G.M.; *Mod. Plast.* **1945**, *11*, 157.
33. Chaimberg, M.; *Ph.D. Thesis*; University of California, Los Angeles, CA; **1989**.
34. Meza, R.; Gargallo, L.; *European Polymer Journal* **1977**, *13*, 235.
35. Scheidegger, A. E.; *The Physics of Flow through Porous Media*; University of Toronto Press: Toronto, Canada, 1960; Part IV, pp 68.

RECEIVED July 29, 1991

Chapter 16

Polyanionic Polymers in Bio- and Mucoadhesive Drug Delivery

Sau-Hung S. Leung[1] and Joseph R. Robinson[2]

[1]Columbia Research Laboratories, Madison, WI 53713
[2]School of Pharmacy, University of Wisconsin, Madison, WI 53706

Bio/Mucoadhesive drug delivery improves intimacy of contact between the drug delivery system and the absorptive surface and increases dosage form residence time. Bio/Mucoadhesive drug delivery systems adhere to the mucin-network and/or mucosal layer and remain in place until they dissolve or in the case of cross-linked polymers until the mucin or tissue layer replaces itself. Adhesive polymers can be neutral, anionic or cationic. On the basis of balance between adhesiveness and toxicity antonic polymers are preferred over neutral and cationic form. Bio/Mucoadhesion begins with establishment of intimate contact between the polymer and substrate. This is followed by bond formation and interpenetration/interdiffusion of polymer and substrate. Any factor that favors intimacy of contact, increases bond formation and enhances physical entanglement, will potentially increase bioadhesive strength.

Bio/Muco adhesion is becoming an important strategy for drug delivery because of the substantial advantages of localization and duration. As a result bioadhesive polymers have recently received considerable attention as platforms for controlled drug delivery. *(1-12)* Mucoadhesive drug delivery systems can localize the medicament in specified regions to improve and enhance both local concentration and systemic bioavailability, as well as promote intimate contact between the formulation and underlying absorbing mucosal layer. Furthermore, mucoadhesives prolong drug delivery system residence time with a corresponding reduction in dosing frequency.

There are four possible interactions between mucoadhesives and glycoproteins on epithelial surfaces: (1) covalent attachment; (2) electrostatic interaction; (3) hydrogen

0097–6156/92/0480–0269$06.00/0
© 1992 American Chemical Society

bonding; and (4) hydrophobic interactions. Aside from covalent attachment e.g., cyanoacrylate- "Super Glue", which is not presently a prominent mechanism for mucoadhesion, the remaining three interactions require intimate contact between polymer and mucin both for optimum adhesion. Polyelectrolyte polymers and mucin both swell in an aqueous medium. The expanded network of the swollen polymer and mucus enhances interdiffusion resulting in an increase in mechanical entanglement and area of contact, with a subsequent increase in interactions between polymer and mucin strands and thus an increase in adhesion.

When traditional formulations are administered to different locations of the body, e.g., ocular, nasal, buccal, vaginal, rectal and gastrointestinal, they are removed from the site of administration by the natural clearing mechanisms of the body. Thus, ocular formulations are diluted by the tear and removed via the drainage system. Nasal formulations are removed, with the aid of blinking, together with the mucous fluid, propelled by cilia toward the throat. Buccal gels or films are washed daily by 0.5 to 2 liters of saliva secreted by the salivary gland. Furthermore, oral formulations are propelled distally toward the large intestine by the motility pattern of the intestine. All mocoadhesive dosage forms must have sufficient adhesive force to resist the normal cleaning mechanism(s) of specific route of administration.

SCREENING METHODS FOR MUCOADHESIVES

Attachment of soluble polymers to the cultured human conjunctival epithelial cell surface was studied by Park and Robinson using a fluorescent probe technique*(2)*. Thus, the lipid bilayer portion of the cell was labeled with a fluorescence probe, pyrene. Adhesive polymers which bind to the cell surface, compress the lipid bilayer and change fluorescence intensity of the probe. A large value of fluorescence change indicates a strong binding potential of the soluble polymer to the cell membrane. It was found that, compared with polycations and neutral polymers, polyanions with high charge density, such as polyacrylic acid, are better bioadhesives. This method, however, cannot be used to measure the interaction of water-insoluble polymers to the cell membrane because of the interference of water-insoluble polymers to fluoresence spectra.

A majority of screening methods are based on determination of the shear or tensile strength of bioadhesion*(13)*. Tensile and shear strength are the vertical and horizontal components of the mucoadhesive strength. The tensile strength was measured by a Wilhelmy plate method*(4)* and a modified tensiometer method*(5,11)*. The Wilhelmy plate method*(4)* was designed by Smart et al. to measure the tensile strength between mucus and water-soluble polymers. The plates were first coated by submerging in a 1% test polymer solution and then oven-dried at 60°C to constant weight. The coated glass plate was then immersed in homogenized mucus solution. The tensile strength of the polymer-mucus interaction was measured after 7 minutes of contact, the shortest time to achieve measurable mucoadhesion. This is a quick screening method for water-soluble polymer, however, the mucin used is in homogenized state and not in natural state, consisting of a sol and a gel phase. The modified tensiometer*(5,11)* was designed to measure the tensile strength of the interaction between test polymers and the mucus layer. In this method a section of rabbit's stomach tissue was mounted onto a lower support, and another section of rabbit's stomach tissue was mounted onto an upper stopper, mucosal surfaces facing each other. A test polymer was placed between the mucin layers in a suitable test medium, e.g., 0.1M pH2 isotonic phosphate buffer. The force of detachment was then measured and the tensile strength determined. The mucoadhesion to natural mucus layer is measured, however, one has to be careful not to disrupt the mucus layer during tissue preparation.

Shear strength represents the horizontal component of mucoadhesion. A dual tensiometer was designed by Leung and Robinson *(5)* to measure the shear strength in mucoadhesion. The combination of tensile and shear strengths in mucoadhesion provides a better understanding of the mucoadhesion process. In this method two modified tensiometers were used.

Peppas and Buri *(6)* have developed a method to study the static and dynamic mucoadhesive properties of polymeric particles. They used a thin channel filled with artificial mucus gel or natural mucus. The particles were preswollen in a mucin solution and placed inside the mucus gel channel. The motion of these particles was generated by a controlled air flow and their motion with time were photographed. The velocities of the particles were determined from the pictures, and the hydrodynamic force required for detachment calculated. In this method, both freshly excised mucus layer or gastrointestinal tissue can be used.

The mucoadhesiveness of micro-size particles was determined by Teng and Ho *(7)* using a falling liquid film system. In this method an excised intestinal segment was spread on a plastic flute and inclined at an angle. A suspension of test particles was allowed to flow down the flute. The concentration of particles entering and leaving the intestinal segment were measured, and the amount of micro-particles retained by the mucous layer determined. By comparing the fraction of retained micro-particles, adhesion of polymer coated microparticles to the intestinal mucosal surface can be compared.

In addition to the above screening methods, an in-situ method using glass spheres coated with polymers to test the mucoadhesiveness of these polymers on rat jejunum or stomach was used by Ranga Rao and Buri *(8)*. The coated glass spheres were placed on the rat jejunum or stomach. The tissues were then rinsed with phosphate buffer or dilute acid. The number of particles remaining after rinsing was determined and used as an index for mucoadhesion.

The gastrointestinal transit of mucoadhesives in rats was studied by Ch'ng et al. *(11)*. Male Sprague-Dawley rats were fasted with free access to water for 48 hours prior to the experiment. A capsule containing radioactive test material or solid control was inserted into the stomach through a 2 to 3 mm opening in the stomach. 4 ml of normal saline was injected into the stomach. The rats were allowed to recover from the anesthesia and sacrificed at selected time intervals. The stomach and small intestine were removed, and the small intestine was cut into segments. The amount of test material and control were determined by measuring the amount of radioactivity or counting the number of particles in each segment of gut.

The adhesiveness of bioadhesive dosage forms was studied by Ishida et al. *(3,14)* and Marvola et al. *(15)*. A "stickiness" test apparatus *(3)* was used to measure the adhesiveness of tablets. In this method mouse peritoneal membrane was mounted on a support by a metal ring. The adhesiveness of the tablet on the peritoneal membrane was measured by a spring balance. A shearing "stickiness" test apparatus*(14)* was also used to measure the adhesiveness of an ointment. A horizontal force was applied to the interacting surfaces, and adhesiveness of the ointment was measured by a spring balance. Furthermore, Marvala et al.*(15)* used the isolated swine esophagi in aerated medium to determine the adhesive tendency of both tablets and capsules to esophagus.

A colloidal gold staining technique was developed by Park*(9)* to study mucoadhesion with the use of mucin-gold staining technique. In this method red colloidal gold particles were stabilized by forming mucin-gold conjugates with the mucin molecules. When the mucin-gold conjugates were added to a mucoadhesive hydrogel, a red color developed in the interacting interface. The mucoadhesive properties of different polymers can then be evaluated by measuring the intensity of the red color developed or by measuring the decrease in concentration of the conjucates.

Radiolabelling technique was used by Bridges et al.*(16)* to evaluate the bioadhesiveness of N-(2-hydroxypropyl) methacrylamide copolymers containing pendent sugar residues or quaternary ammonium groups. In this method intestinal rings were incubated with ^{125}I-labelled copolymers. The intestinal rings were then solubilized in 1 M NaOH. The amount of protein and radioactivity were determined, and amount of polymer captured per unit weight of tissue protein calculated.

Interactions of mucoadhesives with biological substrate surfaces may be studied using a confocal imaging technique. The scanning laser fluorescent confocal microscopy, an improved light microscopy, can be used to view biological structures with a minimal amount of disturbance, and scan optically with little loss in image quality of the interior sections*(17)*. Confocal fluorescence microscopy can image samples within a narrow depth of focus with improved resolution and elimination of out-of-focus noise. Which allows visualization of biological structures that emit or reflect light in thick specimens that are reasonably transparent. Permeation of protein through cornea, a transparent tissue, has been successfully studied using the confocal imaging technique*(18)*. Thus, confocal fluorescence microscopy may be used to study interactions of polymers with the natural mucin network with minimal disturbance to the mucus layer.

THE PROCESS OF BIO/MUCOADHESION

In the bio/mucoadhesive delivery of drugs, three different layers are involved, (Figure 1): (1) the mucoadhesive drug delivery system; (2) the mucus layer; and (3) the epithelial cell layer.

Epithelial Cell Layer. In the epithelial cell layer, the cell membrane is viewed as a two dimensional oriented viscous lipid solution where protein are free to move*(19)*. There are polysaccharids - containing structures on the external surface of cells and are collectively referred to as the glycocalyx*(20)*, which is maintained and synthesized continuously by the underlying cell*(21)* and appears to be partly responsible for the adhesive property of the cell. Most epithelial tissue carry a net negative charge, and the surface of this tissue is a dynamic interface containing regions for all types of non-specific and specific interactions. Adhesion at the epithelial layer can occur when the continuity of the mucus layer is interrupted either naturally, through the sloughing of cells and mucus, chemically by mucolytic agents, or by physical abrasion. The adhesion of certain polymers onto the epithelial layer depends significantly on the interaction between the polymer and specialized structures and/or macromolecules on the epithelial cell surface. The presence of macromolecules surrounding the cell has been well established*(22)*, e.g., fibronectins*(23)*, lectin*(24)*, concanavalin*(24)*, ligatin*(25)*, and polypeptides*(26)*. Attachment of natural and synthetic polymers onto the exposed epithelial layer may serve several different functions: formation of a continuous layer with the mucus network to cover exposed epithelium and thus protect the underlaying cell against physical and chemical insults; carry out some of the functions of natural mucin, e.g., lubrication and tissue hydration; facilitation of the recovery of damaged or diseased cells; and the sustained delivery of medicaments.

Mucus Layer. On the surface of all orifices of the body, with the exception of the ears, there is a continuous layer of mucus. Mucins are synthesized by goblet cells lining the mucosal epithelium layer and by special exocrine glands, such as salivary glands*(27)*. The basic component of all mucus is the mucin glycoproteins, which have strong adhesive properties to each other and other bioadhesive molecules. Mucin glycoproteins bind firmly to the epithelial layer resulting in a continuous and unstirred gel layer over the mucosa. Mucin glycoproteins have oligosaccharide side-chains*(28)*, and terminal L-fucose*(29)* and sialic acids*(30, 31)* groups at periodic intervals. The pKa of sialic acid is 2:6 giving a substantial negative charge to the tissue surface at physiological pH. Mucus is a viscoelastic gel with macromolecules linked together via cross-linkings, disulfide bonds*(32)* and physical entanglement*(33)* to form an infinite network.

Bio/Mucoadhesive. Bio/mucoadhesives include water-soluble, linear macromolecules and water-insoluble, crosslinked macromolecular networks. In a study of the adhesiveness of a series of anionic, cationic and neutral polymer*(2)*, it was found that polyanionic polymers are better bioadhesives than polycationic and neutral polymers when both bioadhesiveness and cellular toxicity are considered. Furthermore, polyanions with carboxyl groups e.g., polyacrylic acid, appear to be better candidates than those with sulfate groups*(2)*.

Bio/mucoadhesives are usually polymers with numerous hydrophilic functional groups that can form hydrogen-bonds, e.g., carboxyl, hydroxyl, amide, and sulfate groups*(34)*. Some contribution of electrostatic interaction to bio/mucoadhesion is also likely, because mucin and mucosal surfaces are negatively charged at pH 7.4*(30, 31)*. When mucoadhesives make contact with aqueous media, they swell and form a gel. The rate and extent of water uptake by the mucoadhesive will depend on the type and number of hydrophilic functional groups in the polymer structure, as well as the pH and ionic strength of the aqueous media. For water-insoluble polymers, e.g., cross-linked polyacrylic acid, the polymer network absorbs water, swells and expands when in contact with an aqueous media. However, the amount of water sorbed, i.e., the degree of hydration was found to increase with an increase in the percent of charged functional groups*(5)*. As the degree of hydration increases, the expanded nature of the polymer network increases, with a corresponding increas in mucoadhesive strength*(5)*. Thus, an increase in openness of the polymer network favors the bio/mucoadhesion process. Furthermore, the extent of equilibrium swelling of cross-linked polyacrylic acid was found to decrease almost linearly with an increase in the concentration of cross-linking agent*(35)*.

In a swollen macromolecular network, the mesh size, ζ, of the effective area for diffusion is the space between the spheres where the chains are confined. Critical hole formation occurs by increased chain flexibility and chain movement. The average mesh size of the network in angstroms can be represented as*(42)*

$$\zeta = \propto_s (2\, n l^2)^{1/2} \tag{1}$$

where ℓ is the bond length, n is the number of links (C-C bonds) between two cross-links, and

$$\propto_s = (V/Vo)^{1/2} \tag{2}$$

V is the volume of the swollen polymer, and Vo is the volume of the unswollen polymer. Mucoadhesive strength was found to increase with an increase in the openness of the polymer and mucin network*(5)*.

Mechanism of Bio/Mucoadhesion. The process of bio/mucoadhesion may be visualized as having two stages: firstly, the development of close contact between the adhesive and substrate; and secondly, interpenetration between the polymer and mucin (glycoprotein) network. At both stages interficial bonds are formed between the functional groups of both networks. Interfacial bonding occurs primarily through secondary bonding, e.g., electrostatic and hydrophobic interactions, hydrogen bonding, and van der Waals intermolecular interactions. For bioadhesives with charged groups, electrostatic interaction and hydrogen bonding appear to be of primary importance.

The development of close contact between the adhesive and substrate depends heavily on the miscibility of the two phases or the wetting ability of one phase to the other. The Dupré equation*(37)* describes the spreading ability of the adhesive to the substrate.

$$W_{AB} = \gamma_A + \gamma_B - \gamma_{AB} \tag{3}$$

Where W_{AB} is the thermodynamic work of adhesion, γ_A and γ_B are the surface tensions, and γ_{AB} is the interfacial tension. The interaction of cultured human endothelial cells with polymeric surfaces of different wettabilities have been studied, and an optimal cell adhesion was found with moderately wettable polymers*(38)*. Cell adhesion within a series of cellulose polymers was also found to increase with an increase in contact angle of the polymer surface*(39)*. Recently a combined spreading coefficient was used to study the mechanism bioadhesion*(40)*. The magnitude of the combined spreading coefficient was found to correlate well with the bioadhesiveness of a series of polymers. Thus, surface properties that effect surface tension and/or the wettabilties between adhesive and substrate have a significant role in bio/mucoadhesion.

Bio/mucoadhesion between an adhesive and substrate involves a hydrated macromolecular network in the adhesive or the substrate or both. At the initial contact and during the establishment of intimate contact of the two phases, interpenetration/interdiffusion occurs between the adhesive and substrate. The ability of the polymer chains to interpenetrate can be approximated by their ability to diffuse. Thus, chain-segment mobility of the polymers and substrates can be related to their viscosities and diffusion coefficient. Over a sufficiently restricted temperature range, the experimental diffusion coefficient, D, shows an exponential temperature dependence of the Arrhenius type*(41)*

$$D = Do \exp(-E/RT) \quad (4)$$

Where the pre-exponential factor Do is a constant and is independent of temperature over a finite temperature range, and E is the experimental activation energy for diffusion of the segment chains. Bioadhesion is a temperature dependent process (Figure 2), and the tensile strength increases with an increase in temperature*(36)*.

The diffusion theory of adhesion was first discussed by Voyutskii,*(43)* where the chains of the adhesive and substrate interpenetrate each other to a sufficient depth and create a semi-permanent adhesive bond. The representative mean diffusional path, S, for macromolecules can be estimated as*(44)*

$$S = (2t\,D)^{1/2} \quad (5)$$

where D is the diffusion coefficient.

The degree of hydration and hence expanded nature of the polymer was found to decrease with a decreasing percent of charged groups*(5)*. The average mesh size of the polymer network correlates linearly with the percent of acrylic acid*(45)*. Thus, the degree of hydration and the average mesh size of the polymer network is controlled by the percent of hydrophilic functional groups. Furthermore, the diffusion coefficient of acrylic polymers was found to depend on the mesh size of the network to the sixth power*(36)*, and the tensile strength was proportional to the average mesh size of the acrylic polymer network to the third power*(36)*. Thus, tensile strength depends significantly on the diffusion coefficient of the adhesive polymers.

The mucoadhesive strength of cross-linked polyacrylic acid was found to depend on the pH of the bathing medium*(12)*. The mucoachesive strength is high at acidic pH of 1 to 2 and drops sharply at around pH 4*(12)*. The sharp decrease in mucoadhesive strength around the pKa of polyacrylic acid (4.75) suggests the significant of the acid form of the carboxylic group in mucoadhesion. Mucoadhesion was also found to depend on the density of carboxyl-group*(35)*. Thus, for mucoadhesion to occur, polymer chains must be able to form hydrogen bonds with the underlying substrate and should be flexible enough to maximize the formation of hydrogen bonds*(35)*.

APPLICATION OF BIO/MUCOADHESIVES

Bioadhesive polymers can be used in a variety of different formulations, e.g., tablet,

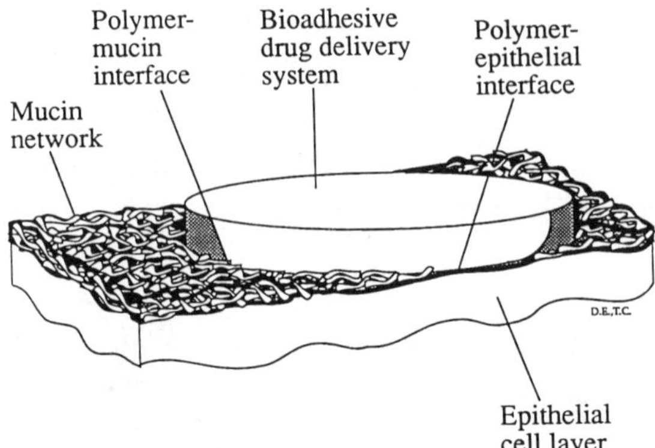

Figure 1: Schematic diagram at the site of bioadhesion.

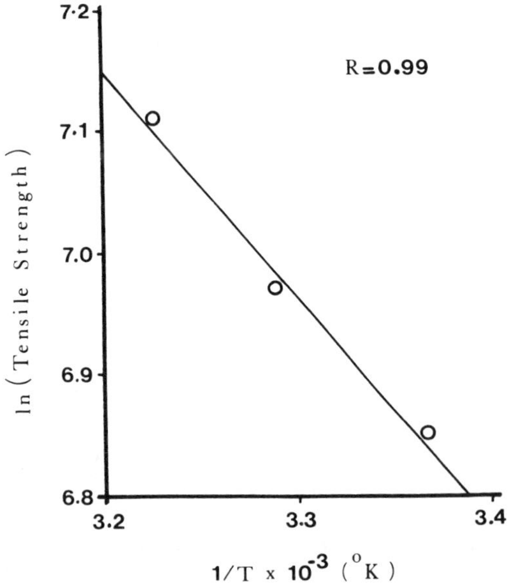

Figure 2: Graph of ln (tensile strength) versus 1/T. (Reproduced with permission from ref. 36. Copyright 1990 Elsevier.)

capsule, minocapsule, film, patch, gel or powder. Bio/mucoadhesive formulations can be used in any route that has a mucus covering, e.g., ocular, nasal, buccal, gastrointestinal, rectal and vaginal.

Ocular route. In the precorneal area physiological pH is 7.3 to 7.4. At the negatively charged corneal surface, there is preferential uptake of cationic liposomes over anionic and neutral liposomes*(46)*. Thus, it appears that some electrostitic interactions are occuring at the corneal surface with possible involvement of the mucus layer. The mucus layers in the cornea and the conjunctuva are 0.02 μm to 0.05 μm*(47)* and 1.4 μm*(48)* thick, respectively. Mucus forms the inner layer of the tear film and is in contact with the microvilli of both the corneal and conjunctival epithelia.

The major problem of ocular drug delivery is the excessive loss of drug via the lacrimal drainage system into the nasopharynx before adequate corneal absorption can occur. Besides nonproductive absorption and protein binding, excessive ocular drainage caused by induced lacrimation is a major problem in ocular drug delivery. The eye is an extremely sensitive organ that responds quickly to external stimuli. Administration of an ocular formulation is likely to stimulate the eye to increase tear secretion with subsequent drug loss. Thus, in formulation of bio/mucoadhesive ocular drug delivery systems, it is essential to carefully adjust the formulation to minimize eye irritation. Typical ocular bioavailability of topically applied drugs is 1 - 10%.

When an ocular mucoadhesive formulation is administered, there will be adhesion of the dosage form to the precorneal tissue. The drug will then dissolve/diffuse/partition into the tear film and release drug into the precorneal area for absorption.

A mucoadhesive ocular dosage form was used in the rabbit to increase the ocular bioavailability of progesterone*(49)*. The mucoadhesive used was cross-linked polyacrylic acid, commonly known as polycarbophil, and the area under the concentration-time curve was 4.2 times greater than a conventional ocular suspension. Similar results were obtained for ocular bioavailability of pilocarpine*(50)*. The mucoadhesives used were hyaluronic acid, polygalactouronic acid, mesoglycan and carboxymethylchitin.

Nasal Route. The human nasal cavity is divided by the median septum into two symmetrical and approximately equal chambers that open into the nasopharynx through two oval openings. These two oval openings, the choanae, measure 2.5 cm vertically and 1.5 cm transversely*(51)*. The nasal cavity is lined with a continuous mucosal layer from the nares to the nasopharynx. The mucosal layer lining the nasal choanae, is the thickest and most vascular*(51)*. Thus, drugs can be easily absorbed via the nasal route and will bypass first-pass metabolism and lumenal degradation associated with the oral route. In terms of nasal drug absorption, it seems that hydrophobic drugs are better absorbed than hydrophilic drugs*(54)*. The nasal mucosa and submucosa are liberally interspersed with goblet cells, as well as numerous mucous and serous glands, to secret mucus that keep the nasal mucosal surface moist*(52)*. The mucus layer consists of 90 to 95% water, 1 to 2% salt and 2 to 3% mucin. Mucus moves at a rate of 5 mm/min toward the throat by the ciliated cells present on the surface of the mucosa*(53)*.

Different mucoadhesive formulations have been developed to deliver drugs nasally. Carbopol 934 and hydroxypropyl cellulose were used in development of powder insulin dosage form*(55)*. The resultant hypoglycemic effect using a 3 u/kg was twice that of intraverrous administration of 0.5 u/kg. When hydroxypropyl cellulose was used as the bioadhesive base, the nasal powder was found to adhere to the nasal mucoas for up to 6 hours after application*(56)*. Mucoadhesive microspheres were prepared using albumin, starch and DEAE-dextran. The clearance of the microspheres was slowed, and contact between microspheres and the mucin/epithelial layer prolonged*(57)*. Furthermore, polyacrylic acid gel was used in formulations of insulin and calcitonin. The absorption of both insulin and calcitonin were found to be enhanced*(58)*.

Buccal Route. The oral mucosa can be divided into non-keratinized, keratinized and highly selective keratinized regions*(61)*. The borders of the lips and dorsal surface of the tongue is highly selective keratinized. The other keratinized regions are the hard plate and the gingiva. The non-keratinized regions consist of the soft palate, the mouth floor, the ventral surface of the tongue, the labial and buccal mucosa. The buccal cavity is a convenient and easily accessible site for bio/mucoadhesive drug delivery. In addition, the buccal route has lower enzymatic activity*(59, 60)* and can by-pass first-pass metabolism and gastric degradation.

Absorption of drug through the buccal mucosa is likely to occur most efficiently through the non-keratinized regions. Thus, an appropriate site of adhesion for mucoadhesive dosage form is on the non-keratinized regions.

The buccal cavity is lined by a relatively thick mucous membrane. It is highly vascularized and approximately 100 sq cm in area*(62)*. A layer of mucus lining the entire buccal cavity and copious water from the salivary gland (1 to 2 liters per day at around pH6) to act as a bathing medium to hydrate the bioadhesive for proper attachment. There are three main sources of saliva: the three pairs of main glands (the parotid, the submaxillary and the sublingual); the minor salivary glands in buccal, palatal and sublingual mucosa; and variety of smaller glands dispersed on the tongue, buccal and sublingual mucosa.

Mucoadhesive buccal tablets which remain in place for 3 hours were prepared using hydrozypropyl cellulose and lactose*(63)*. After placement of the mucoadhesive tablet on the buccal mucosa, the tablet was eroded slowly with time. Similary, a mucoadhesive buccal nitroglycerin tablet was found to have pharmacodynamic effects for up to 5 hours*(64)*. Mucoadhesive gels which have prolonged residence time over solution were investigated using polyacrylic acid*(65, 66)* and polymethylmethacrylate*(67)* as gel-forming polymers.

Mucoadhesive laminated buccal films and patches have also been investigated*(3, 68-70)*. Mono- or multilaminated films can be prepared by using a standard laminating technique, provided attention is given to the usual swelling property of the bioadhesive and thus potential seporation of the rate limiting membrane from the bioadhesive. Different layers can function as an adhesive layer, as a platform for drug delivery or simply as a drug reservoir.

Gastrointestinal Route. The oral route is the most conmonly used route of drug delivery. When a dosage form is administered orally, it will quickly reach the stomach via the esophagus. The dosage form will stay in the stomach for a certain period of time and will subsequently be released into the small intestine. The small intestine is 7 meters long from the pylorus to the ileocecal junction. It is divided into three portions: the duodenum; the jejunum; and the ileum. The mucous membrane of the small intestine is composed of the villi, small projections of cytoplasm, and rich network of capillary blood vessels and lymphatic vessels. Thus, the small intestine is the main absorption site of the majority of drugs. Bioavailability of drugs would be greatly improved if the dosage form can be retained in the stomach or the proximal portion of the small intestine and release medicaments slowly. The effect would be more significant for drugs with a narrow absorption window.

In order to formulate a good mucoadhesive oral drug delivery system, the gastrointestial (GI) motility pattern needs to be considered. GI motility in human and animal follow a cyclic pattern of digestive (fed) mode and the interdigestive (fasted) mode. In the fasted mode the motility pattern is divided into 4 distinct phases*(71)*. Phase I is a quiescent period, phase II has random spikes of electrical activity or intermittent contractions, phase III is a period of regular spike bursts of regular contractions at moximal frequency, and phase IV is the transition period between phase III and phase I. However, in the fed mode there is continuous motility with retropulsive forces approximately 50% the magnitude of the phase III activity. The GI motility in the fasted mode is commonly known as the interdigestive migrating motor complex (IMMC) with an average duration of 90 to 120 minutes. The IMMC is interrupted when food is administered. Phase III in the IMMC is commonly known as the "housekeeper wave". It originates in the foregut and propagates distally to the terminal ileum. The "housekeeper wave" serves as a cleaning mechanism to clear all indigestible materials, including dosage forms, from the stomach and small intestine. Thus, a good oral mucoadhesive drug delivery system needs to overcome the cleaning action of the "housekeeper wave" and remain in the stomach or proximal small intestine. In addition, the fasted stomach has considerable debris consisting of cells and degraded mucus, both of which will coat the bioadhesive and make its attachment to the wall very difficult.

Rectal Route. The rectum is 5 inches in length and is the distal portion of the large intestine. The mucous membrane of the large intestine is smooth and devoid of villi and hence has relatively small surface area for drug absorption. A large number of goblet cells are dispersed in this mucous membrane, and it is responsible for the thick mucus layer covering the surface of the large intestine*(72)*. The primary function of the rectum is water absorption, thus the water content is relatively low and viscosity of the fecal matter is relatively high. In formulation of rectal mucoadhesive dosage form, hydration rate of the adhesive and dosage form, as well as diffusion of released drug needs to be properly controlled. Once the drug molecule reaches the absorbing epithelicum, absorption is primarily via simple diffusion through the lipid membrane*(73)* and is consistent with the pH partition theory*(74)*.

The proximal one-third of the rectum is drained via the superior hemorrhoidal vein to the inferior mesentiric vein and finally to the hepatic portal system. The distal two-thirds of the rectum is drained directly into the general circulation and by-passes the first-pass metabolism. Thus, for rectal drug deliveries, the formulations need to be localized in the lower or middle rectal areas. However, after insertion, a rectal suppository has a tnedency to migrate upward into the middle and upper third of the rectum. Thus, mucoadhesives can be used in rectal dosage forms to slow the upward migration by attaching to the mucin/epithelial layer of the lower two-thirds of the rectum. The more intimate contact of the formulation to the epithelial layer may also enhance the effect of adjuvants, e.g., penetration enhancers, which are present in the formulation. Examples of penetration enhancers used in rectal formulations are sodium salicylate*(75, 76)*, 5-methoxysalicylate*(75)* and enamine*(77)*.

A cylindrical mucoadhesive dosage form consisting of hydroxyethyl methacrylate (HEMA) cross-linked with ethylene glycol dimethacrylate was used for rectal delivery of antipyrine and theophylline*(78)*. The formulation gave a rapid increase in plasma concentration for the first 4 to 5 hours and an almost constant release thereafter. A mucoadhesive, polyacrylic acid gel, was used in rectal delivery of insulin*(79)*, and was shown to promote the rectal absorption of this drug in rats.

Vaginal Route. The uterine cervix consists of the portio vaginalis or exocervix and the endocervix. The exocervix is covered by nonkeratinized, stratified squamous epithelium which protrudes into the vagina. The endocervix consists of a layer of tall, columnar endocervical glands, which are responsible for mucin secretion which covers the surface of the cervix and vagina. During reproductive years the squamous epithelium lining the vagina is thick and statified, and contain a rich supply of glycogen. Glycogen is metabolized by the lactobocillus to produce an acidic environment*(84)* which decreases vaginal infections. The mucosal lining of the vagina is extremely sensitive to estrogen levels. The squamous epithelium responds to both increases or decreases in endogenous and exopenous estrogen level. For post-menopaudal women the endogenous estrogen level decreases, and the cervix shrinks and becomes atrophic with a reduction in mucus production. Furthermore, with atrophy of the squamous epithelium, the glycogen content decreases, with a subsequent increase in vaginal pH and infections.

Cervical mucus contains more sialic acid than does gastric*(80, 81)*, thus the electrostatic interaction between adhesives and cervical mucus may be higher. The visco-elastic character of cervical mucin was found to vary and be dependent on the amount of water*(82)* and local calcium concentration*(83)*. Cervical mucus is well known for its cycle variation in viscosity. When estrogen dominates the mucus appears clear and with low viscoelasticity, which is readily penetrated by sperm. Whereas when progesterone dominates the mucus is thick and with high viscoelasticity, which resists sperm penetration*(85)*.

Hydroxypropyl cellulose and Carbopol were used in a disc-like*(86)* and a stick-like*(87)* vaginal formulation. Once administered to the cervical canal, the formulation swelled and the shape of the dosage form was fairly well maintained*(87)*. The amount of released bleomycin was found to increase with the amount of mucoadhesive incorporated in the formulation*(86)*. A soluble hydroxypropyl cellulose cartridge impregnated with drug was also used for vaginal drug delivery*(88)*. Once placed in the vagina, drug is released over an extended period of time. The hydrated cartridge becomes a highly viscous gel and provides a soothing effect on the vaginal wall. Furthermore, a cross-linked polyacrylic acid, polycarbophil, was used in a mucoadhesive vaginal moisturizer to treat vaginal dryness*(89)*. The hydrated polycarbophil attaches to the vaginal tissue surface for three to four days to alleviate dryness and subsequent irritations. The low pH of the formulation (around pH 3) maintains the desiccated vaginal tissue at an acidic pH and may minimize vaginal infections.

CONCLUSION

Bio/mucoadhesives drug delivery systems can be designed to localize in specified absorption sites with enhanced bioavailability and minimized first-pass metabolism and enzymatic degradation. For each chosen administration route, a unique natural clearing mechanism exists. Once the natural clearing of the mucoadhesive formulation is minimized, design of sustained release dosage form can be successful. Thus, a thorough understanding of the administration route and the cleaning mechanism is crusial in design of mucoadhesive formulation. Currently, there is a lacking of a good understanding of intact mucus layer in the natural undisturbed state and the interaction between polymers and natural mucin network. To this end, the scanning laser fluorescent confocal microscopy may provide some useful information.

REFERENCE:

1. Gurny, R., Meyer, J. M., and Peppas, N. A., *Biomaterial,* 1984, 5, pp. 336.
2. Park, K., and Robinson, J.R., *Int. J. Pharm.*, 1984, 19, pp. 107.
3. Ishida, M., Machida, Y., Nambu, N., and Nagai, T., *Chem. Pharm. Bull.*, 1981, 29, pp. 810.
4. Smart, J. D., Kellaway, I. W., and Worthington, H. E. C., *J. Pharm. Pharmacol.*, 1984, 36, pp. 295.
5. Leung, S. H. S., and Robinson, J. R., *J. Controlled Rel.*, 1988, 5, pp. 223.
6. Peppas, N. A., and Buri, P. A., *J. Controlled Rel.*, 1985, 2, pp. 257.
7. Teng, C. L. C., and Ho, N. F. H., *J. Controlled Rel.*, 1987, 6, pp. 133.
8. Ranga Rao, K. V., and Buri, P., *Int. J. Pharm.*, 1989, 52, pp. 265.
9. Park, K., *Int. J. Pharm.*, 1989, 53, pp. 209.
10. Nagai, T., *Medicinal Res. Revs.*, 1986, 6, pp. 227.
11. Ch'ng, H. S., Park, H., Kelly, P., and Robinson, J. R., *J. Pharm. Sci.*, 1985, 74, pp. 399.
12. Park, H., and Robinson, J. R., *J. Controlled Release,* 1985, 2, pp. 47.
13. Reich, S., Levy, M., Meshores, A., Blumental, M., Yalon, M., Sheets, J. W., and Goldberg, E. P., *J. Biomed. Mater. Res.*, 1984, 18, pp. 737.
14. Ishida, M., Nambu, N., and Nagai, T., *Chem. Pharm. Bull.*, 1983, 31, pp. 1010.
15. Marvola, M., Vahervuo, K., Sothmann, A., Marttila, E., and Rajaniemi, M., *J. Pharm. Sci.*, 1982, 71, pp. 975.
16. Bridges, J.F., Woodley, J.F., Duncan, R., and Kopecek, J., *Int. J. Pharm.*, 1988, 44, pp. 213.

17. White, J.G., Anios, W.B., and Fordham, M., *J. Cell. Biol.,* 1987, 105, pp. 41.
18. Rojanasakul, Y., Ph.D. Thesis, University of Wisconsin-Madison, 1989.
19. Singer, S.J., and Nicolson, G.L., *Science,* 1972, 175, pp. 720.
20. Jones, G.W., *Microbial Interaction,* Reissing, J.L., ed., Chapman and Hall, London, 1977, pp. 139.
21. Ito, I., *Fed. Proc., Fed. Am. Soc. Exp. Biol.,* 1969, 28, pp.12.
22. Bell, G.I., Dembo, M., and Bongrand, P., *Biophys. J.,* 1984, 45, pp. 1051.
23. Pierschbacher, M., Hayman, E.G., and Ruoslahti, E., *Proc. Natl. Acad. Sci. U.S.A.,* 1983, 80, pp. 1224.
24. Edelman, G.M., *Prog. Clin. Biol. Res.,* 1977, 17, pp. 467.
25. Marchase, R.B., Harges, P., and Jakoi, E.R., *Dev. Biol.,* 1981, 86, pp. 250.
26. Bertolotti, R., Rutishauser, U., and Edlemann, G.M., *Proc. Natl. Acad. Sci. U.S.A.,* 1980, 77, pp.4831.
27. Schachter, H., and Williams, D., *AEMB,* 1982, 144, pp. 3.
28. Allen, A., and Garnec, A., *Gut,* 1980, 22, pp. 249.
29. Chantler, E.N., and Scudder, P.R., *Mucus and Mucosa,* Ciba Foundation Symposium 109, Pitman, London, 1984, pp. 180.
30. Gottschalk, A., *The Chemistry and Biology of Sialic Acid and Related Substances,* Cambridge University Press, London, 1960, pp. 1.
31. Jeanloz, R.W., Glycoprotein, Their Campostition, Structure and Function, Gottschalk, A., ed., Elsevier, Amsterdam, 1972, pp. 403.
32. Starkey, B.J., Snary, D., and Allen, A., *Biochem. J.,* 1974, 141, pp. 633.
33. Silberberg, A., and Meyer, F.A., *Mucus in Health and Disease-II. Advances in Experimental Medicine and Biology,* 144, Chantler, E.N., Elder, J.B., and Elstein, M., eds., Plenum Press. New York, 1982, pp. 53.
34. Chen, J.L., and Cyr, G.N., in *Adhesive Biological System,* Manly, R.S. ed., Academic Press, New York and London, 1970, Chap. 10.
35. Park, H., and Robinson, J.R., *Pharmaceutical Res.,* 1987, 4, pp. 457.
36. Leung, S.H.S., and Robinson, J.R., *J. Controlled Rel.,* 1990, 12, pp. 187.
37. Hiemenz, P.C., *Principles of Colloid and Surface Chemistry,* Hiemenz, P.C., ed., Marcel Dekker, New York, 1977, Chap. 6, pp. 209.
38. Van Wachem, P.B., Beugeling, T., Feizen, J., Bantjes, A., Detmers, J.P., and Van Aken, W.G., *Biomaterials,* 1985, 6, pp. 403.
39. Schonhorn, H., and Hansen, R.H., *J. Polym. Sci.,* 1966, B4, pp. 203.
40. Boddé, H.E., Lehr, C.M., deVries, M.E., Bouwstra, J.A., and Junginger, H.E., *5th International Pharmacentical Technology Symposium,* 1990, pp. 108.
41. Peppas, N.A., and Reinhart, C.T., *J. Membr. Sci.,* 1983, 15, pp. 275.
42. Peppas, N.A., and Reinhart, C.T., *J. Membr. Sci.,* 1984, 18, pp. 227.
43. Voyutskii, S.S., *Autohesion and Adhesion of High Polymers,* John Wiley and Sons/Interscience, New York, 1963.
44. Campion, R.P., *J. Adhes.,* 1974, 7, pp. 1.
45. Leung, S.H.S., Ph.D. Thesis, University of Wisconsin-Madison, 1987.
46. Schaeffer, M.E., and Krohn, K.L., *Invest Ophthamol Vis. Sci.,* 1982, 22, pp. 220.

47. Sade, J., Eliezer, N., Silberberg, A., and Nevo, J., *Am. Rev. Resp. Dis.*, 1970, 104, pp. 48.
48. Nichols, B.A., Chiappino, M.L., and Dawson, C.R., *Invest, Ophthalmol. Vis. Sci.*, 1985, 26, pp. 464.
49. Robinson, J.R., and Li, V.H.K., *Recent Advances in Glaucoma,* Ticho, U., David, R., eds., Excerpta Medica, Amsterdam, 1984, pp. 231.
50. Saettone, M.F., Mont, D., Torracca, M.T., Chetoni, P., and Giannaccini, B., *Drug Develop. Ind. Pharm.*, 1989, 15, pp. 2475.
51. Gray, H., *Anatomy of the Human Body,* Clemente, C.D., eds., Lea and Febiger, Philadelphia, 1985, Chap. 15, pp. 1357.
52. Taylor, M., *Laryngoscope,* 1974, 84, pp. 612.
53. Mygind, N., *Nasary Allergy,* Blockwell Scientific, Oxford, 1978, pp. 39.
54. Duchateau, G.S.M.J.E., Zuidema, J., Albers, W.M., and Merkus, W.H.M., *Int. J. Pharm.*, 1986, 34, pp. 131.
55. Nagai, T., Nishimoto, Y., Nambu, N., Suzuki, Y., and Sekine, K., *J. Controlled Rel.*, 1984, 1, pp. 15.
56. Kuroishi, T., Aska, H., and Okamoto, M., *Jpn. Pharmacol. Ther.*, 1984, 27, pp. 4055.
57. Illum, L., Jorgensen, H., Bisgaard, H., Krogsgaard, and Rossing, N., *Int. J. Pharm.*, 1987, 39, pp. 189.
58. Morimoto, K., Morisaka, K., and Kamada, A., *J. Pharm. Pharmacol.*, 1985, 37, pp. 134.
59. Garren, K.W., Topp, E.M., and Repta, A.J., *Pharm. Res.*, 1989, 6, pp. 966.
60. Lee, V.H.L., Dodda, K., Patel, R.M., Hayakawa, E., and Inagaki, K., *Proc. Int. Symp. Controlled Release Bioact. Mater.*, 1987, 14, pp. 23.
61. Jarrett, A., *The Physiology and Pathophyciology of the Skin, Vol. 6.*, Academic Press, London, 1980, pp. 1871.
62. Ho, N.H.F., and Higuchi, W.I., *J. Pharm. Sci.*, 1971, 69, pp. 537.
63. Schor, J.M., Davis, S.S., Nigalaye, A., and Bolton, S., *Drug Dev. Ind. Pharm.*, 1983, 9, pp. 1359.
64. Erb, R.J., *Controlled Release Nitroglycerin in Buccal and Oral Form, Advances in Pharmacotherapy, Vol. 1,* Bussmann, W.D., Dries, R.R., and Wagner, W. eds., S. Karger, Basel, 1982, pp. 35.
65. Ishida, M., Nambu, N., and Nagai, T., *Chem. Pharm. Bull.*, 1982, 30, pp. 980.
66. Ishida, M., Nambu, N., and Nagai, T., *Chem. Pharm. Bull.*, 1983, 31, pp. 4561.
67. Bremecker, K.D., Strempel, H., and Klein, G., *J. Pharm. Sci.*, 1984, 73, pp. 548.
68. Anders, R., and Merkle, H.P., *Int. J. Pharm.*, 1989, 49, pp. 231.
69. Yotsuyanagi, T., Yamamura, K., and Akao, Y., *The Lancet,* 1985, 14, pp. 613.
70. Brook, I.M., Tucker, G.T., Tuckley, C.C., and Boyes, R.N., *J. Controlled Rel.*, 1989, 10, pp. 183.
71. Szurszewski, J.H., *Am. J. Physiol.*, 1969, 217, pp. 1757.

72. Gray, H., *Anatomy of the Human Body,* Clemente, C.D., ed., Lea and Febiger, Philadelphia, 1985, Chap. 15, pp. 1402.
73. Binder, M.J., *Biochem. Biophys. Acta.,* 1970, 219, pp. 503.
74. Muranishi, S., *Methods Findings Exp. Clin. Pharmacol.,* 1984, 6, pp. 763.
75. Nishihata, T., Rutting, J.H., Kamada, A., Higuchi, T., Routh, M., and Caldwell, L., *J. Pharm. Pharmacol.,* 1983, 35, pp. 148.
76. Nishihata, T., Sudoh, M., Inagaki, H., Kamada, A., Yagi, T., Kawamori, R., and Shichiri, M., *Int. J. Pharm.,* 1987, 38, pp. 83.
77. Nishihata, T., Okamura, Y., Kamada, A., Higuchi, T., Yagi, T., Kawamoni, R., and Shichiri, M., *J. Pharm, Pharmacol.,* 1985, 37, pp. 22.
78. de Leede, L.G.J., de Boer, A.G., Portzger, E., Feijen, J., Gind Breimer, D.D., *J. Controlled Rel.,* 1986, 4, pp. 17.
79. Morimoto, K., Kamiya, E., Takeeda, T., Nakamoto, Y., and Morisaka, K., *Int. J. Pharm.,* 1983, 14, pp. 149.
80. Allen, A., *Brit. Med. Bull.,* 1978, 34, pp. 28.
81. Meyer, F.A., Vered, J., and Sharon, N., *Mucus in Health and Disease,* Elstein, M., and Parke, D.V., eds., Plenum Press, New York, 1977, pp. 239.
82. Wolf, D.D., Sokoloski, J., Khan, M.A., and Litt, M., *Fertil. Steril.,* 1977, 28, pp. 53.
83. Creeth, J.M., *Mod. Probl. Paediat.,* 1977, 19, pp. 34.
84. Bergman, A., and Brenner, P.F., *Menopause-Physiology and Pharmacology,* Mishell, D.R. Jr., ed., Year Book Medical Publishers, Inc., Chicago, London, 1987, Chap. 5.
85. Chantler, E., *Adv. Exp. Med. Biol., Vol. 144,* Chantler, E.N., Elder, J.B., and Elstein, M., eds., Plenum Press, New York and London, 1982, pp. 251.
86. Machida, Y., Masuda, H., Fujiyama, N., Ito, S., Iwater, M., and Nagai, T., *Chem. Pharm. Bull.,* 1979, 27, pp.93.
87. Machida, Y., Masuda, H., Fujiyama, N., Iwater, M., and Nagai, T., *Chem. Pharm. Bull.,* 1980, 28, pp.1125.
88. Williams, B.L., U.S. Patent 4,317,447, March 2, 1982.
89. Replens by Columbia Laboratories, Miami, Florida.

RECEIVED August 30, 1991

Chapter 17

pH-Sensitive Hydrogels

Characteristics and Potential in Drug Delivery

Helle Brøndsted[1] and Jindřich Kopeček[1,2]

Departments of [1]Pharmaceutics/CCCD and of [2]Bioengineering, University of Utah, Salt Lake City, UT 84112

The factors influencing the equilibrium degree of swelling of pH-sensitive hydrogels, such as charge and pKa of the ionizable monomer; the crosslinking density and hydrophilicity of the polymer; and pH, ionic strength and composition of the surrounding solution are reviewed in relation to applications in drug delivery. pH-sensitive swelling can be exploited in the development of drug delivery systems, which would release the drug in response to the pH of the surrounding solution. pH-sensitive hydrogels have potential in drug delivery to the gastro-intestinal (GI) tract, due to the pH-variation throughout the GI tract, and for more complex systems where a physiological stimuli results in a change in pH and release of drug. The potential application of pH-sensitive hydrogels in drug delivery is illustrated and an example of new pH-sensitive hydrogels for colon-specific drug delivery is given.

Hydrogels are defined as infinite three-dimensional polymeric networks containing a considerable amount of water, e.g., more than 20% (*1,2*). Due to their high water content, hydrogels possess excellent biocompatible behavior (*3*). Hydrogels have been extensively studied for biomedical applications such as implants (*4*) and soft contact lenses (*3*) Drug delivery systems developed for many different purposes have been based on hydrogels because of their high and, at the same time, controllable permeability to drug molecules (*5*). However, the majority of the studies on hydrogels are related to neutral hydrogels. Hydrogels and their properties have been reviewed by several authors (*2,3,6-9*).

pH-sensitive hydrogels usually contain pendant acidic or basic groups, such as carboxylic acids and primary amines, or strong acid and bases, such as sulfonic acids and quaternary ammonium salts, which change ionization in response to changes in pH, thus changing the properties of the gel (*10*). Among the first ionic hydrogels investigated were gels based on acrylic acid and methacrylic acid (*11,12*). It was observed that the equilibrium degree of swelling of these gels responded to changes in pH. When the pH of the swelling solution was increased the gels showed an increase in swelling. The fact that polyelectrolyte gels change properties in response to pH can be exploited in the development of new drug delivery systems.

0097–6156/92/0480–0285$06.00/0
© 1992 American Chemical Society

Not many such systems have been described so far, even though the swelling properties and permeability of pH-sensitive hydrogels have been studied.

In this chapter we first intend to review the preparation and characteristics of pH-sensitive hydrogels, such as equilibrium degree of swelling and drug permeability. Their properties will be related to applications in drug delivery of low and high molecular weight compounds, and it will become apparent that there exists a great potential in using pH-sensitive hydrogels for the development of drug delivery systems. Then the application of pH-sensitive hydrogels in drug delivery to the colon will be demonstrated using new acidic hydrogels containing enzymatically degradable crosslinks.

Preparation.

Basic preparation techniques of pH-sensitive hydrogels will be discussed along with some classical examples.

Monomers. pH-sensitive hydrogels are hydrogels containing one or more ionic or ionizable monomer in the polymeric backbone or in crosslinks. Copolymerization can be achieved with neutral hydrophilic or hydrophobic monomers. The latter improve the mechanical strength of the usually weak hydrogel. In principle, hydrogels with any desired property can be obtained by varying the amount and type of monomer and crosslinker used. Table I lists some of the monomers used in the preparation of pH-sensitive hydrogels. The monomers may contain weakly acidic and basic groups, such as carboxylic acids and primary or substituted amines, or may contain strong acids and bases such as sulfonic acids and quartenary ammonium salts.

Table I. Monomers used in pH-sensitive Hydrogels

Type	Monomer	pH-sensitive group	Reference
Acidic	Acrylic acid		13,33
	Methacrylic acid	-COOH	10,38,40
	Sodium styrenesulfonate	$-SO_3^-Na^+$	17,18
	Sulfoxyethyl methacrylate	$-SO_3H$	43
Basic	Aminoethyl methacrylate	$-NH_2$	
	N,N-dimethylaminoethyl methacrylate	$-N(CH_3)_2$	39,47
	N,N-diethylaminoethyl methacrylate	$-N(CH_2CH_3)_2$	10,40,70
	Vinylpyridine	[pyridyl ring structure, N]	
	Vinylbenzyl trimethylammonium chloride	$-N(CH_3)_3^+,Cl^-$	17,18

Synthesis of Hydrogels. The hydrogels can be prepared by covalent crosslinking, such as crosslinking copolymerization, crosslinking of polymeric precursors, or by physical crosslinking such as polyelectrolyte complexes where one ionic species is in excess. Ionic hydrogels can also be prepared by hydrolysis of neutral hydrogels, e.g., poly(acrylonitrile) or poly(hydroxyalkyl methacrylate) hydrogels, which results in an acidic hydrogel. Details of these methods and important features of the resulting hydrogels are listed in Table II.

Crosslinking Copolymerization. Crosslinking copoly-merization is performed either in bulk or more usually in the presence of a solvent. In the latter case all monomers and the crosslinking agent are dissolved in a suitable solvent. An initiator, usually a free radical initiator, is added and the polymerization is initiated by the decomposition of the initiator, which can be achieved by elevating temperature or applying UV light. After polymerization the hydrogels should be washed in order to extract all residual monomers and polymers not incorporated into the network (*13*).

Crosslinking of Linear Polymers. Hydrogels may be prepared by crosslinking of linear polymers containing reactive groups with a bifunctional crosslinking agent. The polymerization is usually done in a solvent, and the gel point is reached very quickly (*14,15*).

Interpenetrating Networks. An interpenetrating network consists of two crosslinked networks which are physically entangled. Such a network can be prepared, e.g., from poly(oxyethylene) and poly(acrylic acid) (*16*). The first hydrogel network was composed of crosslinked poly(oxyethylene). Next the membrane was soaked in a mixture of acrylic acid, crosslinker and initiator and polymerized again, resulting in interpenetrating pH-sensitive networks of poly(oxyethylene) and poly(acrylic acid). Interpenetrating networks may also be formed simultaneously using two different polymerizations mechanisms such as step-growth and free radical polymerizations.

Polyelectrolyte Complexes. pH-sensitive hydrogels can be formed from anionic and cationic polymers, where ionic forces hold the polymers together (*17,18*). A pH-sensitive gel can be obtained by using one of the ions in excess. A polyelectrolyte complex composed of poly(vinylbenzyl trimethylammonium chloride) and sodium poly(styrenesulfonate) (Ioplex 101) has been studied with regard to biomedical applications (*2*). A difference between polyelectrolyte complexes and other pH-sensitive gels is that, as the ionic strength increases, the degree of swelling of polyelectrolyte complexes increases, due to interruption of the ionic forces in the gel. Ultimately, the gel will dissolve. As will be discussed later the degree of swelling of other pH-sensitive hydrogels decreases with increasing ionic strength.

Hydrolyzed Neutral Hydrogels. By hydrolysis of already existing groups in neutral hydrogel, such as esters, amides and nitriles, different degrees of ionization can be introduced in the polymeric backbone. The resulting distribution of ionizable groups in the backbone will be different than the distribution after copolymerization of the two corresponding monomers. A pH-sensitive hydrogel was prepared by partial hydrolysis of poly(acrylonitrile) resulting in formation of acidic groups. The poly(acrylonitrile) formed a physical network due to association of blocks of acrylonitrile and blocks of acrylic acid. These hydrogels have a very high water content and very high mechanical strength compared to conventional hydrogels (*19*).

Table II. Methods for Synthesis of Hydrogels

Method	Example	Comments	Ref.
Crosslinking copolymerization	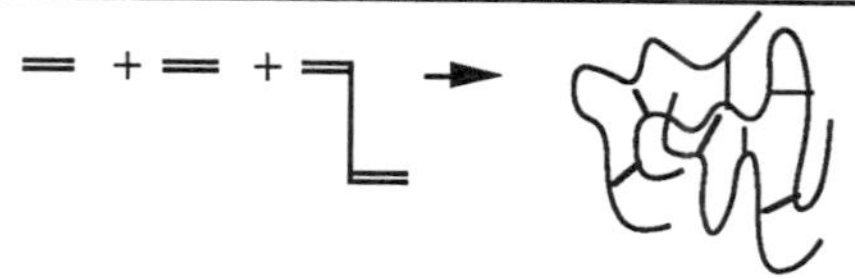	An extensive amount of different monomers are available. Hydrogels with different properties may be obtained by varying the monomers.	10,13,39 p
Crosslinking of polymeric precursors	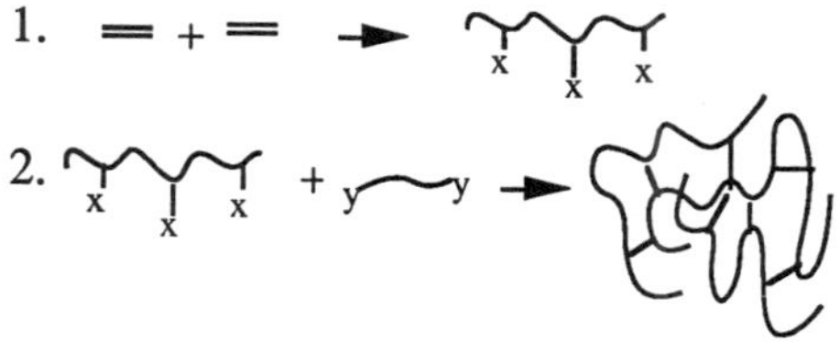 1. x x x 2. x x x + y y	The gel point is reached very fast. Good control of molecular weight of polymer backbone. (Example: x=p-nitrophenylester, y=aliphatic amine)	14,15
Polyelectrolyte complexes	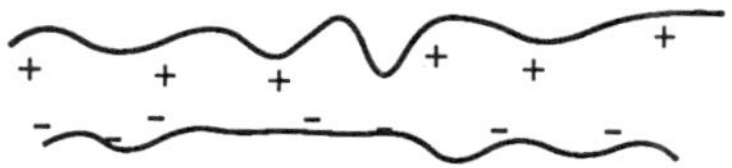	Physical crosslinks (ion-ion interaction). Swells drastically in solutions of high ionic strength.	17,18
Hydrolysed neutral gels	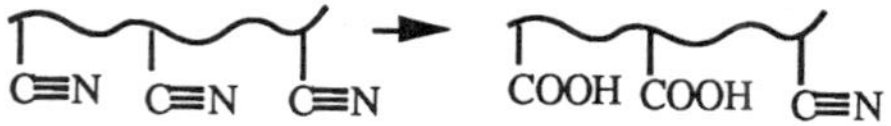 C≡N C≡N C≡N → COOH COOH C≡N	Can vary degree of hydrolysis Block copolymer	19
Interpenetrating networks	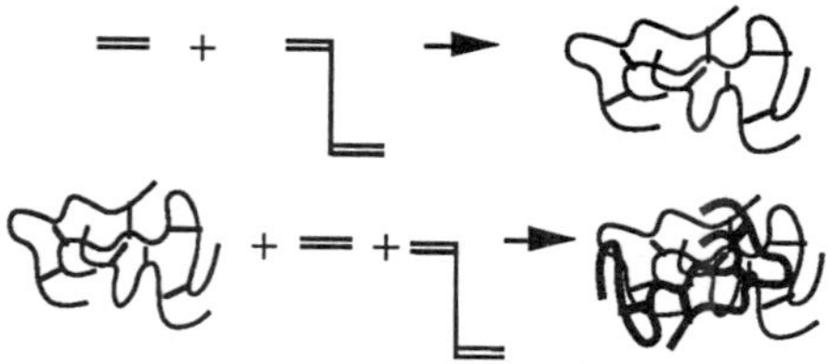	Can be formed in one step using two different polymerization mechanisms, or in two steps polymerizing one network at a time.	16

Equilibrium Degree of Swelling.

Ionized gels often have very high degrees of swelling. A high concentration of ions exists inside the gel due to dissociation of acidic or basic groups in the gel and diffusion of counterions into the gel from the surrounding medium. The high ion concentration will increase water flow into the gel due to osmosis, resulting in an increase in swelling described by the Donnan equilibrium. Another factor contributing to increases in swelling is the interaction and repulsion of charges along the polymer chain. Much literature describes the swelling of ionizable gels; however, the swelling equilibrium is very difficult to model mathematically due to electrostatic interactions along the chain and the often very high swelling of ionic gels. Furthermore, the pKa of the gel, which is a parameter that has to be incorporated into the expression, depends on the degree of ionization which changes with pH. In this section the different approaches taken to model the swelling of ionic gels will be discussed briefly and the factors influencing the degree of swelling of ionic gels will be reviewed.

Theory. The theory for swelling equilibrium for neutral gels was developed by Flory and Rehner and was based on the free energy relationship shown in equation 1 (*20*).

$$0 = \Delta F_1 + \Delta F_2 \quad (1)$$

where ΔF_1 corresponds to the free energy of mixing and ΔF_2 corresponds to the elastic free energy due to expansion. The free energy of mixing favors swelling whereas elastic free energy opposes swelling of the gel by a retractive force. Explicit expressions for ΔF_1 and ΔF_2 have been derived by various research groups (*8,21*).

For ionized networks, equation 1 has been modified to include contributions from the mixing of ions with solvent, ΔF_3 and from interactions of charges along the polymer chain, ΔF_4 (*22*) (equation 2).

$$0 = \Delta F_1 + \Delta F_2 + \Delta F_3 + \Delta F_4 \quad (2)$$

The elastic free energy term ΔF_2 assumes a Gaussian chain length distribution. For high degrees of swelling, this assumption is not valid, and ΔF_2 must include contributions from the non-Gaussian free energy due to the limited extensibility of the chains. Expressions for these terms have been developed by Hasa et al (*22*).

It has been shown that the modulus of elasticity of ionized gels depends on the degree of ionization of the gel and the ionic strength of the solution (*22-26*). The mechanical behavior of ionized gels was described by Hasa et al. by modifying the theory of rubberlike elasticity (equation 3). The modulus of elasticity, G, can be determined by stress-strain measurements in compression or elongation (*22*).

$$f/A=-G(\lambda-\lambda^{-2})=-G^{\circ}(1+\phi_N+\phi_E)(\lambda-\lambda^{-2}) \quad (3)$$

where f/A is applied force per area, G is modulus of elasticity, λ is the ratio of height of deformed sample to height of relaxed sample in compression, ϕ_N and ϕ_E are contributions to modulus due to non-Gaussian behavior (finite extensibility of the chains) and electrostatic interactions along the chain, respectively, G° is the modulus obtained from the theory of rubberlike elasticity for unionized network.

For a highly crosslinked poly(2-hydroxyethyl methacrylate-co-methacrylic acid) hydrogel with low charge density a good correlation between theory and experimental results was observed. However, at very high charge densities, deviation from the theory was found (*23,24*). It was later confirmed that the swelling behavior of ionic gels could be predicted using a theory based on Hasa et al. (*27-29*). The swelling transition was enhanced by an increase in the number of charges on the chain, but decreased with increased concentration of electrolytes and decreased flexibility of the chain (*28*).

Recently, Brannon-Peppas and Peppas derived a model which describes and predicts the swelling behavior of ionized gels. This model is based on three contributions to swelling equilibrium; the contributions from mixing, from the elastic free energy and from ionic interactions. This model differs from the above described model in assuming a Gaussian chain length distribution. These researchers were able to predict the swelling of gels with different pKa as a function of pH and ionic strength of the surrounding medium (*30,31*).

Other research groups have derived models for swelling of ionic gels based on Donnan theory (*32-36*). This theory explains the swelling due to the difference in osmotic pressure of ions inside the gel and in the surrounding medium. This theory is more simple because it does not involve parameters from the elastic free energy of the network. Ricka and Tanaka used the Donnan theory to describe the swelling of poly(acrylamide-co-acrylic acid) networks (*33*). The researchers found that the theory was valid as long as only monovalent ions were involved. Phase transitions in poly(acrylamide) and poly(acrylic acid) interpenetrating networks were recently observed by Ilmain et al. (*37*).

Factors Influencing the Swelling of pH-Sensitive Hydrogels. The equilibrium degree of swelling of pH-sensitive hydrogels is mainly influenced by the charge of the ionic monomer, pKa of the ionizable group, degree of ionization, concentration of ionizable monomer in the network and pH, ionic strength and composition of the swelling solution. Also, factors such as crosslinking density and hydrophilicity/hydrophobicity of the polymer (as for uncharged gels) influences the degree of swelling and the pH-sensitivity, that is, the magnitude of the response to changes in pH. The effects of these factors on the equilibrium degree of swelling are summarized in Table III.

The charge of the ionic monomer effects the pH-sensitivity of the gel. An acidic hydrogel will be ionized at high pH but unionized at low pH, thus, the equilibrium degree of swelling will increase at high pH where the gel is ionized (*38*). A cationic/basic hydrogel has the opposite pH-dependence of swelling (*39*). The specific resistance of hydrogel membranes containing ionogenic groups was related to pH (*40-42*). The hydrogels were based on 2-hydroxyethyl methacrylate; methacrylic acid and N,N-diethylaminoethyl methacrylate were used in order to obtain acidic, basic and ampholytic gels. The dependence of the specific resistance on pH as a function of content of ionogenic groups, crosslinking density and degree of neutralization were evaluated. Figure 1 shows specific resistance as a function of pH for acidic, basic and ampholytic membranes. It was found that the acidic membrane showed maximum specific resistance at pH around 3.5, the basic membrane at pH 10 and the ampholytic membrane around pH 6. Maximum specific resistance was obtained at a pH close to the pKa of the membrane, where the membrane became neutral or at the isoelectric point of the ampholytic membrane (*40*).

pKa was shown to influence the pH-swelling curve. An increase in the pKa of an acidic monomer shifted the curve to a higher pH (*31*). It was shown that the swelling response to pH was very sensitive at a pH close to the pKa of the hydrogel. Strong acid membranes containing sulfooxyethyl methacrylate as the ionogenic monomer were found only to have a small dependence on pH (*43*)

The concentration of ionizable monomer in the hydrogel has been shown to be important for the swelling and pH-sensitivity of the gel. This effect depends on the relative hydrophilicity of the ionizable monomer to the neutral comonomer. It was found that as the amount of methacrylic acid in poly(methacrylic acid-co-2-

Table III. Factors influencing swelling of polyelectrolyte hydrogels

	Factor	Effect	Ref.
Gel properties	Charge of ionizable monomer	Acidic:pH↑⇒ionization↑ Basic:pH↑⇒ionization↓	10,40
	pKa of ionic monomer	pKa↑⇒pH-ionization profile shifts to ↑ pH	31
	Degree of ionization	Ionization↑⇒swelling↑	10,31, 40
	Concentration of ionizable monomer	Concentration↑⇒swelling in ionized state↑	13
	Crosslinking density	Density↑⇒swelling↓	13,38
	Hydrophilicity/ hydrophobicity of polymer backbone	Hydrophilicity↑⇒swelling↑	
Swelling solution	pH	Acidic:pH↑⇒swelling↑ Basic:pH↑⇒swelling↓	13,39
	Ionic strength	Ionic strength↑⇒osmotic pressure inside gel↓⇒swelling↓ (exception:polyelectrolyte complexes)	31,39
	Coion	Usually no change	46
	Counterion	Effect depends on specie (salting-in/salting-out and effect on water structure)	46
	Valency of counterion	Valency↑⇒swelling↓	46,47

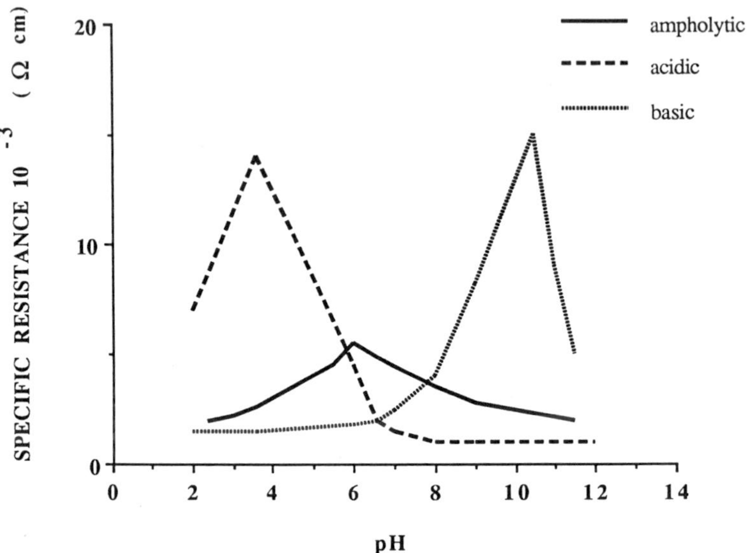

Figure 1. The specific resistance of hydrogel membranes as a function of pH. The membranes are all based on 2-hydroxyethyl methacrylate, containing 2% ethylene dimethacrylate as crosslinking agent. The acidic membrane contains 2.64% methacrylic acid, the basic membrane 2.89% N,N-diethylaminoethyl methacrylate, and the ampholytic membrane 1.60% methacrylic acid and N,N-diethylaminoethyl methacrylate. (Data from ref. 10).

hydroxyethyl acrylate) hydrogels increases, dependence of specific resistance on pH increases (*40*). Using a more hydrophobic monomer such as butylacrylate, the degree of swelling increases as the amount of acrylic acid increases, both at low and high pH (*44*).

Crosslinking density influences pH dependent swelling. An increase in crosslinking density restricts the equilibrium degree of swelling. Investigating the swelling of poly(methacrylic acid-co-2-hydroxyethyl methacrylate) hydrogels with different crosslinking density, it was found that this effect was more pronounced at high pH than at low pH. At high pH the gel was ionized and hydrophobic interactions and hydrogen bonding inside the gel present at low pH were interrupted (*38*).

The influence of structure of the polymer backbone on swelling was investigated using copolymers of methacrylic acid, butyl methacrylate and hydroxyethyl methacrylate (*45*). It was found that the more hydrophilic the polymer backbone, the lower the pH-sensitivity. Siegel found that increasing the hydrophobicity of the more hydrophobic n-alkyl methacrylates would decrease the pH-sensitivity of poly(n-alkyl methacrylate-co-N,N-dimethylaminoethyl methacrylate) hydrogels. In this case the swelling at low pH where the gel is ionized is highly dependent on the hydrophobicity of the neutral monomer (*39*).

Buffer composition and ionic strength effect the swelling of polyelectrolyte hydrogels (*39,46-48*). As ionic strength increases, the swelling decreases due to increased counterion concentration, shielding of charges on the polymer chain and high ion concentration outside the gel. As the concentration of ions outside the gel increases, the concentration of ions inside and outside of the gel will become equal and the osmotic pressure inside the gel will decrease. A buffer containing multivalent counterions decreases the degree of swelling. One multivalent ion is able to neutralize several charges inside the gel. Therefore, the concentration of multivalent counterions inside the gel will be less than monovalent ions; the osmotic pressure and the degree of swelling of the gel will be decreased. Swelling in phosphate buffer was greater than swelling in citrate buffer. The buffer species is not important at high ionic strength; in that region the ionic strength is the only determinant of swelling (*46*).

The type of coion did not influence the equilibrium degree of swelling, since coions do not very easily permeate a swollen gel of the same charge. On the other hand, the counterion species was very important for the swelling equilibrium, even if the valency was the same. For example, the swelling of poly(methyl methacrylate-co-N,N-dimethylaminoethyl methacrylate) hydrogels was much higher with Cl^- present as the counterion than with SCN^- as counterion. The result was explained by the fact that some ions are water structure breakers that increase swelling, whereas others are structure makers and decrease swelling (*46*). However, Refojo (*49*) earlier found that the swelling of neutral poly(2-hydroxyethyl methacrylate) hydrogels was decreased by some electrolytes, e.g., chloride ions, and increased by others, e.g., thiocyanate. The decrease in swelling was explained by the salting-out effect of chloride, decreasing the solubility of the polymer segments, thus decreasing the swelling. In the case of thiocyanate, there is a salting-in effect, increasing the solubility of the polymer segments and the swelling. Thus, the effect of a particular ion on the swelling of a hydrogel is not easy to predict and is connected to factors such as charge on the gel, solubility of the polymer segments in the electrolyte solution, secondary interactions in the gel and effect on water structure. It was found that urea increases the degree of swelling of poly(2-hydroxyethyl methacrylate) hydrogels. The increase in swelling was first believed to be due to the effect on the water structure, since urea appears to break ordered water structure, decreasing hydrophobic effects. As hydrophobic interactions in poly(2-hydroxyethyl methacrylate) hydrogels are broken, the gel swells (*49*). Later, Ratner and Miller proved that the effect of urea on the degree of swelling was due to the interaction of urea with regions in the gel containing hydrogen bonds (*50*).

It has been demonstrated that it is possible to control the swelling of pH-sensitive hydrogels due to a change in pH of the swelling solution. The pH-sensitivity can be controlled by changing the structure of the polymeric backbone, the ionizable monomer and charge density. Ionizable monomer is especially important for pH-dependent swelling.

Drug Permeability.

Drugs, especially hydrophilic low molecular weight drugs, are often assumed to diffuse via the water filled pores in the hydrogel by the so-called "pore-mechanism." Diffusion of solutes in hydrogel membranes has been described by Yasuda et al., who developed the free volume theory (*51,52*). More hydrophobic solutes can permeate via the "partition mechanism," where the solute interacts with and permeates along the polymer matrix. The mechanism of diffusion is dependent on the hydrophilicity/hydrophobicity of polymer, pore size, degree of swelling and size and hydrophilicity of solute (*51,52*).

With respect to ionic gels, it is furthermore necessary to investigate permeability as a function of gel charge, surrounding pH and charge of the drug molecule. A drug with an opposite charge of the hydrogel could act as a counterion, and the release of such a drug would be by ion exchange (*7*). A negatively charged drug molecule may have decreased permeability through a negatively charged hydrogel membrane due to repulsion between like charges.

Kopeček et al. (*10*) investigated the permeability of NaCl and $MgSO_4$ through hydrogel membranes. Four different kinds of membranes were investigated: neutral, acidic containing methacrylic acid, basic containing N,N-diethylaminoethyl methacrylate and ampholytic containing equal amounts of acidic and basic monomer. The dependence of permeability on the content of ionogenic groups, crosslinking density and pH were evaluated. It was found that the permeability was lowest for solely acidic and basic membranes. Total diffusion of the salt was slowed due to repulsion of coion from the membrane. The diffusion was fastest through ampholytic membranes, since these membranes enhance the permeability of both ions. The permeability of neutral membranes fell between that of ampholytic and acidic or basic membranes. The dependence on crosslinking density was found to be strongest in the case of neutral membranes where only pore size and swelling of the membrane were important. For ionized membranes charge interactions were dominant and crosslinking density was not as important. As charge density increased the permeability of ampholytic membranes increased, but was decreased in the case of acidic and basic membranes due to higher repulsion. The permeability of $MgSO_4$ was found to be lower than that of NaCl. For the ampholytic membranes a high dependence of crosslinking density was observed, which was attributed to the higher radius of the hydrated Mg^{2+} and SO_4^{2-} ions as compared to Na^+ and Cl^- ions. As the radius of the solute increases, the permeability becomes more dependent on the pore size (*51*).

Vacík et al. (*53*) found that as for salts the permeability of organic compounds such as urea and creatinine decreased as crosslinking density of the membrane increased. The permeability of these compounds through ampholytic membranes was increased; it was suggested that only molecules which interact with charges on the hydrogel will have increased permeability in ampholytic membranes when compared to neutral membranes.

The permeability of various compounds, up to a molecular weight of 70,000, through hydrolyzed polyacrylonitrile membranes was investigated (*19*). A linear relationship between logarithm of the permeability coefficient and logarithm of the molecular weight up to a molecular weight of about 70,000 was observed. Solutes with a higher molecular weight did not permeate within 72 hours.

As mentioned earlier, the crosslinking density is important for pH-dependent permeability (*54*). Weiss et al. (*54*) synthesized poly(methacrylic acid) membranes with different degrees of crosslinking. Dextran coupled to lissamine-rhodamine B (molecular weight 6000) was used as the permeant. At high pH the permeability was approximately the same for the two membranes. At low pH the membrane with the lower degree of crosslinking had a much higher permeability than the highly crosslinked membrane. The permeability correlated with the degree of swelling of the membranes. When solutes of different molecular weight and charge were investigated, it was determined that the permeability of high molecular weight compounds was more dependent on pH than low molecular weight compounds. As pH increased and the gel became ionized, the degree of swelling increased. The permeability of high molecular weight solutes was more dependent on the degree of swelling, in accordance with the free volume theory (*54*). A molecule possessing a negative charge also had increased permeability through acidic membranes as pH increased. (*54-56*). The results show that diffusion was mainly dependent on swelling and not on the charge density because repulsion of the molecule from the membrane would then be expected.

Models predicting drug release from neutral hydrogels have been developed (*8*). In the case of ionic hydrogels, the models tend to be more complicated (*58-60*). Brannon-Peppas and Peppas developed a model predicting the release from dry poly(methacrylic acid-co-2-hydroxyethyl methacrylate) hydrogels (*58*), whereas the model developed by Kou et al. was based on swollen gels (*60*).

It can be concluded that the actual mechanism of diffusion is dependent on the degree of swelling of the gel, hydrophilicity of polymer backbone, pKa, charge density, pore-size and drug properties, such as size, charge and hydrophilicity.

Approaches to Drug Delivery.

Drug Delivery to the Gastrointestinal Tract. The system developed by Siegel and coworkers, based on a basic poly(methyl methacrylate-co-N,N-diethylaminoethyl methacrylate) hydrogel, which swells at low pH, can be used in the oral delivery of foul tasting drugs (*61*). At the neutral pH of the mouth the gel has a low degree of swelling and drug loaded into the gel will not be released. As pH decreases in the acidic conditions of the stomach, the degree of swelling increases and drug will be released. An acidic gel would have potential for drug delivery of acid labile drugs to the small intestine (*38*).

Recently, pH-sensitive hydrogels were prepared for enteric drug delivery (*62*). The gels were based on N-isopropylacrylamide, acrylic acid and vinyl terminated polydimethylsiloxane. The gels exhibited both pH and temperature sensitive swelling, which allowed fabrication of a hydrogel which was pH-sensitive, mainly at a certain temperature, e.g., 37°C. In-vitro release studies were performed using a model drug, indomethacin, which can cause severe gastric irritation. In 24 hours almost no drug was released at pH 1.4, 37°C, whereas at pH 7.4, 37°C, more than 90% was released during 5 hours

Touitou and Rubinstein (*63*) formulated capsules coated with a poly(acrylic acid) polymer (Eudragit) containing insulin and surfactant, sodium laurate and cetyl alcohol, in arachis oil. This capsule protected the drug in the upper part of the small intestine. As pH increased to pH 7.5 the release of drug increased. It was proposed that this resulted in release of drug in colon. Insulin levels were monitored by blood glucose level depression and a reduction of 45% was measured. Gwinup et al. (*64*) investigated a similar system based on capsules coated with a methacrylic acid copolymer. A weakness of using pH-sensitive release for colonic delivery is that due to fluctuations in pH in the GI tract, release may also take place in the small intestine, and absorption may in that case be decreased due to high enzyme activity in the small intestine.

Mucoadhesives. In drug delivery to the gastrointestinal tract, it is desirable to retain the drug delivery device at the site of action (for a drug with local action) or at the site of absorption (for a drug with systemic action) to assure sufficient time for all of the drug to be released. For this purpose, bioadhesive polymers are being developed which will adhere to the mucin of the GI-tract and increase the residence time. The use of bioadhesives in drug delivery has recently been reviewed by Gupta et al. (*65*). The bioadhesive properties of neutral, acidic and basic polymers were evaluated by Park et al. (*66*); it was found that anionic polymers possessed the best bioadhesive properties.

The use of anionic gels as mucoadhesives has been proposed for the use of prolonging gastric retention time (*67,68*). It was observed that anionic materials with high charge density were good mucoadhesives. A mucoadhesive, polycarbophil, based on acrylic acid, was found to have decreased mucoadhesion with increased crosslinking density due to reduced interpenetration of the mucus. Poly(acrylic acid-co-acrylamide) hydrogels were prepared, and the mucoadhesion in relation to pH, crosslinking density, charge density and degree of swelling was determined. The results suggested the mucoadhesion was favored with high charge density and flexibility of the polymer chains. An increase in crosslinking density decreased the mucoadhesion. The effect of mucus turnover still remains to be investigated.

Pepsin Degradable Hydrogels. Hydrogels were prepared by Shalaby and Park (*69*) based on poly(1-vinyl-2-pyrrolidone) containing albumin crosslinks. The gels swell to a great extent in the stomach which prolongs residence time. The gels have potential for use in drug delivery to the gastrointestinal tract where prolonged gastric residence time and drug release is desired. The albumin crosslinks will be degraded in the presence of pepsin in the stomach resulting in an increased degree of swelling prolonging retention time and release of drug. These gels increased in degree of swelling at pH above 7 due to ionization of the albumin. An even higher swelling at low pH could be obtained if polyelectrolyte gels which swell to a high degree in the stomach were used. Degradation of these gels took place by both surface erosion and bulk degradation.

Self-Regulated Insulin Delivery Systems. Devices which release insulin in response to glucose levels are being developed by several research groups (*47,70-72*). Glucose-sensitive membrane systems composed of a basic polyamine hydrogel membrane and a glucose-oxidase immobilized membrane have been developed by Ishihara et al. (*70,71*) and Albin et al. (*72*). The basic membrane was based on crosslinked N,N-diethylaminoethyl methacrylate and 2-hydroxypropyl methacrylate. As the glucose concentration in contact with the glucose-oxidase immobilized membrane increases, glucose will diffuse into the membrane. There it will be converted to gluconic acid by the action of glucose oxidase, and the pH will decrease, resulting in ionization of the basic groups in the gel an increase in the degree of swelling. Insulin loaded into the gel will then be released. It was shown that the swelling of the membrane was reversible and very sensitive to changes in pH. Insulin release from the polyamine membrane was shown to be controlled by the degree of swelling of the membrane, which was dependent on the pH of the surrounding solution. The release of insulin could be controlled in response to changes in glucose concentration; as the glucose concentration increased, the insulin release was increased. Since it has been shown that the swelling of ionizable hydrogels is very dependent on the pH, composition and ionic strength of the surrounding solution, it is necessary to model the physiological conditions as close as possible to be able to predict the insulin release from this device (*72*).

Siegel et al. (*47*) suggests the use of a self-regulating mechanochemical insulin pump. The pump consists of 3 compartments. Compartment 3 which is in contact with the body fluid consists of a basic hydrogel membrane containing immobilized glucose-oxidase similar to the ones proposed by Ishihara et al. (*70,71*) and Albin et al. (*72*). As the glucose concentration increases, the degree of swelling increases and the gel expands. The gel will push against a diaphragm separating compartment 2 and 3, increasing the pressure in compartment 1 and 2; insulin from compartment 1 will be released through a valve. The advantage of this system compared to the ones where insulin is released from a pH sensitive membrane is that problems concerning aggregation of insulin solutions, adsorption on the hydrogel surface and clogging of the membrane are avoided. In this system stabilized insulin solutions or semi-solid insulin preparations can be used.

Saccharide-Sensitive Hydrogels. Hydrogels containing a immobilized lectin, concanavalin A, were synthesized which exhibit saccharide-sensitive swelling (*73*). Concanavalin A is known to possess binding sites for certain saccharides. It was found that when an ionic saccharide such as dextran sulphate was added to the gel, the degree of swelling of the gel increased due to the negative charges on the dextran molecule. As the ionic surfactant was exchanged with a non-ionic surfactant, the gel collapsed and degree of swelling decreased to initial value. By changing lectins and saccharides used, various swelling profiles can be obtained.

Hydrogels for Drug Delivery to Colon.

Hydrogels containing acidic comonomers and enzymatically degradable azoaromatic crosslinks have recently been developed for the delivery of drugs to the colon (*13*). These crosslinks are degraded by microbial enzymes predominantly present in colon (*74*). Drug delivery to colon is of interest for drugs that are unstable in the upper part of the GI-tract, such as peptide and protein drugs, since the amount of digestive enzymes in colon is drastically reduced. It would also be useful to deliver low molecular weight drugs which are used for treatment of diseases present in colon, such as ulcerative colitis. In the low pH-range of the stomach, the gels have a low degree of swelling and drug loaded into the gels is protected against digestion by enzymes. As the pH increases when the gel passes down the GI-tract, the degree of swelling increases due to increased degree of ionization. Upon arrival in colon the gels have reached a degree of swelling which makes the crosslinks accessible to enzymes or mediators. The gel is then degraded, and the drug released and available for systemic absorption or action in colon. It was found that the enzymatic degradation of crosslinks in hydrogels depends strongly on the degree of swelling of the gels (*75*). A too high swelling would result in the undesired release of drug prior to the arrival of the gel in colon. Therefore, it is necessary to investigate the relationship between pH and swelling of the gel, kinetics of swelling, the pH-profile of the GI-tract, GI-transit time, permeability of drug and rate of degradation.

The hydrogels were synthesized by crosslinking copolymerization,. based on acrylic acid, N-tert-butylacrylamide and N,N-dimethylacrylamide. Two azoaromatic crosslinking agents of different spacer lengths were used, 4,4'-di(methacryloylamino)azobenzene (DMAAB) and 4,4'-di(N-methacryloyl-6-aminohexanoylamino)azobenzene (DMCAAB). The synthesis and characterization of these gels were described previously (*13, 76*).

The degree of swelling at different pH was measured. Figure 2 shows the equilibrium degree of swelling as $1/v_2$ (v_2 is volume fraction of polymer in the swollen state) versus percent acrylic acid in the monomer mixture. As amount of acrylic acid increases, degree of swelling at pH 7.4 increases due to increased ionization of the gel. At pH 2 degree of swelling decreases with amount of acrylic acid, hydrophobic interactions and hydrogen bonding are present.

In Figure 3 the equilibrium degree of swelling as $1/v_2$ versus percent crosslinker in the monomer mixture is shown. As the crosslinking density increases, the swelling is restricted and the pH-sensitivity of the gel decreases.

With respect to applications in drug delivery, the kinetics and reversibility of swelling must also be evaluated. It is often desirable that swelling changes rapidly in response to pH. For "on-off" type delivery systems, the swelling must be reversible. It was showed that, for poly(acrylic acid-co-N,N-dimethylacrylamide-co-N-tert-butylacrylamide) hydrogels, the response to a pH change was very fast and reversible. The swelling after abrupt changes in pH between pH 2 and 7.4 is shown in Figure 4 (*13*).

It was found that the swelling behavior and modulus of poly(acrylic acid-co-N,N-dimethylacrylamide-co-N-tert-butylacrylamide) hydrogels containing 40% acrylic acid could be explained from the theory of rubberlike elasticity for uncharged networks. As the content of acrylic acid in the gels increased to 70%, the swelling behavior and modulus deviated from this theory (*13*). This deviation could be explained in accordance with the theory proposed by Hasa et al (*22*). For lower charge densities the f_N term can be counterbalanced by the f_E term, but it can also be due to ions in the surrounding solution which shield the charges on the hydrogel. As the charge density increases a deviation from the theory will be observed.

Recently the degradability of the gels synthesized by crosslinking copolymerization was studied (*14,77*). The degradation experiments were performed in-vitro using cecum content isolated from rats and in-vivo by implantation of the gels in rat cecum. The degradation was evaluated as an increase in the degree of swelling and as a change in color, since the gels are yellow due to azobonds. After degradation the gels became colorless and disintegrated. As the crosslinks were degraded, the degree of swelling increased. Gels containing different amounts of crosslinker were investigated. Table IV lists the degree of swelling after 2 days of

Table IV. Degree of swelling after degradation of hydrogels in-vitro

% Crosslinker	$1/v_2$ (DMAAB) Day 0	$1/v_2$ (DMAAB) Day 2	$1/v_2$ (DMCAAB) Day 0	$1/v_2$ (DMCAAB) Day 2
0.1	131	317	220	474
0.2 (0.5mm)	90	225	-	-
0.2 (1.0mm)	90	122	129	286
0.5	63	77	87	119
1.0	34	39	39	54

Composition of hydrogels: 40% acrylic acid, 10% N-tert-butylacrylamide, 49-49.9% N,N-dimethylacrylamide and 0.1-1.0% 4,4'-di(methacryloylamino)azobenzene (DMAAB) or 4,4'-di(N-methacryloyl-6-aminohexanoylamino)azobenzene (DMCAAB). Two gels in different thicknesses (0.5 and 1 mm respectively) with 0.2% DMAAB were investigated. Experimental conditions: Rat cecum content was suspended in water, filtered through glass wool and freeze dried. Gel discs were immersed in a cecum content suspension (30 mg cecum content/4 ml potassium phosphate buffer pH 7.4 0.04 M, the suspension was added 1.25 x 10^{-4} M benzyl viologen and 1.25 mg/ml a-D-glucose.), bubbled with nitrogen for 5 min and kept in a shaking water bath at 37°C, 50 strokes/min for 2 days. After degradation the gels were washed and swollen in 0.04 M potassium phosphate buffer pH 7.4 (*14,77*).

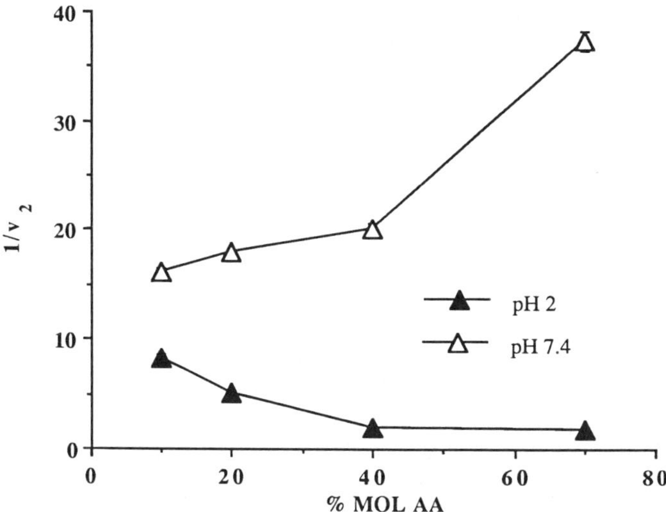

Figure 2. The equilibrium degree of swelling shown as $1/v_2$, where v_2 is volume fraction of polymer in swollen state, for poly(N,N-dimethylacrylamide-co-acrylic acid-co-N-tert-butylacrylamide) gels as a function of %mol acrylic acid. (Reproduced with permission from ref. 13. Copyright 1991, Butterworth-Heinemann Ltd.)

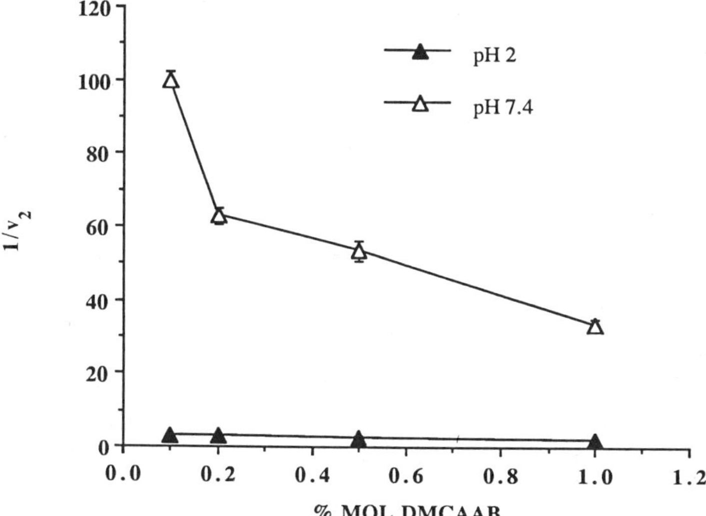

Figure 3. The equilibrium degree of swelling shown as $1/v_2$, where v_2 is volume fraction of polymer in swollen state, for poly(N,N-dimethylacrylamide-co-acrylic acid-co-N-tert-butylacrylamide) gels as a function of %mol crosslinking agent (4,4'-di(N-methacryloyl-6-aminohexanoylamino)azo-benzene.

degradation of different gels. It was found that, as crosslinking density increased, the degradability decreased due to a decrease in the degree of swelling. Two crosslinking agents of different lengths were evaluated. When the crosslinking density was the same, the gel containing the long crosslinker appeared to be degraded faster. The rate of disintegration of the gel also depended on the geometry of the gel; a thin gel was degraded faster than a thick gel. The degradation appeared to take place by surface erosion as well as bulk degradation.

In-vivo degradation was investigated by implantation into the stomach or cecum (Table V). It was found that the degree of swelling of gels implanted into the cecum increased, indicating the gels were degraded. Furthermore, the dry weight of the gels decreased, suggesting that the gel dissolved. The gels implanted into the stomach did not change in degree of swelling. A good correlation between in-vivo and in-vitro degradability was found.

Table V. Degree of swelling after degradation of hydrogels in-vivo

Implantation	$1/v_2$			% dry weight		
site	Day 0	Day 2	Day 6	Day 0	Day 2	Day 6
Stomach	90	94	86	100	103	98
Cecum	90	130	205	100	83	68

Composition of hydrogels: 40% acrylic acid, 10% N-tert-butylacrylamide, 49.8% N,N-dimethylacrylamide and 0.2% 4,4'-di(meth-acryloylamino)azobenzene (DMAAB). Experimental conditions: Gel discs were preswollen in potassium phosphate buffer pH 7.4 0.04 M and placed in nylon bags with small mesh size. The bags were implanted into the stomach or cecum of rats. After 2 and 6 days the gels were recovered, washed and swollen in 0.04M potassium phosphate buffer pH 7.4 (Data from ref. 14).

Acidic hydrogels containing degradable crosslinks were also synthesized by the crosslinking of polymeric precursors (*14,15*). These gels appeared to be degraded faster than gels formed by crosslinking copolymerization (*14,77*). For example, gels prepared by crosslinking of polymeric precursors were completely dissolved after 2 days of implantation in rat cecum. Several factors could contribute to this phenomenon. There is the possibility of side reactions during copolymerization, which results in non-degradable crosslinks, but more probably, differences in the structure of the networks, such as spacing of crosslinks along the polymer backbone, influence the rate of degradation. Another factor is that, the molecular weight of the polymer main chain is much higher for gels synthesized by crosslinking copolymerization than for gels synthesized by crosslinking of polymeric precursors; the higher molecular weight would result in a more entangled network for copolymerized gels.

The permeability of insulin as a model drug through poly(acrylic acid-co-N,N-dimethylacrylamide-co-N-tert-butylacrylamide) hydrogel membranes was investigated at pH 2 and pH 7.4. Membranes with different crosslinking density and content of acrylic acid were studied. The permeability was much higher at pH 7.4 than at pH 2 correlating with the degree of swelling of the hydrogel membrane (*13*). Figure 5 shows the natural logarithm of the permeability coefficient as a function of $1/(1-v_2)$. A non-linear relationship was obtained. Since both the membrane and insulin change ionization between pH 2 and pH 7.4, the properties and interactions between membrane and solute will change. The free volume theory applies to small molecules and its use for insulin is only an approximation. It can be seen from Figure 5 that the permeability is mainly dependent on the degree of swelling of the

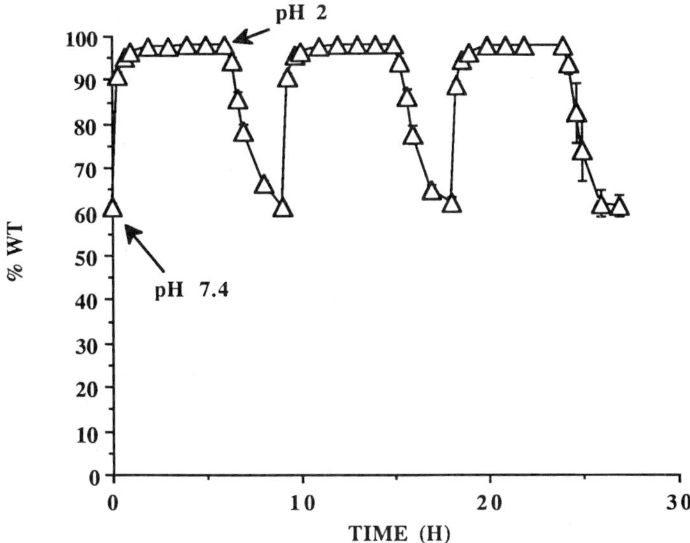

Figure 4. Weight percent of water in the equilibrium swollen state (%wt) for poly(N,N-dimethylacrylamide-co-acrylic acid-co-N-tert-butylacrylamide) gels as a function of time after repeated abrupt changes of pH between pH 2 to pH 7.4. (Adapted from ref. 13.)

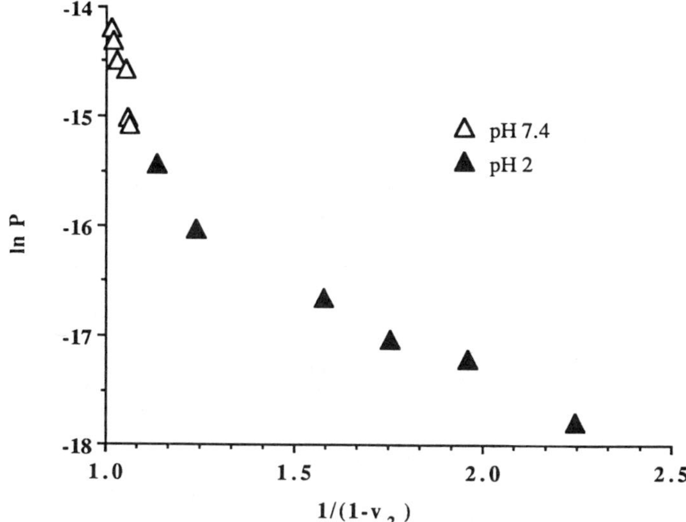

Figure 5. The natural logarithm of the permeability coefficient P (cm^2/s) of insulin through poly(N,N-dimethylacrylamide-co-acrylic acid-co-N-tert-butylacrylamide) gels as a function of $1/(1-v_2)$, where v_2 is volume fraction of polymer in swollen state. (Reproduced with permission from ref. 13. Copyright 1991, Butterworth-Heinemann Ltd.)

membrane and not on charge density. This correlates with the results obtained by other research groups (*54,57*).

The degradability of these gels combined with the pH-sensitivity suggests that the studied gels are suitable for drug delivery to the colon; however, more work is needed in order to optimize the system.

Conclusions.

Factors influencing the swelling and drug permeability of pH-sensitive hydrogels have been discussed. By varying the degree of crosslinking, hydrophobicity/hydrophilicity of the polymer, charge density, monomer ionizability and thickness of the hydrogel, we have illustrated that practically any desired swelling profile can be obtained. In addition to pH, swelling of pH-sensitive hydrogels is influenced by ionic strength and counterions in the swelling medium. The permeability of pH-sensitive hydrogels is mainly dependent on the degree of swelling of the gel, even though the charge of gel and drug may also be important. Hydrogels with pH-sensitivity have potential for use in drug delivery where a target site may have a different pH than its surroundings or where pH fluctuations occur due to pathologic situations or other stimuli. For oral delivery, pH-sensitive hydrogels are useful because of the pH variation throughout the GI-tract. The latter must to be taken into account in designing a drug delivery device. pH-sensitive hydrogels also have potential in the development of new "intelligent" drug delivery systems where a stimulus induces a change in pH and results in release of the drug, such as glucose sensitive membranes resulting in the release of insulin. It was shown how it is possible to formulate hydrogels which have potential for drug delivery to the colon by combining the pH-sensitive properties with enzyme degradable crosslinks.

Acknowledgments.

This research was partially supported by Theratech, Inc. and Insutech, Inc., Salt Lake City, Utah, USA and Hisamitsu Pharmaceutical Co., Tosu, Saga, Japan.

Literature Cited.

1. Wichterle, O.; Lim, D. *Nature* **1960**, *185* , 117
2. Ratner, B.D.; Hoffman, A.S. In *Hydrogels for Medical and Related Applications*; Andrade, J.D., Ed.; ACS Symposium Series 31; American Chemical Society; Washington, D.C., 1976; pp.1-36
3. Wichterle, O. In *Encyl. Polym. Sci. and Technol.*; Mark, H.F.; Gaylord, N.G. Ed.; 1971, 15; pp. 273-291
4. Voldřich, Z.; Tománek, Z.; Vacǐk, J.; Kopeček, J. *J. Biomed. Mater. Res.*, **1975**, *9*, 675
5. Mack, E.J.; Okano, T.; Kim, S.W. In *Hydrogels in Medicine and Pharmacy,*; Peppas, N.A. Ed.; CRC Press, Boca Raton, Fla., 1987, vol III; pp. 65-93
6. Gehrke, S.H.; Lee, P.I. In *Specialized Drug Delivery Systems: Manufacturing and Production Technology,* ; Tyle, P. Ed.; Marcel Dekker, inc., N.Y., 1990, vol 41; pp. 333-392
7. Schacht, E.H. In *Recent Advances in Drug Delivery Systems*; Anderson, J.M.; Kim, S.W. Ed.; Plenum Press, N.Y., 1984; pp. 259-278
8. *Hydrogels in Medicine and Pharmacy*; Peppas, N.A., Ed.; CRC Press, Boca Raton, Fla., 1987; Vol. I-III
9. Kůdela, V. In *Encycl. Polym. Sci. Eng.*; Kroschwitz, J.I. Ed.; John Wiley, N.Y., 1987; pp. 783-807
10. Kopeček, J.; Vacǐk, J.; Lǐm, D. *J. Polym. Sci.* **1971**, *9*, 2801

11. Katchalsky, A.; Michaeli, I. *J. Polym. Sci.*, **1955**, *15* , 69
12. Michaeli, I.; Katchalsky, A. *J. Polym. Sci.*, **1957**, *23*, 683
13. Brøndsted, H.; Kopeček, J., *Biomaterials*, in press
14. Kopeček, J.; Kopečková, P.; Brøndsted, H.; Rahti, R.; Říhová, B.; Yeh, P.-Y.; Ikesue, K. *J. Controlled Rel.*, in press
15. Yeh, P.-Y.; Kopečková, P.; Kopeček, J. to be published
16. Nishi, S.; Kotaka, T. *Macromolecules*, **1986**, *19*, 978
17. Michaels, A.S.; Miekka, R.G. *J. Phys. Chem.*, **1961**, *65*, 1765
18. Michaels, A.S. *Ind Eng Chem.*, **1965**, *57*, 32
19. Dabrovska, L.; Praus, R.; Stoy, V.; Vacík, J. *J. Biomed. Mater. Res.*, **1978**, *12*, 591
20. Flory, P.; Rehner, J. *J. Chem. Phys.*, **1943**, *11*, 521
21. Flory, P., *Principles of Polymer Chemistry*; Cornell University Press, London, 1969; pp.576-593
22. Hasa, J.; Ilavský, M.; Dušek, K. *J. Polym. Sci.*, **1975**, *13*, 253
23. Hasa, J.; Ilavský, M. *J. Polym. Sci.*, **1975**, *13*, 263
24. Ilavský, M.; Dušek, K.; Vacík, J.; Kopeček, J. *J. Appl. Polym. Sci.*, **1979**, *23*, 2073
25. Kienzle-Sterzer, C.A.; Rodriguez-Sanchez, D.; Karalekas, D.; Rha, C. *Macromolecules*, **1982**, *15*, 631
26. Oppermann, W. *Ang. Makromol. Chem.*, **1984**, *123/124*, 229
27. Ilavský, M. *Polymer*, **1981**, *22*, 1687
28. Ilavský, M. *Macromolecules*, **1982**, *15*, 782
29. Vasheghani-Farahani, E.; Vera, J.H.; Cooper, D.; Weber, M.E. Ind. Eng. *Chem. Res.*, **1990**, *29*, 554
30. Brannon-Peppas, L.; Peppas, N.A. *Polym. Bull.*, **1988**, *20*, 285
31. Brannon-Peppas, L.; Peppas, N.A. *Chem. Eng. Sci.*, **1991**, *46*, 715
32. Tanaka, T. *Scientific American*, **1981**, *244*, 125
33. Ricka, J.; Tanaka, T. *Macromolecules*, **1984**, *17*, 2916
34. Nakano, Y.; Seida, Y.; Uchida, M.; Yamamoto, S. *J. Chem. Eng. Japan*, **1990**, *23*, 574
35. Hooper, H.H.; Baker, J.P.; Blanch, H.W.; Prausnitz, J.M. *Macromolecules*, **1990**, *23*, 1096
36. Beltran, S,; Baker, J.P.; Hooper, H.H.; Blanch, H.W.; Prausnitz, J.M. *Macromolecules*, **1991**, *24*, 549
37. Ilmain, F.; Tanaka, T.; Kokufuta, E. *Nature*, **1991**, *349*, 400
38. Kou, J.H.; Amidon, G.L.; Lee, P.I. *Pharm. Res.*, **1988**, *5*, 592
39. Siegel, R.A.; Firestone, B.A. *Macromolecules*, **1988**, *21*, 3254
40. Vacík, J.; Kopeček, J. *J. Appl. Polym. Sci.*, **1975**, *19*, 3029
41. Vacík, J.; Kůdela, V.; Kopeček, J. *Europ. Polym. J.*, **1975**, *11*, 331
42. Kůdela, V.; Vacík, J.; Kopeček, J. *Europ. Polym. J.*, **1977**, *13*, 811
43. Kůdela, V.; Vacík, J.; Kopeček, J. *J. Membr. Sci.*, **1980**, *6*, 123
44. Přádný, M.; Kopeček, J. *Makromol. Chem.*, **1990**, *191*, 1887
45. Kirstein, D.; Braselmann, H.; Vacík, J.; Kopeček, J. *Biotech. Bioeng.*, **1985**, *27*, 1382
46. Siegel, R.A.; Firestone, B.A.; Johannes, I.; Cornejo, J. *Polym. Prepr.*, **1991**, *31*, 231
47. Siegel, R.A. In *Pulsed and Self-regulated Drug Delivery*; Kost, J. Ed.; CRC Press, Boca Raton, Fla., 1990; pp.129-157
48. Kwon, G.; Bae, Y.H.; Kim, S.W.; Cremers, H.; Feijen, J. *J. Controlled Rel.*, in press, **1991**
49. Refojo, M.F. *J. Polym. Sci., Part A-1*, **1967**, *5*, 3103
50. Ratner, B.D.; Miller, I.F. *J. Polym. Sci., Part A-1*, **1972**, *10*, 2425

51. Yasuda, H.; Lamaze, C.E.; Ikenberry, L.D. *Makromol. Chem.*, **1968**, *118*, 19
52. Yasuda, H.; Ikenberry, L.D.; Lamaze, C.E. *Makromol. Chem.*, **1969**, *125*, 108
53. Vacík, J.; Czaková, M.; Exner, J.; Kopeček, J. *Collection Czechoslov. Chem. Commun.*, **1977**, *42*, 2786
54. Weiss, A.M.; Grodzinsky, A.J.; Yarmush, M.L. *AIChE Symposium Series*, **1986**, *82*, 85
55. Grodzinsky, A.J.; Grimshaw, P.E. In *Pulsed and Self-regulated Drug Delivery*; Kost, J. Ed.; CRC Press, Boca Raton, Fla., 1990; pp.47-64
56. Grimshaw, P.E.; Grodzinsky, A.J.; Yarmush, M.L.; Yarmush, D.M. *Chem. Eng. Sci.*, **1990**, *45*, 2917
57. Shatayeva, L.K.; Samsonov, G.V.; Vacík, J.; Kopeček, J.; Kálal, J. *J. Appl. Polym. Sci.*, **1979**, *23*, 2245
58. Brannon-Peppas, L.; Peppas, N.A. *J. Controlled Rel.*, **1989**, *8*, 267
59. Gehrke, S.H.; Cussler, E.L. *Chem. Eng. Sci.*, **1989**, *44*, 559
60. Kou, J.H.; Fleisher, D.; Amidon, G.L. *J. Controlled Rel.*, **1990**, *12*, 241
61. Siegel, R.A.; Falamarzian, M.; Firestone, B.A.; Moxley, B.C. *J. Controlled Rel.*, **1988**, *8*, 179
62. Dong, L.-C.; Hoffman, A. S. *J. Controlled Release*, **1991**, *15*, 141
63. Touitou, E.; Rubinstein, A. *Intern. J. Pharm.*, **1986**, *30*, 95
64. Gwinup, G.; Elias, A. N.; Domurat, E. S. *Gen. Pharmac.*, **1991**, *22*, 243
65. Gupta, P.K.; Leung, S.-H.; Robinson, J.R. In *Bioadhesive Drug Delivery Systems*; Lenaerts, V., Gurny, R., Ed.; CRC Press, Boca Raton, Fla, 1990; pp. 65-92
66. Park, K.; Chung, H.S.; Robinson, J.R. In *Recent Advances in Drug Delivery Systems*; Anderson, J.M., Kim, S.W., Ed.; Plenum Press, New York, 1984; pp. 163-183
67. Park, H.; Robinson, J.R. *Pharm. Res.*, **1987**, *4*, 457
68. Robinson, J.R.; Longer, M.A.; Veillard, M. Annals N.Y. *Academy Sci.*. **1987**, *507*, 307
69. Shalaby, W.S.W.; Park, K. *Pharm. Res.*, **1990**, *7*, 816
70. Ishihara, K.; Kobayashi, M.; Ishimaru, N.; Shinohara, I. *Polym. J.*, **1984**, *8*, 625
71. Ishihara, K.; Matsui, K. *J. Polym. Sci.*, **1986**, *24*, 413
72. Albin, G.; Horbett, T.A.; Ratner, B.D. In *Pulsed and Self-regulated Drug Delivery*; Kost, J. Ed.; CRC Press, Boca Raton, Fla., 1990; pp.159-185
73. Kokufata, E.; Zhang, Y.-Q.; Tanaka, T. *Nature,* **1991**, *351,* 302
74. Saffran, M.; Kumar, G.S.; Savariar, C.; Burnham, J.C. *Science,* **1986**, *233*, 1081
75. Ulbrich, K.; Strohalm, J.; Kopeček, J. *Biomaterials*, **1982**, *3*, 150
76. Brøndsted, H. *Hydrogels for colon-specific peptide drug delivery*, Ph. D. Dissertation, University of Utah, 1991
77. Brøndsted, H.; Kopeček, J. to be published

RECEIVED September 19, 1991

Author Index

Affiliation Index

Subject Index

Production: Paula M. Befard
Indexing: Colleen Stamm
Acquisition: A. Maureen Rouhi
Cover design: Sue Schafer

Printed and bound by Maple Press, York, PA

Highlights from ACS Books

Good Laboratory Practices: An Agrochemical Perspective
Edited by Willa Y. Garner and Maureen S. Barge
ACS Symposium Series No. 369; 168 pp; clothbound, ISBN 0–8412–1480–8

Silent Spring Revisited
Edited by Gino J. Marco, Robert M. Hollingworth, and William Durham
214 pp; clothbound, ISBN 0–8412–0980–4; paperback, ISBN 0–8412–0981–2

Insecticides of Plant Origin
Edited by J. T. Arnason, B. J. R. Philogène, and Peter Morand
ACS Symposium Series No. 387; 214 pp; clothbound, ISBN 0–8412–1569–3

Chemistry and Crime: From Sherlock Holmes to Today's Courtroom
Edited by Samuel M. Gerber
135 pp; clothbound, ISBN 0–8412–0784–4; paperback, ISBN 0–8412–0785–2

Handbook of Chemical Property Estimation Methods
By Warren J. Lyman, William F. Reehl, and David H. Rosenblatt
960 pp; clothbound, ISBN 0–8412–1761–0

The Beilstein Online Database: Implementation, Content, and Retrieval
Edited by Stephen R. Heller
ACS Symposium Series No. 436; 168 pp; clothbound, ISBN 0–8412–1862–5

Materials for Nonlinear Optics: Chemical Perspectives
Edited by Seth R. Marder, John E. Sohn, and Galen D. Stucky
ACS Symposium Series No. 455; 750 pp; clothbound; ISBN 0–8412–1939–7

Polymer Characterization:
Physical Property, Spectroscopic, and Chromatographic Methods
Edited by Clara D. Craver and Theodore Provder
Advances in Chemistry No. 227; 512 pp; clothbound, ISBN 0–8412–1651–7

From Caveman to Chemist: Circumstances and Achievements
By Hugh W. Salzberg
300 pp; clothbound, ISBN 0–8412–1786–6; paperback, ISBN 0–8412–1787–4

The Green Flame: Surviving Government Secrecy
By Andrew Dequasie
300 pp; clothbound, ISBN 0–8412–1857–9

For further information and a free catalog of ACS books, contact:
American Chemical Society
Distribution Office, Department 225
1155 16th Street, NW, Washington, DC 20036
Telephone 800–227–5558

Bestsellers from ACS Books

The ACS Style Guide: A Manual for Authors and Editors
Edited by Janet S. Dodd
264 pp; clothbound, ISBN 0–8412–0917–0; paperback, ISBN 0–8412–0943–X

Chemical Activities and Chemical Activities: Teacher Edition
By Christie L. Borgford and Lee R. Summerlin
330 pp; spiralbound, ISBN 0–8412–1417–4; teacher ed. ISBN 0–8412–1416–6

Chemical Demonstrations: A Sourcebook for Teachers,
Volumes 1 and 2, Second Edition
Volume 1 by Lee R. Summerlin and James L. Ealy, Jr.;
Vol. 1, 198 pp; spiralbound, ISBN 0–8412–1481–6;
Volume 2 by Lee R. Summerlin, Christie L. Borgford, and Julie B. Ealy
Vol. 2, 234 pp; spiralbound, ISBN 0–8412–1535–9

Writing the Laboratory Notebook
By Howard M. Kanare
145 pp; clothbound, ISBN 0–8412–0906–5; paperback, ISBN 0–8412–0933–2

Developing a Chemical Hygiene Plan
By Jay A. Young, Warren K. Kingsley, and George H. Wahl, Jr.
paperback, ISBN 0–8412–1876–5

Introduction to Microwave Sample Preparation: Theory and Practice
Edited by H. M. Kingston and Lois B. Jassie
263 pp; clothbound, ISBN 0–8412–1450–6

Principles of Environmental Sampling
Edited by Lawrence H. Keith
ACS Professional Reference Book; 458 pp;
clothbound; ISBN 0–8412–1173–6; paperback, ISBN 0–8412–1437–9

Biotechnology and Materials Science: Chemistry for the Future
Edited by Mary L. Good (Jacqueline K. Barton, Associate Editor)
135 pp; clothbound, ISBN 0–8412–1472–7; paperback, ISBN 0–8412–1473–5

Personal Computers for Scientists: A Byte at a Time
By Glenn I. Ouchi
276 pp; clothbound, ISBN 0–8412–1000–4; paperback, ISBN 0–8412–1001–2

Polymers in Aqueous Media: Performance Through Association
Edited by J. Edward Glass
Advances in Chemistry Series 223; 575 pp;
clothbound, ISBN 0–8412–1548–0

For further information and a free catalog of ACS books, contact:
American Chemical Society
Distribution Office, Department 225
1155 16th Street, NW, Washington, DC 20036
Telephone 800–227–5558